Ecology of Industrial Pollution

Written for researchers and practitioners in environmental pollution, management and ecology, this interdisciplinary account explores the ecological issues associated with industrial pollution to provide a complete picture of this important environmental problem from cause to effect to solution.

Bringing together diverse viewpoints from academia and environmental agencies and regulators, the contributors cover such topics as biological resources of mining areas, biomonitoring of freshwater and marine ecosystems and risk assessment of contaminated land in order to explore important questions such as: What are the effects of pollutants on functional ecology and ecosystems? Do current monitoring techniques accurately signal the extent of industrial pollution? Does existing policy provide a coherent and practicable approach? Case studies from throughout the world illustrate major themes and provide valuable insights into the positive and negative effects of industrial pollution, the provision of appropriate monitoring schemes and the design of remediation and restoration strategies.

LESLEY C. BATTY is a Lecturer in Environmental Science at the University of Birmingham, UK. Her research focuses on ecological aspects of industrial pollution, particularly on the use of plants in remediation and the effects of mining activities on the environment. She is a council member of the British Ecological Society and has acted as a consultant to several government agencies and engineering companies within the UK.

KEVIN B. HALLBERG is a Research Fellow in Environmental Microbiology at Bangor University, UK. His research focuses on the microbial ecology of extremophilic microorganisms, particularly the environmentally and industrially useful acidophiles. He has acted as consultant for international mining companies as well as governmental regulatory agencies including the International Atomic Energy Agency.

Ecological Reviews

Ecological Reviews will publish books at the cutting edge of modern ecology, providing a forum for volumes that discuss topics that are focal points of current activity and likely long-term importance to the progress of the field. The series will be an invaluable source of ideas and inspiration for ecologists at all levels from graduate students to more-established researchers and professionals. The series will be developed jointly by the British Ecological Society and Cambridge University Press and will encompass the Society's Symposia as appropriate.

Biotic Interactions in the Tropics: Their Role in the Maintenance of Species Diversity
Edited by David F. R. P. Burslem, Michelle A. Pinard and Sue E. Hartley

Biological Diversity and Function in Soils
Edited by Richard Bardgett, Michael Usher and David Hopkins

Island Colonization: The Origin and Development of Island Communities
By Ian Thornton
Edited by Tim New

Scaling Biodiversity
Edited by David Storch, Pablo Margnet and James Brown

Body Size: The Structure and Function of Aquatic Ecosystems
Edited by Alan G. Hildrew, David G. Raffaelli and Ronni Edmonds-Brown

Speciation and Patterns of Diversity
Edited by Roger Butlin, Jon Bridle and Dolph Schluter

Ecology of Industrial Pollution

Edited by

LESLEY C. BATTY
University of Birmingham

KEVIN B. HALLBERG
Bangor University

CAMBRIDGE
UNIVERSITY PRESS

CAMBRIDGE
UNIVERSITY PRESS

University Printing House, Cambridge CB2 8BS, United Kingdom

One Liberty Plaza, 20th Floor, New York, NY 10006, USA

477 Williamstown Road, Port Melbourne, VIC 3207, Australia

314-321, 3rd Floor, Plot 3, Splendor Forum, Jasola District Centre, New Delhi - 110025, India

103 Penang Road, #05-06/07, Visioncrest Commercial, Singapore 238467

Cambridge University Press is part of the University of Cambridge.

It furthers the University's mission by disseminating knowledge in the pursuit of education, learning and research at the highest international levels of excellence.

www.cambridge.org
Information on this title: www.cambridge.org/9780521514460

First published 2010

A catalogue record for this publication is available from the British Library

Library of Congress Cataloging in Publication data
Ecology of industrial pollution / edited by Lesley C. Batty, Kevin B. Hallberg.
    p. cm. – (Ecological reviews)
  ISBN 978-0-521-51446-0 (hardback) – ISBN 978-0-521-73038-9 (pbk.)
    1. Environmental monitoring. 2. Pollution–Environmental aspects. 3. Pollution–Measurement. 4. Pollution prevention. I. Batty, Lesley C. II. Hallberg, Kevin B.
III. Title. IV. Series.
  QH541.15.M64E26 2010
  577.27–dc22

                                    2009042292

ISBN 978-0-521-51446-0 Hardback
ISBN 978-0-521-73038-9 Paperback

Contents

The colour plate section is placed between pages 308 and 309.

Contributors

DANIELLE K. ASHTON
Environment Agency, Wallingford,
United Kingdom

MONTSERRAT AULADELL
School of Geography,
Earth and Environmental Science,
University of Birmingham,
Edgbaston, United Kingdom

ALAN J. M. BAKER
School of Botany,
University of Melbourne,
Australia

DECLAN BARRACLOUGH
Environment Agency,
Solihull, United Kingdom

LESLEY C. BATTY
School of Geography,
Earth and Environmental Science,
University of Birmingham,
Edgbaston, United Kingdom

RACHEL BENSTEAD
Environment Agency,
Wallingford, United Kingdom

ALISTAIR BOXALL
EcoChemistry Team,
University of York/Central Science

Laboratory, York,
United Kingdom

PAUL R. BRADFORD
Environment Agency, Solihull,
United Kingdom

IAN T. BURKE
Institute of Geological Sciences,
School of Earth and Environment,
University of Leeds,
United Kingdom

JOHN DAVY-BOWKER
Centre for Ecology and Hydrology,
Wallingford, United Kingdom

ROWLAND ELEY
Environment Agency, Bristol,
United Kingdom

ANTONY VAN DER ENT
B-WARE Research Centre,
Radboud Universiteit,
Nijmegen,
The Netherlands

WILFRIED H. O. ERNST
Institute of Ecological Science,
Vrije Universiteit,
Amsterdam, The Netherlands

ALASTAIR J. D. FERGUSON
Environment Agency,
Bristol, United Kingdom

TIM GANNICLIFFE
Natural England,
Wigan, United Kingdom

ANDREA GEISSLER
Williamson Centre for Molecular
Environmental Science,
School of Earth, Atmospheric and
Environmental Sciences,
University of Manchester,
United Kingdom

PETER GELL
School of Science & Engineering,
The University of Ballarat,
Australia

ROSANNA GINOCCHIO
Centro de Investigación Minera y
Metalúrgica (CIMM),
Chile

ALASTAIR GRANT
School of Environmental Sciences,
University of East Anglia,
Norwich, United Kingdom

KEVIN B. HALLBERG
School of Biological Sciences,
College of Natural Sciences,
Bangor University,
United Kingdom

KEITH HENDRY
Faculty of Life Sciences,
University of Manchester,
United Kingdom

SHELLEY R. HOWARD
Environment Agency,
Bristol, United Kingdom

J. IWAN JONES
Centre for Ecology and Hydrology,
Wallingford, United Kingdom

KEN KILLHAM
School of Biological Sciences,
University of Aberdeen,
United Kingdom

TERRY E. L. LANGFORD
Centre for Environmental Sciences,
School of Civil Engineering and
Environment,
University of Southampton,
United Kingdom

FRANCIS R. LIVENS
Centre for Radiochemistry Research,
Department of Chemistry,
University of Manchester,
United Kingdom

JONATHAN R. LLOYD
Centre for Radiochemistry Research,
Department of Chemistry,
University of Manchester,
United Kingdom

FRANÇOIS MALAISSE
Laboratoire d'Ecologie, Faculté
Universitaire des Sciences
Agronomiques de Gemblaux, Belgium

LORRAINE MALTBY
Department of Animal and Plant
Sciences, University of Sheffield,
United Kingdom

KATHERINE MORRIS
Institute of Geological Sciences,
School of Earth and Environment,
University of Leeds,
United Kingdom

JOHN F. MURPHY
Centre for Ecology and Hydrology,
Wallingford, United Kingdom

DAVID OTTEWELL
Environment Agency,
Bristol, United Kingdom

ACHIM PAETZOLD
Catchment Science Centre,
University of Sheffield,
United Kingdom

JAMES L. PRETTY
Centre for Ecology and Hydrology,
Wallingford, United Kingdom

OLE WILLIAM PURVIS
Department of Botany, Natural
History Museum,
London, United Kingdom

STEPHEN ROAST
Environment Agency,
Exeter, United Kingdom

JON SADLER
School of Geography, Earth and
Environmental Science,
University of Birmingham,
Edgbaston, United Kingdom

MICHAEL JAY SADOWSKY
Department of Soil, Water and
Climate; and the BioTechnology

Institute, University of Minnesota,
St Paul, MN, USA

SONJA SELENSKA-POBELL
Institute of Radiochemistry,
Forschungszentrum Dresden-
Rossendorf,
Dresden, Germany

PETER J. SHAW
Centre for Environmental Sciences,
School of Civil Engineering and
Environment, University of
Southampton, United Kingdom

MARK TIBBETT
Centre for Land Rehabilitation,
School of Earth and Environment,
University of Western Australia,
Crawley, Australia

PHILIP WARREN
Department of Animal and Plant
Sciences, University of Sheffield,
United Kingdom

RACHEL J. WATERFALL
APEM Ltd., Stockport,
United Kingdom

KEITH N. WHITE
Faculty of Life Sciences, University of
Manchester, United Kingdom

PAUL WHITEHOUSE
Environment Agency, Wallingford,
United Kingdom

ADRIAN E. WILLIAMS
APEM Ltd., Stockport,
United Kingdom

Preface

This volume reflects the content of the Symposium on the Ecology of Industrial Pollution which was held in Birmingham, UK, from 7 to 8 April 2008. The principal aim of this symposium and the associated volume was to provide an interdisciplinary analysis of a particular environmental problem that has historically affected many parts of the world and continues to do so. One of the key approaches was to obtain inputs from academics, enforcers and practitioners in order to provide a balanced view of the issue and to identify any potential areas of disagreement! Second, we recognised that, although many of the individual components of pollutants are dealt with in meetings and publications, there is little interaction between researchers who deal with the 'pure science' issues and those who deal with more technological aspects, nor between microbiologists, botanists and zoologists. By bringing these normally disparate areas together, we hoped to provide new research areas and more importantly exchange of existing knowledge and experiences.

In our introduction we try to provide an overview of the key subject areas covered within the volume and identify what we consider to be the principal questions that were generated from discussions at the meeting. The remainder of the book is not subdivided but instead the chapters are arranged into what we hope to be a logical progression from the main ecological impacts through monitoring techniques and finishing with ecological remediation technologies and system recovery. The final chapter picks up on the developments in recent years of the idea of ecosystem services and applies this to industrial pollution to provide a new possibility for management of such systems. One area that we must recognise was not covered in this volume is that of climate change (undoubtedly a consequence of industrial activity). We felt that this was outside the scope of this meeting and necessitates a volume to itself.

We hope that the readers find the material as stimulating and exciting as did the attendees of the meeting.

Acknowledgements

We would like the opportunity to thank the British Ecological Society for their support in the meeting and preparation of the volume, key members of staff being Hazel Norman, Lindsay Haddon, Richard English and Hefin Jones. We also recognise the input of Dr Adam Jarvis in the early stage of the symposium and volume preparation. We would also like to thank the symposium speakers who provided their chapters for this book. Without their commitment to the project, we would never have made it past the first hurdle!

CHAPTER ONE

Consequences of living in an industrial world

LESLEY C. BATTY AND KEVIN B. HALLBERG

Introduction

One of the first questions that faced us when preparing this introductory chapter was 'what do we mean by industry?' In modern terms, one often refers to the industrial revolution that began in the latter half of the eighteenth century, which circumscribes the change from an agriculturally based economy to one dominated by manufacturing. However, industrial processes have a history far longer than this, and can be traced back to the Bronze Age and even before, particularly the extraction of minerals. We could also consider agriculture to be an industry as it is the extraction of raw resources albeit in a rather different form. Therefore when we refer to industry, we are actually considering a very wide range of processes and activities. Common to all these, however, is the fact that the production of goods from raw resources creates by-products that can pollute the environment and adversely affect ecosystems.

The industrial pollutants produced and their impacts are potentially as varied as the sources from which they derive, and there has been extensive research into specific effects of individual contaminants on specific organisms or communities. The problem with this approach is that the resulting view is one that can be rather blinkered. It is becoming increasingly clear that, rather than simply causing deterioration of ecosystems, contaminated sites may well be sources of biodiversity. Organisms living on such sites can show great genetic adaptation and may prove useful in the remediation of other contaminated sites. In addition, the limitations of ecological monitoring have potentially caused problems in the assessment of impacts, and the detachment of research into remediation from that of ecotoxicology has resulted in inappropriate application of technologies and poor results in terms of restoration or remediation. Within this volume, we have tried to select a number of different types of industry in order to illustrate these general themes. It should be noted that, although many industrial processes release(d) carbon dioxide and other greenhouse gases into the atmosphere, we

Ecology of Industrial Pollution, eds. Lesley C. Batty and Kevin B. Hallberg. Published by Cambridge University Press. © British Ecological Society 2010.

have chosen not to cover this topic within this volume. Climate change is linked to many different causes (not just industry) and the discussions surrounding this area are sufficiently complex to merit their own substantial volume. However, it may be said that many of the key processes that result in impacts on ecological systems and the limits to restoration ecology could be extremely important in predicting responses and adaptation to climate change.

Impacts of industry

It is interesting to note the changing attitudes to industrial activity over the years as nations develop. A good example of this is the activities associated with the Parys Mountain copper mine on Anglesey, UK. In letters written in the late eighteenth century by a professor studying the mining around the area, it is reported that the copper is sent to factories in Flintshire where '…the view of the valley…is particularly charming because the cotton spinning mills are lit up from top to bottom and reflected in the ponds…' and '… I think this is the prettiest valley that I have ever seen…. Poverty and misery are not to be found' (Rothwell 2007). In contrast, Greenly (1919), when commenting on the mining area itself reports that the '…higher and central portions are of the most utter desolation imaginable'. It is easy to forget that the industrial revolution led to great improvements in the wealth of many nations and did indeed have positive impacts on human health (as well as the more widely reported negative impacts). However, over the last decades the negative view of industry as a major source of contamination to the environment has increased in volume, particularly in recent years with the issue of climate change. If we examine the effects of industry upon the ecological environment, then typically reports are of reduced biodiversity either through the direct toxic effects of pollutants or through indirect effects on habitat quality and food webs. In this volume, Purvis et al. (Chapter 3) and Batty et al. (Chapter 4) provide examples of this as a result of air and water pollution, respectively.

It is the recognition of these negative impacts of industrial pollutants on the environment that led to significant advances in the protection of both human health and environmental health from the effects of pollution since the industrial revolution. If we take air pollution as an example, then we can see that over time changes in legislation and critically the Clean Air Act of 1956 have acted to both reduce the incidences of respiratory illness and to significantly reduce the concentrations of sulphur in the air, resulting in re-establishment of clean air ecology (particularly lichens) in many previously affected areas (Chapter 3). Equivalent legislative changes for freshwaters and soils have also been implemented with the focus mainly on the protection of human health. However, in recent years there has been a change in emphasis within developed countries to also consider ecological health, as the importance of the function of ecosystems in the health of the human environment has been recognised. The EU Water Framework Directive (WFD) is a key example where it is not simply the chemical environment that is considered, but a healthy ecological status is also a key objective.

Through changes in legislation there has gradually been an improvement in the quality of many environmental compartments within the landscape. However, in many parts of the world, legislation is underdeveloped, or not sufficiently implemented, resulting in continuing threats to the ecological health of the receiving environment. In addition continuing advances in technology result in new pollutants such as those from the nanotechnology and pharmaceutical industries (Chapter 5), and critically the effects of these on organisms are largely unknown.

The long history of global industrial activity has led to the accumulation of contaminants within the environment, especially where the pollutants in question are persistent. It is therefore extremely important that, when considering the impacts of industry, we also recognise the potential contribution of these historical sources to the impacts on current ecological communities. Floodplain deposits, in-river sediments and ancient mine workings are a few examples where old contamination can still affect present day ecosystems (Chapter 2, Chapter 13 and Chapter 14, respectively).

It is a clear ecological concept that environmental heterogeneity within a landscape can drive biological diversity (e.g., see Hutchings *et al.* 2000), but this can also be true within polluted environments (Chapter 2). The presence of highly metalliferous soils upon naturally occurring outcrops of metal ores has led to the adaptation of a number of species (particularly plants, lichens and bacteria) to these conditions, and indeed some may only survive where metal concentrations are high. The exploitation of these resources by man has led to these communities being extremely rare, but they can often survive and proliferate on abandoned sites. The lead rakes of the Peak District are a prime example of this, where calminarian grasslands can be found (Barnatt & Penny 2004). However, changing environmental legislation, the decline of industrial activity and potentially also environmental change now pose a significant threat to these highly biodiverse areas. For example, the requirement to meet the objectives of the Water Framework Directive may result in the removal or remediation of metal sources, such as spoil heaps from the environment. A number of different threats to these unique communities are identified by Baker *et al.* (Chapter 2).

More modern remains of industry are also proving to be valuable as refuges for threatened species. The move from a manufacturing based economy to one based largely on service industry within many countries in the developed world has led to large areas of so-called brownfield land, which are often contaminated by a mixture of pollutants. The lack of human access, together with particular environmental conditions has allowed the colonisation of these areas with a range of species. Although this has not been dealt with in this volume, readers are directed to Chapter 3 of Natural England's Report on the State of the Natural Environment (2008).

Monitoring ecological response to pollution

If we are to determine the impacts of pollutants upon ecosystems and their recovery following remediation, it is absolutely essential that we have a robust method of monitoring. There have been significant advances in the methods of monitoring freshwater systems, details of which are provided in Jones *et al.* (Chapter 6); however, it is clear that there are several challenges to be met. The first of these is to define what is meant by the term 'reference condition'. It is a term used in many key pieces of legislation (including the WFD) to assess an ecological community in relation to the community that is expected to be present based on 'reference conditions'. However, due to the extent of human impact both on a temporal and spatial scale it is difficult to find a 'real' example of this, or to model one. It is the general consensus that, when tackling pollution within the environment, the aim is not to attain a reference condition that reflects pre-industrial conditions (Chapter 6) but rather to achieve high quality and sustainable water resources. However, as Gell points out (Chapter 8), it may be preferable to have pre-industrial baseline targets due to future risks from pollutants (in the form of sediments) on the longevity of a system (current evaluation of status shows little divergence from a standard reference condition). The relative value of monitoring against some ideal ecological community is a concept that is clearly questionable.

There has been far less progress in the ecological monitoring of either land or marine environments in terms of ecological impacts. It is suggested that, rather than using a similar system used for freshwaters where the physical environment is very different, for marine environments sediment toxicity testing to monitor lethal and more importantly sub-lethal effects would be more beneficial (Chapter 7). This idea that sub-lethal effects may be of critical importance is also highlighted in other systems where there is a need for new toxicity tests for emerging contaminants due to the inappropriate nature of existing tests that lack subtle endpoints (Chapter 5). The assessment of land contamination is notoriously difficult due to the extremely heterogeneous nature of soils which strongly control the bioavailability of contaminants. There has been some attempt to assess contamination using chemical approaches (production of Soil Guideline Values for a number of contaminants), but the limitations of these are extensive, and there is little if any link to the ecology of the area. Recent work has made progress in providing a much clearer link between land contamination and associated communities using a risk-based approach (Chapter 9), and the success of this approach will be monitored in forthcoming years.

Remediation and ecological recovery

We now return to the previous question of 'reference conditions'. Whenever the remediation of a contaminated site and its associated ecological recovery is considered, a 'target' must be defined by which the remediation activity can be deemed a success. However, as we have previously highlighted, there is a question as to where this point should lie. Is the aim to return an environment to its pre-contaminated conditions, or alternatively should the target be a particular community composition, such as the presence of a rare species, or more generally a sustainable water resource? It appears that returning a particular habitat to its pre-industrial conditions is an unrealistic goal for two main reasons. The first of these is a lack of good quality data that provide a detailed characterisation of the abiotic and biotic components of any environment. Although there has been some attempt to use palaeoecological data (Chapter 8), this is limited in many environments due to preservation, and the lack of consistency in methods and data analysis has been highlighted (Chapter 6). The second is that, even where more recent data are available, it is evident that the community does not return to its previous state even when given sufficient time. Tibbett, Williams *et al.*, Langford *et al.*, Batty *et al.* and Purvis (Chapters 15, 14, 13, 4 and 3, respectively) all report that, despite improvements in the physico-chemical environment, either through a decline in industry or active remediation activities, the community does not return to its pre-contaminated state (or other target condition). This constraint to recovery is probably due to the lack of sources of colonising organisms, lack of physical habitat (although chemical conditions may improve), the impact of other pollutants in the environment (other than those directly targeted by remediation) and transfer of industrial pollutants from long-term sources (such as sediments) not tackled by remediation. In addition, ecological function may continue to be impaired as a result of the changes in community structure, although there is little information on the causes of ecological dysfunction in recovering communities (Chapter 15).

The presence of adapted organisms on industrially contaminated sites provides a potential opportunity to exploit these organisms for either the stabilisation of such sites or active remediation. The potential for using metallophyte plants in the remediation of metalliferous soils has been postulated for a number of years and successfully applied in some cases; however, this potential is rather under-exploited due to a lack of knowledge of the mechanisms of adaptation and metabolic and genetic responses to pollutants (Chapter 2). Rather more progress has been made in the use of bacteria in remediation activities, particularly where the land is contaminated by organic pollutants or radionuclides (Chapters 12 and 11, respectively). Advances in knowledge of the genetics involved in adaptation of microorganisms and mechanisms of action in remediation processes (Chapter 10) provides great potential for these organisms to be applied in many situations, particularly where there are problems of mixed contaminants.

Conclusions

This volume provides an overview of the impacts of industrial pollution, ways of monitoring and remediation and recovery of such systems. It is clear that, although in many countries the main polluting industries have now declined or in fact ceased, the legacy of industrial activity continues to affect ecological communities, and changes in industrial processes now provide new potentially harmful substances within the environment. The need therefore to understand the links between the contaminating substances and the ecological responses both on an individual organism and whole ecosystem level is vital. Only when this is achieved will we be able to appreciate the full extent of the impacts, provide appropriate monitoring schemes and design remediation strategies that best tackle the specific problems. A number of key questions arise within this volume:

How important is pollution in driving diversification in communities?
To what extent do polluted areas constitute a valuable habitat for rare species?
What are the limitations of resilience and/or functional redundancy within an impacted community?
Is biomonitoring effective and accurate in assessing the extent of contamination within an environment?
Are 'reference conditions' helpful in either monitoring or restoration/remediation?
Can adapted organisms effectively be used in remediation technologies and how can this potential be maximised?

Although significant progress has been made in all these areas, there is clearly the need for further research in order to create more integrated and sustainable management of our industrial areas (Chapter 16).

References

Barnatt, J. and Penny, R. (2004) *The Lead Legacy. The Prospects for the Peak District's Lead Mining Heritage*. Peak District National Park Authority, Buxton, UK.

Greenly, E. (1919) *Geology of Anglesey Vol 11*. Memoirs of the Geological Survey of England and Wales, London.

Hutchings, M.J., John, E.A. and Stewart, A.J.A. (2000) *The Ecological Consequences of Environmental Heterogeneity*. Blackwell Science Ltd, Oxford, UK.

Rothwell, N. (2007) *Parys Mountain and the Lentin Letters*. Awlwch Industrial Heritage Trust, Anglesey, UK.

CHAPTER TWO

Metallophytes: the unique biological resource, its ecology and conservational status in Europe, central Africa and Latin America

ALAN J. M. BAKER, WILFRIED H. O. ERNST,
ANTONY VAN DER ENT, FRANÇOIS MALAISSE
AND ROSANNA GINOCCHIO

Introduction

Metalliferous soils provide very restrictive habitats for plants due to phytotoxicity, resulting in severe selection pressures. Species comprising heavy-metal plant communities are genetically altered ecotypes with specific tolerances to, e.g., cadmium, copper, lead, nickel, zinc and arsenic, adapted through micro-evolutionary processes. Evolution of metal tolerance takes place at each specific site (Ernst 2006). A high degree of metal tolerance depends on the bioavailable fraction of the metal(loids) in the soil and the type of mineralization. At extremely high soil metal concentrations, especially on polymetallic soils, even metal-tolerant genotypes are not able to evolve extreme tolerances to several heavy metals simultaneously. Adapted genotypes are the result of the Darwinian natural selection of metal-tolerant individuals selected from surrounding non-metalliferous populations (Antonovics *et al.* 1971; Baker 1987; Ernst 2006). Such selection can lead ultimately to speciation and the evolution of endemic taxa. Heavy-metal tolerance was first reported by Prat (1934) in *Silene dioica* and demonstrated experimentally in grasses by Bradshaw and co-workers in *Agrostis* spp. and by Wilkins in *Festuca ovina* in the late 1950s and 1960s (see Antonovics *et al.* 1971) and from the early 1950s onwards in the herb *Silene vulgaris* by Baumeister and co-workers (see Ernst 1974). Metal-tolerant plants avoid intoxication by an excess of heavy metals by means of special cellular mechanisms, as long as the soil metal levels do not exceed the levels of metal tolerance (Ernst 1974; Ernst *et al.* 2004). They can thus thrive on soils that are too toxic for non-adapted species and ecotypes. These unique plants with an ability to tolerate metal toxicities and survive and reproduce on metalliferous soils are called *metallophytes*.

Ecology of Industrial Pollution, eds. Lesley C. Batty and Kevin B. Hallberg. Published by Cambridge University Press. © British Ecological Society 2010.

(a)

(b)

Figure 2.1. Metallophyte vegetation on ancient lead-mining sites in the UK. (a) Sparse cover of *Agrostis capilliaris* and *Silene uniflora* on acidic wastes at Goginan lead mine, central Wales; (b) Continuous metallophyte turf colonising superficial mine workings at Gang mines, near Matlock, Peak District. The calcareous substrate here and mosaic of metal contamination levels produce a rich assemblage of metallophytes including *Minuartia verna* in the most metal-contaminated areas. Photos: A. J. M. Baker. See colour plate section.

Heavy-metal sites and their vegetation in Europe
Evolution and distribution of metallophytes

After the last Quaternary Ice Age, forest developed on nearly all soils in Europe, except on those with extreme climatic or edaphic conditions. In the latter group are soils with elevated concentrations of heavy metals, too toxic for trees. In such situations, shadow-sensitive xerophytes were able to survive when they had the genetic advantages in metal tolerance (Ernst *et al.* 1992). Heavy-metal-tolerant vegetation was originally restricted to natural outcrops of metal ores, scattered as a relic of the Late Glacial epoch over Europe. Most of these habitats were destroyed or modified by mining activities from the Bronze Age onwards. However, metal mining has considerably enlarged the potential habitat range by creating further areas of metal-contaminated soils (Ernst 1990; Ernst *et al.* 2004). In Europe, sparsely distributed sites with metal-enriched soils form residual sanctuaries for metallophyte communities. Most sites are disconnected spatially and are of very limited extent. The UK has many sites in Wales (Davies & Roberts 1978), the Peak District (Barnatt & Penny 2004) and the North Pennines, and some isolated sites in Cornwall and in the Mendips (Ernst 1974; JNCC 2002). The central part of Germany is well-known for its heavy-metal vegetation (Schubert 1953, 1954; Ernst 1964, 1974; Becker *et al.* 2007). Alluvial heavy-metal vegetation occurs along the rivers Innerste and Oker in the Harz Mountains. In the Mansfeld area, several hundreds of large Cu-Pb-Zn-mine spoil heaps are scattered with metal-lophyte communities (Schubert 1953; Ernst & Nelissen 2000). In the European Alps in Austria, Slovenia and Italy, in the French Pyrenees, and several small sites are known in the Spanish Picos de Europa. The most studied and extensive communities are those of the three-border area of Belgium, the Netherlands and Germany, the Harz Mountains area and the Pennine orefield in the UK. Metallophyte vegetation makes up an important component of the biodiversity of Europe (Whiting *et al.* 2004).

Thalius (1588) was the first to recognise a relationship between the plant *Minuartia verna* and heavy-metal-enriched soils in the Harz Mountains, Germany. Subsequently, the association of the plant with lead-mine wastes in the Pennine orefield, UK, gave rise to its local name 'leadwort'. Schulz (1912) speculated that *M. verna* is in fact a glacial relict species surviving on heavy-metal soils as an isolated population; this was later confirmed by genetic analysis (Baumbach 2005). Libbert (1930) then defined the *Armerietum halleri* as a plant association specific to metalliferous soils, and the *Violetum calaminariae* was described from the Breiniger Berg near Aachen by Schwickerath in 1931. Plant associations specific to metal-enriched soils were thus recognised.

Types of heavy-metal sites

The history of metal sites determines the species composition of the vegetation. Three types of heavy-metal vegetation can be distinguished on syntaxonomy and on their occurrence: primary, secondary and tertiary.

Primary sites
Primary sites are those with metallophytes where elevated concentrations of metals are due to natural mineralisation or ore outcropping, and not that which is anthropogenically influenced. Primary sites in Europe are therefore extremely rare today and mostly found as very patchy small sites in Central Europe, in the Pyrenees and in the Alps (Ernst 1974). Virgin sites like those in tropical woodlands and rainforests (Duvigneaud 1958; Brooks *et al.* 1985) are virtually non-existent, although many of the African sites are also threatened by mining activities (Leteinturier *et al.* 1999). Besides a high concentration of metals like zinc, lead, cadmium or copper in soil, heavy-metal vegetation types are characterised by a low nutrient availability. Hence, these plant communities are of very low productivity.

Secondary sites
Almost all primary metal-enriched sites in Europe have been anthropologically influenced by mining activities. These secondary sites result from mining activities, e.g., disturbed primary sites, spoil and slag heaps, ore processing and concentration (beneficiation) areas. The distinction between primary and secondary is often difficult to elaborate especially with ancient sites. Early mining has diminished most primary occurrences of metallophytes. From the Bronze Age to the late Middle Ages mining had a relatively low impact on the local environment. Metallophytes occurred locally on primary sites, and superficial mining created secondary habitats. Both habitat types were ecologically very similar. At that time mining was restricted to areas with metals outcropping. After the Middle Ages, much larger secondary habitats were created, often far away from areas with primary habitats, by deep underground mining or by metal refining on site. Exceptionally high concentrations of metals in soils at primary habitats result from weathering of natural mineralisation on well-developed soils. Modern secondary habitats, however, have a totally different substratum; mining has created soils with altered metal composition, depleted phosphorus and organic matter concentrations and low water retention capacity. Besides evolving metal tolerance, plants growing on these wastes were co-selected for tolerance to P-deficiency, resistance to drought and an ability to grow on loose substrates (Ernst 2000). This has affected the edaphic conditions and is a major cause of differences between primary and early secondary habitats.

Tertiary sites
Tertiary metal vegetation types can be subdivided into those communities whose genesis is a result of atmospheric deposition in the vicinity of metal smelters or alluvial deposition of metal-enriched substrates by sedimentation

in river floodplains and on raised riverbanks. Tertiary atmospheric habitats originate by an input of a surplus of metals in a non-metal-enriched environment by industrial emissions (Baumbach *et al.* 2007) often far way from primary sites supporting metallophyte populations. They are often strongly influenced by acidification (by co-emission of sulphur oxides) whose effects are stronger than those of metals in soil. Species occurring at such sites have been selected from the local non-metal-enriched environment. These sites are frequently species-poor, e.g., monocultures of those grass species which have the ability to rapidly evolve metal tolerances such as: *Agrostis stolonifera* at the copper refinery at Prescot, England (Wu *et al.* 1975); *A. capillaris* at the Cd/Zn smelter at Budel, the Netherlands (Dueck *et al.* 1984); and *Agropyron repens* at the copper smelter at Legnica, Poland (Brej 1998). Sometimes metallophytes have arrived at smelter sites with the ores: an example is the moss *Scopelophila cataractae* in Wales and in the Netherlands (Corley & Perry 1985; Sotiaux *et al.* 1987). An unintentional introduction of *Armeria maritima* subsp. *halleri* into the Littfeld area (Germany) may have been caused by mine workers when moving from the Harz area to new mining sites (Ernst 1974). Such 'transport endemism' (Antonovics *et al.* 1971) has probably been a major reason for the extended local distribution of metallophytes, such as *Thlaspi caerulescens* and *Minuartia verna* in the Pennine orefield, UK. Frequent visits by botanists may be the reason for the import of *T. caerulescens* to the Overpelt Zn/Cd smelter site in Belgium and to its extended distribution in the Peak District, UK. Revegetation of tailings with poplar trees in the Auby smelter area in France was not successful; therefore, in the 1920s and in the 1950s *Arabidopsis halleri* and *Armeria maritima* subsp. *halleri* were introduced from Central European calaminarian grassland (Dahmani-Müller *et al.* 2000), and still show a good performance on the metal-contaminated soils around the Auby smelter (Bert *et al.* 2000).

Tertiary alluvial habitats are more of a natural kind and are generally species-rich, because they originate as a result of metal loadings to well-developed soils in riverine systems, often close to primary and early secondary sites (Van der Ent 2007). Downstream of mining activities, riverbanks have been flooded with metal-enriched materials and seeds of metallophytes since the Middle Ages in the Tyne valley, England (Macklin & Smith 1990), in the Innerste and Oker valley in Germany (Libbert 1930; Ernst 1974; Ernst *et al.* 2004) and in the Geul valley in the Netherlands (Kurris & Pagnier 1925). Due to leaching of heavy metals from the surface soils, the survival of this alluvial heavy-metal vegetation type depends on irregular metal replenishment by incidental riverbank flooding, such as in the Tyne valley in 1986 (Rodwell *et al.* 2007), and in the Innerste and Oker Valley in 1969 (Ernst 1974) and 2007 (Klein & Niemann 2007). These heavy-metal-enriched sediments not only affect agricultural crops in other parts of the riverbank lands (Von Hodenberg & Finck 1975), but also transfer propagules from metallophytes into the adjacent agricultural fields (Ernst 1974). In the Harz area, large dams have been constructed to avoid regular riverbank flooding to the detriment of maintaining the alluvial heavy-metal vegetation (Ernst *et al.* 2004). The alluvial tertiary sites of the Geul valley can be species-rich, but are extremely prone to eutrophication (Van der Ent 2007, 2008); here they originate from metal-ore dressing facilities.

Classification of European metallophyte vegetation

The heavy-metal content of soil is one of the most important edaphic factors determining vegetation composition. Heavy-metal toxicity of the soil, as well as low nutrient status, poorly developed soil structure and often water-restricted conditions maintain open vegetation, retarding succession. Many sites also harbour important populations of rare bryophytes, lichens and insects in addition to metallophytes. Being immobile, plants can survive only by adapting their physiological processes, and because metal tolerance is so specific, eco-types of plants are restricted to individual sites, so-called 'local endemism'.

Heavy-metal plant communities of Europe are grouped within the vegetation order of *Violetalia calaminariae*. Ernst (1974, 1976) allocated alpine heavy-metal vegetation to the vegetation alliance *Galio anisophylli-Minuartion vernae* with *Galium anisophyllum, Poa alpina, Euphrasia salisburgensis* and *Dianthus sylvestris* and in the Italian and Austrian Alps the hyperaccumulator *Thlaspi rotundifolium* subsp. *cepaeifolium* and with the endemic *Viola dubyana*. In Western-central and Western Europe heavy-metal vegetation belongs to the alliance *Thlaspion calaminariae* with *Arabidopsis* (*Cardaminopsis*) *halleri* in addition to *T. caerulescens* (see also Brown 2001), and in Central Europe to the alliance *Armerion halleri* (see also Dierschke & Becker 2008). The heavy-metal vegetation types in the Eastern Alps of Austria, Italy and Slovenia are included within the *Thlaspion rotundifolii* (Punz & Mucina 1997), although in Slovenia *T. rotundifolium* is substituted by *T. praecox* (Vogel-Mikuš *et al.* 2007). In Scandinavia, *Lychnis alpina* is a marker species for metallophyte vegetation (Ernst 1974, 1990; Brooks & Crooks 1979; Nordal *et al.* 1999). The *Violetum calaminariae* can be subdivided geographically in eastern and western areas with the blue flowering zinc violet (*Viola guestphalica*) in the *Violetum calaminariae westfalicum* at Blankenrode (Germany) its only site in the world, and the yellow flowering *Viola lutea* subsp. *calaminaria* in the *Violetum calaminariae rhenanicum*. In the British Isles, Rodwell *et al.* (2007) allocate the metallophyte vegetation as *Festuca-Minuartia* community to the calaminarian grassland of the *Violetalia calaminariae*.

Classification of metallophytes

The following classification of metallophytes is adapted from Lambinon and Auquier (1963):

(a)   (b)

Figure 2.2. (a) *Viola calaminaria, Festuca ovina* subsp. *guestphalica* and *Thlaspi caerulescens* at the ancient mining site of Schmalgraf, Belgium. (b) *Armeria maritima s.l., Viola calaminaria* and *Thlaspi caerulescens* at what was one of the richest metallophyte habitats of northwestern Europe at Rabotrath, Belgium. Most of the metallophyte communities have disappeared in the last 4 years since the land was taken into agricultural production. Photos: A. Van der Ent. See colour plate section.

1. Metallophytes
    a. Obligate metallophytes
    b. Facultative metallophytes

2. Associate species
    a. Metal-tolerant species
    b. Non-metal-tolerant species

(1a) Obligate metallophytes: species with an exceptional tolerance to heavy metals in soils as well as a dependence upon the occurrence of these metals in soil. Some are also hyper-metal-accumulators ('hyperaccumulators'). They are not found outside this narrow ecological amplitude within the same phytogeographical area. These species are local endemics with sometimes a large geographical distribution. Examples are: *Alyssum pintodasilvae* (Dudley 1986), *Viola guestphalica* and *V. lutea* subsp. *calaminaria* (Hildebrandt *et al.* 2006; Bizoux & Mahy 2007).

(1b) Facultative metallophytes: genotypes or ecotypes/subspecies of common species with a specific tolerance to metals. They also occur in distinct non-metal-enriched phytogeographical areas. The highly specialised ecotype, sub-species or genotype is dependent on the occurrence of specific metals in the soil. Examples are: *Armeria maritima s.l.* (Baumbach & Hellwig 2007; Baumbach & Schubert 2008), *Minuartia verna s.l., Silene vulgaris* (Ernst 1974) and *Thlaspi caerulescens s.l.* (Koch *et al.* 1998).

(2a) Associated metal-tolerant species: matrix species that are associated with the related plant association with a large ecological amplitude. They are either called 'pseudo-metallophytes' or 'accompanying species' of the true metallo-phyte vegetation. These species are moderately tolerant of heavy metals in soil, but not dependent on their presence. Examples of such species which are both common and have a wide geographic distribution are: *Achillea millefolium, Campanula rotundifolia, Euphrasia* spp., *Plantago lanceolata, Polygala vulgaris, Ranunculus acris, Rumex acetosella, Thymus pulegioides, Agrostis capillaris, Holcus lanatus* and *Phragmites australis*.

(2b) Associated non-metal-tolerant species from related associations, with little or no metal tolerance, the so-called 'indifferent' or 'accidental' species: these are usually weedy species, often annuals, showing neither vigour nor persistence on metalliferous soils.

Metallophytes can occur as a mosaic of patches in other vegetation classes: especially in nutrient-poor grasslands. The once very extensive alluvial tertiary metal vegetation in the Geul valley of the Netherlands and Belgium, for example, is a mixture with the association of *Festucetum-Thymetum serpilli*, char-acteristic for sandy soils, within the class *Koelerio-Corynephoretea* (Weeda *et al.* 2002), whereas the calaminarian grassland grows on clay soil on the riverbank (cf. Ernst 1978).

Ecophysiology of metallophytes

Up to now, no investigations have detected any specific metabolites in metal-tolerant ecotypes. Metal tolerances are due to differential gene activities which are up- or down-regulating enzymes. In the case of metal uptake into the roots, there is a down-regulation of the high-affinity phosphate transporter in arsenic-tolerant plants (Macnair & Cumbes 1987) or the elevated expression of Zn transporter genes (Assunção *et al.* 2001). Once the metal is in the cell, metal-tolerant plants have modified the activity or the metal affinity of enzymes in such a way that a surplus of heavy-metal ions is rapidly removed from the plant cell metabolism to prevent physiological damage. These processes are metal-specific (Ernst *et al.* 1992, 2008; Clemens 2001). Examples are the over-expression of the metallothionein gene *MT2b* in Cu-tolerant ecotypes of *Silene vulgaris* and *S. paradoxa* (Van Hoof *et al.* 2001; Mengoni *et al.* 2003), *cis*-regulatory changes and triplication of the heavy-metal ATPase gene *HMA4* in the Zn-hyperaccumulator

Arabidopsis halleri (Hanikenne *et al.* 2008), the over-production of histidine in Ni-tolerant *Alyssum* spp. (Krämer *et al.* 1996) and the enhanced phytochelatin synthesis in As-tolerant *Holcus lanatus* (Bleeker *et al.* 2006). A high rate of Cd and Zn translocation from roots to shoots is essential for metal hyperaccumulation and differs between ecotypes of *Thlaspi caerulescens* (Xing *et al.* 2008). Within the leaves, metals have to be allocated to different cell types, showing a preference for epidermal cells (Chardonnens *et al.* 1999). Finally, there is a restricted metal transport into the seeds (Ernst 1974), so that the young seedling is not already loaded with metals. All these different aspects of metal metabolism can explain that an exposure of metal-tolerant plants to metals results in the modification of hundreds of enzymes, as evidenced by transcriptomes and proteomes (Tuomainen *et al.* 2006; Weber *et al.* 2006; Hammond *et al.* 2006). These multiple reaction patterns indicate that there is still a long way ahead for understanding all aspects of metal tolerance mechanisms (Clemens *et al.* 2002).

As established by Mendelian genetics, the number of genes necessary for tolerance to cadmium, copper and zinc per se are two for each element, with many modifiers determining the degree of metal tolerance (Bröker 1963; Macnair *et al.* 1993; Schat *et al.* 1996; Bert *et al.* 2003). In addition to any prevailing metal toxicities, metallophytes have also to adapt to other extreme chemical and physical soil factors (Baker 1987), such as dry soils, by structurally enhanced proline levels (Schat *et al.* 1997), differences in calcium status (Zhao *et al.* 2002), iron availability (Lombi *et al.* 2002) and sulphur supply (needed to synthesise adequate amounts of metal-binding compounds) (Ernst *et al.* 2008). The low availability of the major nutrients nitrogen and phosphorus (Ernst 1974), characteristic of open oligotrophic environments, requires metal-tolerant plants to evolve a high degree of major nutrient efficiency, especially on secondary and most tertiary metal-enriched sites. Heavy-metal-resistant ecotypes do not occur on non-polluted soils. As most species in ecosystems with moderately vegetated soils, metallophytes are sensitive to shade (Schubert 1953; Kakes 1980). The most shade-sensitive species is *Minuartia verna,* already disappearing during vegetation succession on metalliferous soils (Ernst 1964, 1974, 1976). Examples of populations of *Thlaspi caerulescens* which have managed to maintain stable populations on small metal-enriched, shadowed patches in woodlands are at Aberllyn zinc mine (N. Wales) and at the Silberberg near Osnabrück (Germany). Mechanisms of adaptation are energy-expensive, and plants that can tolerate high concentrations of heavy metals are thus weak competitors (Wilson 1988; Ernst *et al.* 1992).

Most species of heavy-metal plant communities have a symbiosis with arbuscular mycorrhizal (AM) fungi, which by binding metals in the fungal cells prevents the host from damage (Griffioen *et al.* 1994; Pawlowska *et al.* 1996; Hildebrandt *et al.* 1999; Tonin *et al.* 2001; Turnau & Mesjasz-Przybylowicz 2003; Whitfield *et al.* 2004). Colonisation by AM fungi is almost absent in *M. verna,*

Silene paradoxa, S. vulgaris (Caryophyllaceae), *Alyssum* spp., *Arabidopsis halleri, Biscutella laevigata, Cochlearia pyrenaica* and all *Thlaspi* spp. (Brassicaceae) on heavy-metal-enriched soils (Regvar *et al.* 2003). Metalliferous soils are extremely restricted habitats, posing a strong Darwinian challenge to candidates for survival. This combination of intense selection with restricted location promotes microevolution and speciation processes on sites (Antonovics *et al.* 1971). Metallophytes are typically endemic to their native metalliferous sites and, as a result, have a very restricted geographical distribution (Baker & Brooks 1989).

Conservation and management of metallophyte communities in Europe

Threats

A species can be considered rare when it meets one or more of the following three criteria: restricted geographical distribution; a habitat with restricted ecological conditions; or small population size (Olivieri & Vitalis 2001), though rarity in itself does not constitute a threat of extinction. Soils with elevated metal concentrations, however, are extremely fragmented and dispersed habitats in Europe, often forming small geographically isolated 'islands' in areas of background vegetation with non-elevated metal concentrations (Baker & Proctor 1990). Because of their restricted geographical distribution and very limited ecological amplitude, metallophytes are prone to extinction due to habitat destruction. This results in genetic drift, demographic stochasticity and inbreeding (Bizoux *et al.* 2004). Rare endemic metallophyte species or ecotypes are, therefore, priority targets in biodiversity conservation programmes.

Heavy-metal vegetation is a fragile community type and frequently faces a severe threat of extinction leading to absolute rather than local extinction of metallophytes. The main threats to metallophyte vegetation include: (1) agricultural reclamation, application of herbicides, and application of fertilisers and lime, causing eutrophication in the nutrient-poor metallophyte communities; (2) vegetation succession due to lack of active ecological management; (3) soil remediation enforced by regulatory agencies and (4) site destruction due to mining, gravel extraction, landscape development or tree-planting.

Decline

The designation of metallophyte habitats as 'wastelands' or 'derelict land' in need of rehabilitation has been largely responsible for the disappearance of most metallophyte communities in the last decades. Metallophyte habitats are variously regarded as ecologically degraded sites, derelict brownfields, environmental problems, wastelands and pollution threats. This has resulted in an unambiguous European-wide effort for site rehabilitation and remediation. Sites were either made suitable for agriculture, levelled, used as waste dumps or for gravel and aggregate production, or built upon with industrial developments.

Remediation and land reclamation of metalliferous mining sites is often in direct conflict with conservation efforts (Johnson 1978). In Natura 2000, calaminarian grasslands are considered as Special Areas of Conservation under the Code 6130.

Regulatory drivers seem strongly biased to classify sites as either 'clean' or 'polluted'. Instead of considering metallophyte habitats 'polluted', which implies negative value and determines ontological consequences (namely rehabilitation to clean background conditions), sites could be considered as 'metal-enriched'. Metallophyte habitats present metal-enriched islands in a sea of background concentrations of metals. Landscape heterogeneity with environmental gradients, even in what is considered a pollution scenario, drives biological diversity. The intrinsic quality of metal-enriched sites enables the development of endemic metallophyte communities. Many sites have been destroyed on the assumption that chemically and physically hostile environments are biologically insignificant (Johnson 1978).

Habitats of metallophytes are in conflict with common existing perceptions of naturalness. The influence of humans in pristine undisturbed habitats is generally considered negative in ecocentric nature visions (Keulartz 2005). Restoration ecology is tailored to deal with these alterations (i.e., mining demands) of the environment to the original natural situation. Anthropocentric nature visions, on the other hand, consider metallophyte habitats as industrial wasteland. Metaphors like 'ecosystem health' subscribe rehabilitation to chemically, physically and ecologically degraded systems. It is generally anticipated in nature policy that strongly modified environments do not possess significant natural and biodiversity values (Lenders *et al.* 1997).

In former decades and centuries metallophyte habitats are considered at best valueless (and ignored), causing environmental problems and were seen as an industrial blemish on the landscape. Many sites have been efficiently eliminated from the landscape in the last decades (Smith 1979). Due to changed awareness, they are currently protected in Europe by the Habitat Directive. Most mined areas have a long history of mining and the evidence is the remains of the former mining industry that shaped the landscape. These relics (including metallophyte communities) are part of the heritage value of a specific area. Eliminating these elements from the landscape, even if they are considered scars, cleans a landscape of its past. Today most landscapes are now heavily human influenced, and the discrepancy in policy between natural and anthropogenic genesis of a site is virtually non-existent in the ecology of metallophyte communities. Moreover, history can add to natural values. Besides land reclamation and remediation, metallophyte communities face the same fate as most other nutrient-poor communities such as chalk grasslands in the last century which depended on extensive traditional agriculture. European metallophyte communities have experienced a large-scale decline due to modern intensive agriculture and its application of fertilisers and lime. Some specific examples can be cited.

Currently, the large landscape characteristic spoils heaps of the Mansfeld area in Germany face complete destruction of the metallophyte habitat. Tailings continue to be removed for use as road construction materials. With the increasing prices for copper, these historic spoil heaps are scheduled for reworking. At other sites, spoil heaps dating back to the Middle Ages and today lying amidst agricultural fields (Schubert 1953; Ernst 1974) have been removed, but the soil underneath the spoil heaps, often related to copper shales, is still so highly metal-enriched that the agricultural crops (wheat, sugar beet) are very chlorotic and yield poorly. In the Harz Mountains, the materials from many tailings areas were used for highway construction, still visible by the metallophytes on the verges of the Göttingen-Kassel highway (Germany). Secondary sites in the Stolberg and Eschweiler area in Germany were used as landfills for waste, and have subsequently disappeared. Another tailings site was revegetated, but unfortunately used as a children's playground, resulting in symptoms of Cd and Pb toxicity in those children consuming leaves of *Rumex acetosa,* a plant species with a high accumulation capacity (MAGS 1975). The *locus classicus* for the *Violetum calaminariae*, the Breiniger Berg near Aachen is now partly overgrown by *Pinus sylvestris* trees.

In the UK Peak District, over 75% of all remnants of the lead mining industry (especially rakes and surface works with metallophyte communities) have disappeared, mainly due to agricultural improvement of pasture in the last two centuries (Barnatt & Penny 2004).

Plombières in Belgium was a 30 ha ancient secondary site well-known for its assemblage of rare species (Simon 1978). Only 5 ha remain today after a large-scale remediation project involving surface capping with 'clean' soil in 1996. Casino Weiher Halde at Kelmis, also in Belgium, is the site of the former Altenberg mine, worked since Roman times. The site has largely disappeared due to the building of shopping premises. After construction of houses on the site in 2007, the last remaining part is now designated as a nature reserve. Based upon a 1962 mapping and a site visit in 2006, it is estimated that of the former 5 ha heavy-metal vegetation, only *c.* 1 ha remains, a loss of around 80% (Ernst unpublished data).

The once extensive tertiary alluvial metallophyte vegetation along the Geul River of the Netherlands and Belgium covering over 18 km on both sides of the border have nearly completely disappeared due to intensive agriculture. Over 99% has disappeared since 1925 and only 0.5 ha now remains (Van de Riet *et al.* 2005). The best-developed site was destroyed by the construction of a trailer park in the late 1970s. At nearby Rabotrath well-developed tertiary metallophyte vegetation could be found up until 2005, when these meadows were also put under agricultural practice, and have since diminished substantially.

(a)

(b)

Figure 2.3. (a) Casino Weiher Halde, Kelmis, Belgium. At this site, the former Altenberg mine which operated from Roman times until its closure in 1882. Despite legal protection, the site is now dominated by shopping lots and industrial premises. On the small fragment that remains, *Silene vulgaris, Minuartia verna* subsp. *hercynica, Festuca ovina* subsp. *guestphalica, Viola calaminaria, Armeria maritima s.l.* and *Thlaspi caerulescens* occur abundantly. (b) Spoil heaps east of Hettstedt, Germany, from the mining period 1780–1815 now set in an agricultural landscape bear only a sparse vegetation cover mostly of highly specialised ecotypes of *Silene vulgaris*. In the background, the conical tailings tip is from the twentieth century; there is only a cover of ruderal vegetation at its base due to the hostile edaphic conditions. Photos: A. Van der Ent.

In Belgium, the total coverage of the association *Violetum calaminariae* encompasses only 38 ha, almost exclusively of the secondary type (Graitson *et al.* 2003). The quality of the majority of these sites has deteriorated in the last decades due to lack of management (Van der Ent 2007). Succession has resulted in many sites being overgrown by shrubs and trees. The impact of fertilizers and atmospheric deposition of nitrogen on nutrient-poor grasslands has accelerated this process.

Immediate action in pragmatic site management is imperative to protect remaining metallophyte vegetation in Europe; this must be given priority in nature conservation. It is clearly possible to regenerate secondary and tertiary sites as Raskin (2003) and Van der Ent (2008) have shown. Urgent action towards protecting metallophytes is also necessary because of the ever-increasing threat of extinction and the rapid decline in the number of sites. Today almost nothing is left of the former distribution compared to 50 years ago.

Site management

Depending on metal concentrations in the soil, metallophytes can thrive on primary sites for thousands of years, on secondary sites for perhaps hundreds of years, and on tertiary sites for less after the cessation of mining activities. Metallophytes on secondary and tertiary sites are relics of local historic land use and dependent upon site dynamics. To sustain the metallophyte communities, ecological management is necessary. Without management, these plant communities face extinction due to vegetation succession and substrate attenuation.

Most sites are small-scale, but are habitats for highly endangered plants and can therefore be classified as especially valuable for nature conservation (Pardey 2002). Conservation management of metallophytes is strongly linked to maintaining sites with high metal concentrations in the soil. To conserve and develop secondary sites, disturbance and mixing of topsoil/subsoil is necessary to sustain concentrations of metals high enough to hinder the succession of grasses and herbs. Because metallophytes are generally of very low productivity and, hence, uncompetitive, eutrophication remains a serious threat. Site management includes mowing and removal of hay, and/or removal of the top layer of soil to reduce nutrient loadings (especially P) in the system and retard vegetation succession.

Environmental legal protection within Europe

In contrast to former times, metallophytes are protected by legislation Europe-wide. This protection strengthens conservation and can contribute to the restoration of metallophyte vegetation. Under the EU Habitats Directive Annex I (Fauna-Flora-Habitat), heavy-metal vegetation is coded as 'Calaminarian grasslands of the order *Violetalia calaminariae*' under Code 6130. This also includes assemblages of metal-tolerant lower plants on mine waste, even if higher plants are absent. EUNIS Habitat Classification coded 'Heavy-metal grassland' E1.B2a, and Natura 2000 coded as 34.221 with 92/43/CEE I non-priority protection. EUNIS, the European Environment Agency Biodiversity Database, lists per country the Netherlands 1, Belgium 6, Germany 35, Italy 3, France 3, Spain 1 and the UK 23 sites, with 'calaminarian grasslands of the order *Violetalia calaminariae*' under Code 6130.

In Germany, five Federal States designate legally protected biotopes under §30 of the Federal Nature Conservation Act. In the State of Nordrhein Westfalen (NRW), which hosts important metallophytes sites, calaminarian grasslands are protected as a 'Protected Area' following §30 Bundesnaturschutzgesetz and as §62 Landschaftsgesetz NRW, Nature Reserve (NSG) and/or Protected Landscape Elements (LB). There is also the 'Naturschutz-Rahmenkonzeption Galmeifluren NRW' (Pardey *et al.* 1999) ('Concept for conservation of heavy-metal vegetation'), which is an important instrument for efforts to conserve metalliferous vegetation types. In the Netherlands, the remaining tertiary metal vegetation was protected as a nature reserve in 1954, the first site in the world where an industrially contaminated site was protected by law. In Belgium, with its extensive metallophyte sites in the Walloon region, most are cited in CORINE (Inventaire des sites d'importance majeure pour la conservation de la nature dans la Communauté européenne), in ISIWAL (Inventaire des Sites Wallons d'un très grand intérêt biologique) and in SGIB (Sites de Grand Intérêt Biologique). Most species of heavy-metal vegetation in Germany, the Netherlands and Belgium are listed on the National Red Lists and are protected by the National Species Protection Acts. In all three countries, many sites are also being included in the Natura 2000-network, and additionally in the Netherlands a 5-year national research programme and restoration plan is currently being implemented for the conservation of metallophyte vegetation (Van de Riet *et al.* 2005).

Often conflicting with the legal protection of metallophyte species are National Soil Protection Acts and Environmental Hygiene Acts, which require 'remediation measures' for heavily contaminated soils, and thus their habitat destruction. Further, NGOs have for decades been demanding that the pollution of the environment be reduced, similarly supporting metallophyte habitat destruction.

Metallophyte communities are not only ecological interesting entities, they also amount to a cultural-historical account, an archaeological record. The restoration and conservation of historic metal-mining sites can be seen as a practical measure within the spirit of the European Conference on the Conservation of Archaeological Heritage (European Convention on the Protection of the Archaeological Heritage; Valleta, 16I 1992). Most governmental bodies distinguish between 'natural' occurrences (primary sites) and man-made (secondary) sites, like mining spoil heaps; the more natural are given priority. Because of their rarity and ecological status, primary sites are important but historical interest can add value to anthropogenic sites. Discussion as to whether sites are natural or anthropogenic has centred less on ecology than on the demands of legislation, which designate heritage status based upon industrial archaeological values.

Heavy-metal vegetation in Central Africa

In Central Africa, there are a lot of sites enriched with cobalt (Duvigneaud 1959), copper (Duvigneaud & Denaeyer-De Smet 1963; Wild 1968; Brooks *et al.* 1992a) and nickel (Wild 1970) and only a few soils have high levels of lead and zinc (Ernst 1972). The greatest interest has concentrated on the flora on the cobalt and copper outcrops in Katanga (Brooks *et al.* 1985).

The copper-cobalt metallophytes of Katanga

The copper-cobalt flora of Katanga (Democratic Republic of Congo) is without doubt the richest described to date globally in terms of numbers of endemic metallophytes. Our knowledge of this flora has emerged from alternating periods of active exploration with periods of total inactivity. This progression has been, and still is, linked to efforts on several fronts but primarily from a knowledge of the existence of Cu-Co outcrops and site accessibility. Subsequent collection of plant materials (and later also soils), taxonomic works, a few systematic studies and biogeochemical investigations have followed.

Historical perspective

Five broad periods of progress may be recognised (Leteinturier 2002; Leteinturier & Malaisse 2002). Although local mining activities in Katanga commenced as early as the fourteenth century (De Plaen *et al.* 1982), the first plant collections from Cu-Co sites were made by Rogers probably around 1910, and certainly in 1914 at the Étoile du Congo mine. Other collections in this first period were made by Burt-Davy (1919), Rogers (1920), Robert (1921), Robyns (1926), Quarré (1937, 1939) and by priests from Saint François-de-Sales (1939). Only one paper (Robyns 1932) was published in this period; it provides a list of the first 45 Katangan copper metallophytes recorded. From 1940 to 1953 further collections were few and erratic, and included those by Hoffman (1946) and Duvigneaud (1948). A second period extended from 1954 to 1963 with independent plant collections from the Cu-Co outcrops by Duvigneaud and his collaborators and by Schmitz. Duvigneaud collected at least 3704 voucher specimens, and the number of Cu-Co metallophytes was raised from 72 (Duvigneaud 1958) to 218 (Duvigneaud & Denaeyer-De Smet 1963). Other collections during this period are rare, but include those of Symoens (1956–1963), Plancke (1958), Bamps (1960) and Ledocte (1960). From 1963 until 1978, few occasional collections were made (Evrard & Léonard in 1968, Lisowski 1968–1971, Bercovitz 1971, Breyne 1977 and Pauwels 1978). A third period of botanical exploration started in 1978 with great attention devoted to Cu-Co sites by Malaisse and Brooks. This period extended until 1986. Nearly 2000 voucher specimens were collected, resulting in 17 publications (see, e.g., Malaisse *et al.* 1983). Other investigations were made by Shewry *et al.* (1979), Wechuyzen, and a diverse group of botanists from the Belgian National Botanic Gardens. The fourth period stretched from 1987 to 2002; sporadic collections of plants were made and 14 publications appeared. Two surveys are to be highlighted: one in the Tenke-Fungurume area by Malaisse, Dikumbwa, Kisimba and Muzinga, and the other on copper sites in southcentral Africa, including Katanga by Leteinturier, whose PhD thesis (Leteinturier 2002) greatly advanced our knowledge of the Cu-Co metallophytes, listing 548 taxa.

Since August 2003, the fifth period, a new Congolese policy was implemented permitting 1644 mining titles in Katanga for Cu-Co sites to about 200 different mining societies. Their activities have to comply with the requirements of the applicable laws of the Democratic Republic of the Congo (DRC), notably Mining Law 007/2002 and Decree 037/2003. Mining companies are required to carry out an Environmental and Social Impact Assessment (ESIA), which in turn results in a Biological Diversity Action Plan (BDAP). Botanical surveys are part of the BDAP but are very rarely undertaken (<5% of the mining titles), and very fanciful BDAPs have been submitted and accepted by the Department of Environment of the Ministry of Mines. Examples consist of surveys giving only poor comments on woodlands but nothing regarding mineralised areas that were never visited. Moreover Latin names of plants are inaccurate and frequently misspelt. Such a situation results in complete destruction of Cu-Co flora and vegetation from mining activities conducted by most of the so-called 'mining societies'. Moreover, some sites delivered for mining have not commenced any operational activities. This has provided a unique opportunity for more than 50 000 (maybe even 100 000) men and boys (aged 6 years and above) to be involved in illegal mining. These 'miners' collect heterogenite, a cobalt oxide [CoO(OH)], which is sent to South Africa without any importation tax at the border, made possible through a Congolese-Libanese network. About 30 trucks each with 300 sacks of heterogenite move daily to the south. In some places, small villages of 300 people involved in illegal mining have been established. This artisanal activity is of great importance for the survival of these small communities, but surface injury to the steppe-savanna vegetation may reach near total destruction. Heaps of sterile rocks and shallow pits produce a lunar desert landscape devoid of all plant cover.

The current state of knowledge

The actual number of Cu-Co ore bodies in Katanga is far above that documented in 1960. If 150 sites have received some botanical investigation (from one to more than a thousand vouchers collected), at least some 150 others have never been explored; some of these sites may have already been totally destroyed. Since 2005, an important plant collecting initiative has taken place. Some 5 000 voucher specimens have been collected in less than 4 years, about 39% of the total collected since 1914. However, these collections presently need to be carefully checked for naming and classified regarding Cu-Co metallophyte biodiversity. At least 15% of previous collections are in the same situation, awaiting assignment of correct scientific names. It has been estimated that more than 700 higher plant species exist on the Katangan Cu-Co outcrops, but the real figure could in fact be much higher. How many species new to science are in these collections? To what extent are they copper-cobalt endemics? Leteinturier (2002) already highlighted the time extending between a collection and the description of a new taxon, a few years even up to 50. Several families and genera of the flora of southcentral Africa (Flore d'Afrique centrale, *Flora Zambesiaca*, Flora of Tropical East Africa) are in need of revision, and this adds a further complication. A good example is given by the genus *Basananthe* Peyr. (Passifloraceae). Revision of the material hosted in the National Botanic Garden of Belgium (BR) has allowed the description of six new species, all from Katanga, including two species restricted to Cu-Co sites (Robyns 1995; Malaisse & Bamps 2005). On the other hand, a re-evaluation of the copper indicator *Haumaniastrum katangense* (Lamiaceae) comes to the conclusion that this species is more widespread on soils with low Cu levels than on those with high ones (Choo *et al.* 1996). The Zambian copper flower *Becium homblei* (Lamiaceae) has lost specific rank; it is now considered nothing more than a Cu-tolerant ecotype of *Becium centrali-africanum* (Sebald 1988), the species mostly occurring on low-Cu soils.

Data concerning heavy-metal tolerance and accumulation in the Cu-Co flora have also to be reviewed thoroughly. Very few experimental studies under controlled conditions have been performed to date (see, e.g., Baker *et al.* 1983), and there is an urgent need to confirm the putative status of the 35 Cu-, and 36 Co-hyperaccumulators listed by Reeves and Baker (2000). Fine superficial dusts tightly adhering to dried herbarium specimens may account for the variable and possibly spurious apparent hyperaccumulation values published in early biogeochemical studies (Brooks *et al.* 1987). Several experimental studies on metal uptake and localization in the metallophytes of Katanga are currently under way.

Conservation and management of the Cu-Co metallophyte flora

Conservation of the Co-Cu metallophyte diversity in relation to current and future mining activities is a major challenge. A few mining groups have, however, recognised the issues and developed excellent scientifically based conservation strategies. The operations at Luiswishi and Tilwezembe mines are good examples (Fig. 2.4).

(a)

(b)

Figure 2.4. Luiswishi copper mine, Katanga. (a) Remnant metallophyte vegetation surrounded by active mining operations. (b) Aerial view of the mine site in 2007, showing the small conservation area. Photos: F. Malaisse.

The flora and vegetation of Luiswishi had been studied previously (Malaisse *et al.* 1999). Seventy-one taxa were listed in 1997. A new survey took place in 2001 to prospect for further metallophytes and discovered a new, unpublished taxon of *Chlorophytum* (Antheriaceae). Further surveys took place in 2006 and 2007, by which time only 10% of the site was unmined. These surveys confirmed the presence of 42 of the 71 taxa previously recorded, as well as 11 species reported for the first time. It is clear that recent mining activities have destroyed most of the metallophyte vegetation at this important mine site (Fig. 2.4a). Of the 29 species that have disappeared, three are of major concern; they were known from only two other sites (Étolie and Sokoroshe mines) where they are endangered. However, recently a small area of the site has been preserved and protected from mining (Fig. 2.4b). Unfortunately, this area does not support the new species. A similar study was made at Tilwezembe, near the Lupula river where Malaisse and co-workers have been allowed to carry out botanical surveys in two different seasons. A total diversity of 73 higher plants has been recorded there. As at Luiswishi, mining activities have progressed rapidly and only a very reduced rocky area is conserved.

Our knowledge of natural revegetation processes at Cu-Co metalliferous mine workings in Katanga has only received preliminary study (Leteinturier *et al.* 1999). A set of nine ecological conditions has been suggested, based upon both the heavy-metal content of soil and its state of hydration. Taxa have been identified for each condition recognised, with one indicator species suggested. Research on the cultivation of these taxa is urgently needed in order to recommend appropriate revegetation strategies to sustain the metallophyte flora. One very positive initiative commenced in 2004 through a PhD programme at Lubumbashi University established in cooperation with Belgian universities (Brussels and Gembloux). It is designed to produce Congolese specialists in Katangan phytogeochemistry able to take over conservation programmes in the mining companies.

Status of metallophytes in Latin America

Knowledge of tropical and sub-tropical metallophyte distribution and ecology lies far behind that for temperate taxa, especially those in Europe. Few metal-tolerant and metal hyperaccumulator plants have been reported in Latin America in comparison to other areas of the world, such as North America, Oceania, Asia, Europe and Africa (Brooks 1998). A total of approximately 172 plant species have been described in the literature for the Region as either metal-tolerant (30 species) or hyperaccumulators (142 species; Ginocchio & Baker 2004), a very low number when compared to the high plant diversity described for the Region (Cincotta *et al.* 2000). Most of these plants are nickel-tolerant and hyperaccumulator plants (89%) as most studies pertain to serpentine areas in Brazil, Cuba, Dominican Republic and Venezuela, followed by copper (5%) and arsenic (3%) (Baker & Brooks 1988; Ginocchio & Baker 2004). Investigations over ultramafic rocks in Argentina and Paraguay (Reeves & Baker 2000) have also been performed but they have not revealed further metal-tolerant plants.

Latin America is a potential area where metallophytes could be found, not only due to the high and unique plant diversity but also to the presence of a high number of ore deposits (e.g., gold, silver, copper, iron and lead) and metal-enriched areas such as those near abandoned tailings dumps and metal smelters, due to historical mining operations under no environmental regulations. Many areas in Latin America are major centres of plant diversity, not only because of species-rich tropical forests but also because of many geographical areas where a diverse and unique endemic flora exists. For example, 8 of the 25 hotspot areas defined for their high biodiversity in the world are located in Latin America (Myers *et al.* 2000). However, this diversity is still poorly evaluated, studied and protected, including metallophytes. If a high and unique plant diversity has co-existed and evolved in ore-rich environments, it is reasonable to think that metallophytes may have evolved in the Region, and thus, it is necessary to start reconnaissance work, before possible extinction from metal mining activities, notably strip-mining. Current information on metallophytes in the Region has derived from three main sources: scientific research performed by botanists and plant ecologists, geobotanical surveys performed by geologists and mine engineers and traditional knowledge from small-scale artesanal miners.

Information from scientific research

Although scientific research on metallophytes in the Region is limited, a number of plants have been described in the literature (see review: Ginocchio & Baker 2004). Recent evidence indicates that three As-hyperaccumulating plants (*Bidens cynapiifolia* (Asteraceae), *Paspalum racemosum* and *P. tuberosum* (Poaceae)) and one possible Cu-tolerant plant (*Bidens cynapiifolia*) grow near a copper mine in the northern Peruvian Andes (Bech *et al.* 1997); 11 Ni-hyperaccumulating plants in the serpentine flora of the Goiás State in Brazil (Brooks *et al.* 1990, 1992b); 106 Ni- and one Cu-hyperaccumulating plants in serpentine soils of Cuba (Brooks *et al.* 1990, 1992b). One Se-hyperaccumulating tree ('monkey nut tree', *Lecythis ollaria* (Lecythidaceae)) has been reported in Venezuela (Aronow & Kerdel-Vegas 1965); three Zn-tolerant plants and some possible Pb- and As-tolerant plants have been found in Ecuador (Bech *et al.* 2001, 2002); and some Cu-tolerant ecotypes of wide-spread plant species have been discovered in Chile near a copper smelter (*Mimulus luteus* var. *variegatus* (Scrophulariaceae)) (Ginocchio *et al.* 2002), on tailings sands (*Mullinum spinosum* (Apiaceae)) (Bech *et al.* 2002) and near copper mines (*Nolana divaricata* (Solanaceae)) (Ruelle 1995), *Cenchrus echinatus* (Poaceae) and *Erigeron berterianum* (Asteraceae) (Bech *et al.* 2002). Investigations over ultramafic rocks in Argentina and Paraguay (Reeves & Baker 2000) and heavily metal-polluted soils near a copper smelter in central Chile (Ginocchio 1997, 1999, 2000) have also been performed by local and international scientists but they have not revealed further metal-tolerant plants.

In the last decade, however, increasing research has occurred in the Region in order to identify metallophytes. For example, scientific explorations performed in two different areas of northcentral Chile have recently resulted in an important number of new descriptions of copper-tolerant plants native and endemic to the country. The first study was carried out in a large area of northcentral Chile that has a semi-arid Mediterranean climate type (the Coquimbo Region). Due to historic copper/gold mining a large number, 395 (SERNAGEOMIN 1989, 1990), of abandoned tailings storage facilities (TSF) are scattered throughout the area. Furthermore, metal-polluted soils and natural mineralised areas are also quite common inside the area. The results showed that 76 abandoned TSFs have been colonised by 106 local plants, 71% being native and endemic to Chile. In laboratory tests, 33 species were found to be copper-tolerant. A second study was performed in the Yerba Loca Natural Sanctuary (YLNS), a well-known high-alpine valley in central Chile, for its historic surface water anomalies. The YLNS (33° S 60° W) is located *c.* 60 km east of the city of Santiago and west from the south edge of the Río Blanco-Los Bronces-Yerba Loca Cu-Mo deposit. A large porphyry Cu deposit ($>40 \, km^2$) with secondary formation of tourmaline and Cu-Zn-Mo sulphides thus exists at high elevations of the basin. Mineral deposits have long influenced surface water quality of streams in the area which broadly differs from dilute waters described for the high Andes in central Chile, particularly in terms of pH, sulphate content and mineral concentrations. For example, the main stream along the YLNS has acidic pH (4.1–5.3) and high sulphate ($>150 \, mg \, L^{-1}$) and metal content waters (3.6–9.1 mg Cu L^{-1} and 0.2 mg Zn L^{-1}). These marked gradients in surface water chemistry may have an important role in structuring plant communities at the YLNS, particularly in meadows, as acidic and metal-rich waters are highly toxic to most plant species, thus resulting in tertiary heavy-metal vegetation. In harsh water quality environments only a limited suite of species is adapted to survive and reproduce on high-alpine meadows at the YLNS. Indeed, *Festuca purpurascens* (Poaceae), *Gaultheria caespitosa* (Ericaceae), *Calamagrostis chrysostachya* (Poaceae) and *Empetrum rubrum* (Empetraceae) are abundant in habitats with acidic and metal-rich waters, while *Carex macloviana* (Cyperaceae), *Patosia clandestine* (Juncaceae) and *Erigeron andicola* (Asteraceae) are abundant in habitats with diluted waters. In the YLNS, a total of 30 potential metallophytes have already been identified but further laboratory testing for metal tolerance is under way.

Information from geobotanical surveys

Large-scale mining in Latin America has employed traditional methods of exploration for minerals based on geology (rock colouration), radiometrics, photogeology, thermal analysis, geochemistry and satellite imaging. Thus geobotanical methods of mineral exploration, concerned with the detection of subsurface mineralisation by an interpretation of its vegetative cover or 'indicator plants' (Brooks 1998) are rarely reported in the literature of the Region. Two of the few geobotanical surveys published in the Region are those of Viladevall *et al.* (1994) in Bolivia and Fernández-Turiel *et al.* (1994) in Argentina. Viladevall *et al.* (1994) suggested that *Baccharis incarum* (Asteraceae) and *Fabiana densa* (Solanaceae) are good shrubs or 'tholas' to be used as regional metal indicator plants in geobotanical surveys for Au, As, Sb and other metals in the altiplanic areas of Bolivia, as their leaves are indicators of the metal contents in the subsoil. In the Puna belt of Argentina, however, these species grow on many soil types (Bonaventura *et al.* 1995). Although these plants cannot be classified as hyperaccumulator plants as they only reached a maximum of $540 \, \mathrm{mg \, kg^{-1}}$ Sb in their leaves, a value below the criteria of $>1000 \, \mathrm{mg \, kg^{-1}}$ for Sb-hyperaccumulating plants, they may have metallophyte status. Fernández-Turiel *et al.* (1994) suggested that *Prosopis alba* (Mimosaceae) and *Larrea divaricata* (Caesalpiniaceae) growing near an old smelter in the Sierra Pampeanas in Argentina had two to six times more Sr, Cd, Bi, Zn, Ni, Li and Cu than the same plants growing on unpolluted soils. They reported that all the shrubs studied had the same pattern of metal accumulation in above-ground structures as metal levels increased in soils, with the exception of *P. alba* and *P. nigra* which accumulated more Zn than the other plants growing in the same soils (700 vs. $200 \, \mathrm{mg \, kg^{-1}}$ Zn in ash), a characteristic of metallophytes. Furthermore, with the exception of serpentine floras described in Cuba and Brazil, there are no reports of unusual locations where metallophytes dominate.

Information from artesanal small-scale miners

Small-scale mining activities reveal additional, albeit unconfirmed, data on metal ore indicating plants, as artesanal miners searched for metal ores using simple biogeochemical methods of exploration based on rock colouration and the associated plants. For example, when the Chilean endemic shrub *Gymnophyton robustum* (Apiaceae) grows on green-coloured rocks, miners are sure that a copper ore is present in the subsoil. Although this information has not been gathered extensively and reported formally in the literature, it may represent an important source of information to commence a search for further metallophytes, as artesanal mining is common in the Region.

Environmental threats from metal mining in Latin America

A lack of environmental regulations in most countries of the Region until recent decades has allowed metal mining to result in serious and diverse environmental problems threatening local vegetation in general and metallophytes in particular. Although the situation has changed more recently due to the establishment of environmental laws and regulations in some countries, such as Peru and Chile, others have followed a tendency to review or relax the environmental standards governing mining to encourage investment, trade liberalization, technological change, cross-border merges and acquisitions, increase influence from large corporations and investor pressure (WWF International & IUCN 1999). Despite these problems, some progress is being made at policy, regulatory and technical levels by Latin American govern-ments, mining industries and international actions to protect biodiversity such as the Convention on Biological Biodiversity. Latin America has attracted the majority of the world's investments in mining and so new laws and regulations are not robust enough to prevent present and future threats to metallophytes that may be discovered in the Region, mainly due to habitat loss.

Five mining hotspots have been identified in the world in 1999, one being located in the Guyanan and Andean regions of Latin America and another at the Pacific Rim (WWF International & IUCN 1999). Many are coincident with hotspots for biodiversity (Myers *et al.* 2000), and vegetation has already been affected by mining operations or is under heavy pressure by metal mining, particularly due to badly planned and managed mining operations. For example, it is interesting to note that many of the Chilean vascular plants, particularly perennial herbs and shrubs, especially metallophytes, have a very limited distributional range (Arroyo & Cavieres 1997; Villagrán & Hinojosa 1997), and thus they are highly vulnerable to extinction if present and future mining are not adequately regulated.

A major initiative by local scientists, governmental agencies and mining companies is needed to promote the determination of metallophytes that may exist in arid and semi-arid areas of Latin America through extensive geobotanical exploration, not only on mineralised areas but also on aban-doned tailings dumps or other metal-enriched areas. Metallophytes are key resources for the minerals industry as they can be used for rehabilitation of mined areas and massive mine wastes (i.e., phytostabilisation), such as tailings storage facilities. Their proven and potential use in mine rehabilitation can help drive conservational efforts as they normally thrive on mine sites that are worked, thus improving environmental sustainability of the mining industry beyond the limited regulatory framework (Whiting *et al.* 2004; Ginocchio *et al.* 2007).

Research initiatives

Metallophytes are of proven special scientific interest, and metallophyte research is carried out at many universities and government research instit-utes worldwide. In Europe, collaborative research programmes have been facili-tated by COST (Co-Opération Scientifique et Technologique) Action 837 ('Plant biotechnology for the decontamination of waters and sites contaminated by organic pollutants and metals'), COST Action 859 ('Phytotechnologies to promote sustainable land use and improve food safety of the EU'), by the Research and Development project PHYTAC ('Development of systems to improve phytoremediation of metal contaminated soils through improved phytoaccumulation') and by the EU Research Training Network METALHOME ('Molecular mechanisms of metal homeostasis in higher plants'), both within the EU Framework V programme. Other very active research centres are in the USA at, for example, Cornell and Purdue Universities, and USDA-ARS, and recently in China (Lou et al. 2004; Deng et al. 2007; Ke et al. 2007; Xiong et al. 2008). Almost all current research focuses on the biochemistry, physiology and genetics of mechanisms for metal adaptation, and on mycorrhizal symbioses, root-associated microbes and metal tolerance (Whiting et al. 2004), plant-animal interactions (Ernst 1987; Boyd & Martens 1994; Pollard & Baker 1997; Huitson & Macnair 2003; Noret et al. 2007), phytoremediation (Baker et al. 2000; Chaney et al. 2000; van Ginneken et al. 2007; Wieshammer et al. 2007), aspects of the bioavailability of metals in soil and ecotoxicology predominantly in the context of risk assessment and regulatory and legislative aspects of soil contamination.

Causes of the rapid decline in the vitality of metallophyte vegetation are well-known for secondary and tertiary sites, as mentioned above. Science lacks coherent insight into the exact measures for its restoration. For conservation and restoration efforts it is important to set up programmes with experimental restoration measures with scientific research focused on the interaction of micro-organisms and metallophytes. Furthermore, research into the geographic distribution, ecological amplitude and niche differentiation of metallophytes, and the impact of ecological management and habitat alteration on metallophyte vegetation, is necessary to facilitate conservation and to develop and manage sites in the future.

Action towards conserving the global metallophyte resource base is imperative, because many species are under threat of extinction from the current quest for base metals and the mining boom. The extent of this unique resource, and its potential in future phytotechnologies is unknown (Whiting et al. 2004) but clearly represents a great asset in the care of the minerals industry. This has been identified as a priority area in the Mining, Minerals and Sustainable Development (MMSD) Project of the Global Mining Initiative (IEED 2002), but positive responses from the minerals industry have to date been slow. The potential importance of mine sites for biodiversity has, however, long been recognised (Johnson 1978; Smith 1979, Whiting et al. 2004; Batty 2005; Baker & Whiting 2008), but guidelines for its conservation and management have only recently been formalized (ICMM 2006).

The European Heavy Metal Ecology Network (EHMEN) was started in 2006 to promote research collaboration and increased insight into the biodiversity, ecology and biogeochemistry of metallophyte vegetation in order to facilitate conservation efforts and restoration ecology in Europe. EHMEN has organised conferences and field visits, the first of which was held at Kelmis, Belgium, in 2006. Initiatives like this are also needed on a global scale to document and research this rich source of plant biodiversity and to document and conserve the important biotechnological metallophyte resources (Whiting *et al.* 2004; Baker & Whiting 2008). The authors of this chapter strongly support the need for further biogeochemical exploration particularly in parts of the world other than those highlighted in this review, notably much of southeast Asia, and the establishment of a global database of metallophytes.

Acknowledgements

We dedicate this chapter to the memory of Professor Tony Bradshaw, FRS, a former President of the British Ecological Society and pioneer in metallophyte research and whose vision in recognising the nature and scale of evolutionary adaptation of plants in extreme habitats has led to the discipline of restoration ecology. He sadly passed away on 21 August 2008, during the preparation of this chapter.

References

Antonovics, J., Bradshaw, A. D. and Turner, R. G. (1971) Heavy metal tolerance in plants. *Advances in Ecological Research* **7**, 1–85.

Aronow, L. and Kerdel-Vegas, F. (1965) Seleno-cystathionine, a pharmacologically active factor in the seeds of *Lecythis ollaria*. *Nature* **205**, 1185–1186.

Arroyo, M. T. K. and Cavieres, L. (1997) The Mediterranean-type climate flora of central Chile. What do we know and how we can assure its protection. *Noticiero de Biología* **5**, 48–55.

Assunção, A. G. L., Da Costa Martins, P., De Folter, S., Vooijs, R., Schat, H. and Aarts, M. G. M. (2001) Elevated expression of metal transporter genes in three accessions of the metal hyperaccumulator *Thlaspi caerulescens*. *Plant, Cell and Environment* **24**, 217–226.

Baker, A. J. M. (1987) Metal tolerance. *New Phytologist*. **106** (Suppl.), 93–111.

Baker, A. J. M. and Brooks, R. R. (1988) Botanical exploration for minerals in the humid tropics. *Journal of Biogeography*. **15**, 221–229.

Baker, A. J. M. and Brooks, R. R. (1989) Terrestrial higher plants that hyperaccumulate metallic elements – a review of their distribution, ecology and phytochemistry. *Biorecovery* **1**, 81–126.

Baker, A. J. M. and Proctor, J. (1990) The influence of cadmium, copper, lead, and zinc on the distribution and evolution of metallophytes in the British Isles. *Plant Systematics and Evolution* **173**, 91–108.

Baker, A. J. M. and Whiting, S. N. (2008) Metallophytes – a unique biodiversity and biotechnological resource in the care of the minerals industry. In: *Proceedings of the Third International Seminar on Mine Closure*, 14–17 October 2008, Johannesburg, South Africa. (eds. A. Fourie, M. Tibbett, I. Weiersbye and P. Dye), pp. 13–20. Australian Centre for Geomechanics, Nedlands, Western Australia.

Baker, A. J. M., Brooks, R. R., Pease, A. J. and Malaisse, F. (1983) Studies on copper and cobalt tolerance in three closely related taxa within the genus *Silene* L. (Caryophyllaceae) from Zaïre. *Plant and Soil* **73**, 377–385.

Baker, A. J. M., McGrath, S. P., Reeves, R. D. and Smith, J. A. C. (2000) Metal hyperaccumulator plants: a review of the ecology and physiology of a biological resource for phytoremediation of metal-polluted soils.

In: *Phytoremediation of Contaminated Soil and Water* (eds. N. Terry and G. S. Bañuelos), pp. 85–107. Lewis/CRC Press Inc, Boca Raton.

Barnatt, J. and Penny, R. (2004) *The Lead Legacy. The Prospects for the Peak District's Lead Mining Heritage.* Peak District National Park Authority; Buxton, UK.

Batty, L. C. (2005) The potential importance of mine sites for biodiversity. *Mine Water and the Environment* **24**, 101–103.

Baumbach, H. (2005) Genetic differentiation of Central European heavy metal ecotypes of *Silene vulgaris*, *Minuartia verna* and *Armeria maritima* in consideration of biogeographical, mining historical and physiological aspects (in German). *Dissertationes Botanicae* **398**, 1–128.

Baumbach, H. and Hellwig, F. H. (2007) Genetic differentiation of metallicolous and non-metallicolous *Armeria maritima* (Mill.) Willd. taxa (Plumbaginaceae) in Central Europe. *Plant Systematics and Evolution* **269**, 245–258.

Baumbach, H. and Schubert, R. (2008) New taxonomic perception of the characteristic species of the heavy metal vegetation and possible consequences for nature conservation of metal-enriched sites (in German). *Feddes Repertorium* **119**, 543–555.

Baumbach, H., Volkmann, H. K. M. and Wolkersdorfer, C. (2007) Heavy metal vegetation on smelter dust at the Weinberg near Hettstedt-Burgörner (Mansfelder Region). Results of centuries of emission and a demand for nature conservation (in German). *Hercynia N. F.* **14**, 87–109.

Bech, J., Poschenrieder, C., Llugany, M. *et al.* (1997) Arsenic and heavy metal contamination of soil and vegetation around a copper mine in Northern Peru. *The Science of the Total Environment* **203**, 83–91.

Bech, J., Poschenrieder, C., Barceló, J. and Lansac, A. (2001) Heavy metal and arsenic accumulation in selected plants species around a silver mine in Ecuador. *Proceedings of the 6th International Conference on the*

Biogeochemistry of Trace Elements (ICOBTE), Guelph, Canada. p. 393.

Bech, J., Poschenrieder, C., Barceló, J. and Lansac, A. (2002) Plants from mine spoils in the South American area as potential sources of germplasm for phytoremediation technologies. *Acta Biotechnologica* **22**, 5–11.

Becker, T., Brändel, M. and Dierschke, H. (2007) Dry grassland on heavy metal-enriched and non-metal-enriched soils of the Bottendorf hills in Thuringia (in German). *Tuexenia* **27**, 255–285.

Bert, V., Macnair, M. R., De Laguerie, F., Saumitou-Laprade, P. and Petit, D. (2000) Zinc tolerance and accumulation in metallicolous and nonmetallicolous populations of *Arabidopsis halleri* (Brassicaceae). *New Phytologist* **146**, 225–233.

Bert, V., Meerts, P., Saumitou-Laprade, P., Salis, P., Gruber, W. and Verbruggen, N. (2003) Genetic basis of Cd tolerance and hyperaccumulation in *Arabidopsis halleri*. *Plant and Soil* **249**, 9–18.

Bizoux, J. P. and Mahy, G. (2007) Within-population genetic structure and clonal diversity of a threatended endemic metallophyte, *Viola calaminaria* (Violaceae). *American Journal of Botany* **94**, 887–895.

Bizoux, J. P., Brevers, F., Meerts, P., Graitson, E. and Mahy, G. (2004) Ecology and conservation of Belgian populations of *Viola calaminaria*, a metallophyte with a restricted geographical distribution. *Belgian Journal of Botany* **137**, 91–104.

Bleeker, P. M., Hakvoort, H. W. J., Bliek, M., Souer, E. and Schat, H. (2006) Enhanced arsenate reduction by a CDC25-like tyrosine phosphatase explains increased phytochelatin accumulation in arsenate-tolerant *Holcus lanatus*. *Plant Journal* **45**, 917–929.

Bonaventura, S. M., Tecchi, R. and Vignata, D. (1995) The vegetation of the Puna Belt at laguna de Pozuelos Biosphere Reserve in northwest Argentina. *Plant Ecology* **119**, 23–31.

Boyd, R. S. and Martens, S. N. (1994) Nickel hyperaccumulated by *Thlaspi montanum* var.

montanum is acutely toxic to an insect herbivore. *Oikos* **70**, 21–25.

Brej, T. (1998) Heavy metal tolerance in *Agropyron repens* (L.) P. Beauv. populations from the Legnica copper smelter area, Lower Silesia. *Acta Societatis Botanicorum Poloniae* **67**, 325–333.

Bröker, W. (1963) Genetic-physiological investigations of zinc tolerance in *Silene inflata Sm.* (in German). *Flora* **153**, 122–156.

Brooks, R. R. (1998) Geobotany and hyperaccumulators. In: *Plants that Hyperaccumulate Heavy Metals: their Role in Phytoremediation, Microbiology, Archaeology, Mineral Exploration and Phytomining* (ed. R. R. Brooks), pp. 55–94. CAB International, Oxon, UK.

Brooks, R. R. and Crooks, R. R. (1979) Studies on the uptake of heavy metals by the Scandinavian 'kisplanten' *Lychnis alpina* and *Silene dioica*. *Plant and Soil* **54**, 491–496.

Brooks, R. R., Baker, A. J. M. and Malaisse, F. (1992a) Copper flowers. *National Geographic Research and Exploration* **8**, 338–351.

Brooks, R. R., Malaisse, F. and Empain, A. (1985) *The Heavy Metal-tolerant Flora of Southcentral Africa. A Multidisciplinary Approach.* A. A. Balkema, Rotterdam.

Brooks, R. R., Naidu, S. M., Malaisse, F. and Lee, J. (1987) The elemental content of metallophytes from the copper/cobalt deposits of Central Africa. *Bulletin de la Société Royale de Botanique de Belgique* **119**, 179–191.

Brooks, R. R., Reeves, R. D. and Baker, A. J. M. (1992b) The serpentine vegetation of the Goiás State, Brazil. In: *The Vegetation of Ultramafic (Serpentine) Soils* (eds. A. J. M. Baker, J. Proctor and R. D. Reeves), pp. 67–81. Intercept Ltd, Andover, UK.

Brooks, R. R., Reeves, R. D., Baker, A. J. M., Rizzo, J. A. and Díaz-Ferreira, H. (1990) The Brazilian serpentine plant expedition (BRASPEX), 1988. *National Geographic Research* **6**, 205–219.

Brown, G. (2001) The heavy-metal vegetation of north-western mainland Europe. *Botanische Jahrbücher für Systematik* **123**, 63–110.

Chaney, R. L., Li, Y. M., Brown, S. L. *et al.* (2000) Improving metal hyperaccumulator wild plants to develop commercial phytoextraction systems; approaches and progress. In: *Phytoremediation of Contaminated Soils and Water* (eds. N. Terry and G. S. Bañuelos), pp. 131–160. CRC Press, Boca Raton.

Chardonnens, A. N., Ten Bookum, W. M., Vellinga, S., Schat, H., Verkleij, J. A. C. and Ernst, W. H. O. (1999) Allocation patterns of zinc and cadmium in heavy metal tolerant and sensistive *Silene vulgaris*. *Journal of Plant Physiology* **155**, 778–787.

Choo, F., Paton, A. and Brooks, R. R. (1996) A re-evaluation of *Haumaniastrum* species as geobotanical indicators of copper and cobalt. *Journal of Geochemical Exploration* **56**, 37–45.

Cincotta, R. P., Wisnewski, J. and Engelman, R. (2000) Human population in the biodiversity hotspots. *Nature* **404**, 990–992.

Clemens, S. (2001) Molecular mechanisms of plant metal tolerance and homeostasis. *Planta* **212**, 475–486.

Clemens, S., Palmgren, M. G. and Krämer, U. (2002) A long way ahead: understanding and engineering plant metal accumulation. *Trends in Plants Science* **7**, 309–317.

Corley, M. F. V. and Perry, A. R. (1985) *Scopelophila cataractae* (Mitt.) Broth. in South Wales, new to Europe. *Journal of Bryology* **13**, 323–328.

Dahmani-Müller, H., van Oort, E., Gélic, B. and Balabane, M. (2000) Strategies of heavy metal uptake by three plant species growing near a smelter site. *Environmental Pollution* **109**, 231–238.

Davies, B. E. and Roberts, L. J. (1978) The distribution of heavy metal contaminated soils in north-east Clwyd, Wales. *Water, Air and Soil Pollution* **9**, 507–518.

Deng, D. M., Shu, W. S., Zhang, J. *et al.* (2007) Zinc and cadmium accumulation and tolerance in populations of *Sedum alfredii*. *Environmental Pollution* **147**, 381–386.

De Plaen, G., Malaisse, F. and Brooks, R. R. (1982) The copper flowers of Central Africa and their significance for prospecting and archaeology. *Endeavour, NS* **6**, 72–77.

Dierschke, H. and Becker, T. (2008) The heavy metal vegetation in the Harz – arrangement, ecological conditions, and syntaxonomic classification (in German). *Tuexenia* **28**, 185–227.

Dudley, T. R. (1986) A new nickelophilous species of *Alyssum* (Cruciferae) from Portugal, *Alyssum pintodasilvae. Feddes Repertorium* **97**, 139–142.

Dueck, T. A., Ernst, W. H. O., Faber, J. and Pasman, F. (1984) Heavy metal emission and genetic constitution of plant populations in the vicinity of two metal emission sources. *Angewandte Botanik* **58**, 47–59.

Duvigneaud, P. (1958) The vegetation of Katanga and its metalliferous soils (in French). *Bulletin de la Société Royale de Botanique de Belgique* **90**, 127–286.

Duvigneaud, P. (1959) Cobaltophytes in Upper Katanga (in French). *Bulletin de la Société Royale de Botanique de Belgique* **91**, 111–134.

Duvigneaud, P. and Denaeyer-De Smet, S. (1963) Copper and the vegetation of Katanga (in French). *Bulletin de la Société Royale de Botanique de Belgique* **96**, 92–231.

Ernst, W. H. O. (1964) Ecological and phytosociological investigations of heavy metal plant communities in Central Europe and the Alpine Mountains (in German). Unpublished PhD Thesis, Westfälische Wilhelms-Universität Münster, Germany.

Ernst, W. H. O (1972) Ecophysiological studies on heavy metal plants in South Central Africa. *Kirkia* **8**, 125–145.

Ernst, W. H. O. (1974) *Heavy Metal Vegetation of the World* (in German). Geobotanica Selecta, Band V. Gustav Fischer Verlag, Stuttgart.

Ernst, W. H. O (1976) *Violetea calaminariae.* In: *Prodomus of the European Plant Communities* Vol. 3 (ed. R. Tüxen), pp. 1–132. J. Cramer, Vaduz.

Ernst, W. H. O. (1978) Ecological borderline between *Violetum calaminariae* and *Gentiano-Koelerietum* (in German). *Berichte der Deutschen Botanischen Gesellschaft* **89**, 381–390.

Ernst, W. H. O. (1987) Population differentiation in grassland vegetation. In: *Disturbance in Grasslands* (eds. J. Van Andel, J. Bakker and R. W. Snaydon), pp. 213–236. Junk Publishers, Dordrecht.

Ernst, W. H. O. (1990) Mine vegetation in Europe. In: *Heavy Metal Tolerance in Plants: Evolutionary Aspects* (ed. A. J. Shaw), pp. 21–37. CRC Press, Boca Raton.

Ernst, W. H. O. (2000) Evolution of metal hyperaccumulation and the phytoremediation hype. *New Phytologist* **146**, 357–358.

Ernst, W. H. O. (2006) Evolution of metal tolerance in higher plants. *Forest, Snow Landscape Research* **80**, 251–274.

Ernst, W. H. O. and Nelissen H. J. M. (2000) Life-cycle phases of a zinc- and cadmium-resistant ecotype of *Silene vulgaris* in risk assessment of polymetallic soils. *Environmental Pollution* **107**, 329–338.

Ernst, W. H. O., Knolle, F., Kratz, S. and Schnug, E. (2004) Aspects of ecotoxicology of heavy metals in the Harz region – a guided excursion. *Landbauforschung Völkenrode* **54**, 53–71.

Ernst, W. H. O., Krauss, G. J., Verkleij, J. A. C. and Wesenberg, D. (2008) Interaction of heavy metals with the sulphur metabolism in angiosperms from an ecological point of view. *Plant, Cell and Environment* **31**, 123–143.

Ernst, W. H. O., Verkleij, J. A. C. and Schat, H. (1992) Metal tolerance in plants. *Acta Botanica Neerlandica* **41**, 229–248.

European Convention on the Protection of the Archaeological Heritage (Revised) Valetta, 16.I. 1992, Council of Europe.

EU Habitats Directive Annex I (Fauna-Flora-Habitat), 1992, European Union.

EUNIS (European Environment Agency) database website at: http://eunis.eea.europa.eu/habitats-factsheet.jsp?tab=0&idHabitat=10113.

Fernández-Turiel, J. L., Rossi, J. N., Aceñolaza, P. et al. (1994) Environmental issues of biogeochemical studies at the Famantina range, La Rioja, Argentina (in Spanish).

Proceedings 7° Congreso Geológico Chileno, Concepción, Chile. pp. 613–617.

Ginocchio, R. (1997) Applicability of time-spatial vegetation distribution models to terrestrial polluted ecosystems (in Spanish). Unpublished PhD Thesis, P. Universidad Católica de Chile, Santiago, Chile.

Ginocchio, R. (1999) Copper tolerance testing on plant species growing near a copper smelter in central Chile. *Proceedings of the 5th International Conference on the Biogeochemistry of Trace Elements (ICOBTE)*, Vienna, Austria. pp. 1156–1157.

Ginocchio, R. (2000) Effects of a copper smelter on a grassland community in the Puchuncaví Valley, Chile. *Chemosphere* **41**, 15–23.

Ginocchio, R. and Baker, A. J. M. (2004) Metallophytes in Latin America: a remarkable biological genetic resource scarcely known and studied in the region. *Revista Chilena de Historia Natural* **77**, 185–194.

Ginocchio, R., Toro, I., Schnepf, D. and Macnair, M. R. (2002) Copper tolerance in populations of *Mimulus luteus var. variegatus* exposed and non-exposed to copper pollution. *Geochemistry: Exploration, Environment, Analysis* **2**, 151–156.

Ginocchio, R., Santibáñez, C., León-Lobos, P., Brown, S. and Baker, A. J. M. (2007) Sustainable rehabilitation of copper mine tailings in Chile through phytostabilization: more than plants. *Proceedings of the 2nd International Conference Mine Closure 2007*, Santiago, Chile.

Graitson, E., Melin, E. and Goffin, M. (2003) Listing and characterisation of calaminarian sites in the Walloon region (in French). Société Publique d'Aide à la Qualité de l'Environnement G.I.R.E.A.: Université de Liège.

Griffioen, W. A. J., Ietswaart, J. H. and Ernst, W. H. O. (1994) Mycorrhizal infection of *Agrostis capillaris* on a copper-contaminated soil. *Plant and Soil* **158**, 83–89.

Hammond, J. P., Bowen, H. C., White, J. P. *et al.* (2006) A comparison of the *Thlaspi caerulescens* and *Thlaspi arvense* shoot

transcriptomes. *New Phytologist* **170**, 239–260.

Hanikenne, M., Talke, I. N., Haydon, M. J. *et al.* (2008) Evolution of metal hyperaccumulation required cis-regulatory changes and triplication of *HMA4*. *Nature* **453**, 391–395.

Hildebrandt, U., Hoef-Emden, K., Backhausen, S. *et al.* (2006) The rare, endemic zinc violets of Central Europe originate from *Viola lutea* Huds. *Plant Systematics and Evolution* **257**, 205–222.

Hildebrandt, U., Kaldorf, M. and Bothe, H. (1999) The zinc violet and its colonisation by arbuscular mycorrhizal fungi. *Journal of Plant Physiology* **154**, 709–717.

Huitson, S. B. and Macnair, M. R. (2003) Does zinc protect the zinc hyperaccumulator *Arabidopsis halleri* from herbivory by snails? *New Phytologist* **159**, 453–459.

International Council on Mining and Metals (ICMM). (2006) *Good Practice Guidance for Mining and Biodiversity*. ICMM, London. 142 p.

International Institute for Environment and Development (IEED) and World Business Council for Sustainable Development (WBCSD). (2002) *Breaking New Ground: Mining, Minerals and Sustainable Development* (Report of the MMSD Project). Earthscan, London. 441 pp.

Joint Nature Conservation Committee (JNCC). (2002) Habitat account – natural and semi-natural grassland formation. 6130 Calaminarian grasslands of the *Violetalia calaminariae*. http//www.jncc.gov.uk/ ProtectedSites/SACselection/habitat.

Johnson, M. S. (1978) Land reclamation and the botanical significance of some former mining and manufacturing sites in Britain. *Environmental Conservation* **5**, 223–228.

Kakes, P. (1980) *Genecological Investigations on Zinc Plants*. Unpublished PhD Thesis, Universiteit van Amsterdam, Amsterdam.

Ke, W., Xiong, Z. T., Chen, S. and Chen, J. (2007) Effects of copper and mineral nutrition on growth, copper accumulation and mineral element uptake in two *Rumex japonicus* populations from a copper mine and an uncontaminated field site. *Environmental and Experimental Botany* **59**, 59–67.

Keulartz, J. (2005) *Operating at the Borderline. A Pragmatic View on Nature and Environment* (in Dutch). Damon, Budel, the Netherlands.

Klein, M. and Niemann, H. (2007) After the flood comes mud removal. *Hannoversche Allgemeine Zeitung*, 24 August 2007.

Koch, M., Mummenhoff, K. and Hurka, H. (1998) Systematics and evolutionary history of heavy metal tolerant *Thlaspi caerulescens* in Western Europe – evidence from genetic studies based on isozyme analysis. *Biochemical Systematics and Ecology* **26**, 823–828.

Krämer, U., Cotter-Howells, J. D., Charnock, J. M., Baker, A. J. M. and Smith, J. A. C. (1996) Free histidine as a metal chelator in plants that accumulate nickel. *Nature* **379**, 635–638.

Kurris, F. and Pagnier, J. (1925) Botanical-chemical investigation of the zinc vegetation at Epen (in Dutch). *Natuurhistorisch Maandblad* **14**, 86–89.

Lambinon, J. and Auquier, P. (1963) Flora and vegetation of calaminarian soils in the northern Walloon region and the western Rhineland. Chorological types and ecological groups (in French). *Natura Mosana* **16**, 113–130.

Lenders, H. J. R., Leuven, R. S. E. W., Nienhuis, P. H. and Schoof, D. J. W. (1997) *Nature Management and Development* (in Dutch). Boom, Meppel.

Leteinturier, B. (2002) Evaluation of the phytocenotic potential of the Southern-central African copper outcrops with a view to phytoremediation of sites polluted by mining activity (in French). Unpublished PhD Thesis, Agricultural University Gembloux, Belgium.

Leteinturier, B. and Malaisse, F. (2002) On the tracks of botanical collectors on copper outcrops of South Central Africa (in French). *Systematics and Geography of Plants* **71**, 133–163.

Leteinturier, B., Baker, A. J. M. and Malaisse, F. (1999) Early stages of natural revegetation of metalliferous mine workings in South Central Africa: a preliminary survey. *Biotechnologie, Agronomie, Société et Environment* **3**, 28–41.

Libbert, W. (1930) The vegetation of the Fallstein region (in German). *Beihefte zu den Jahresberichten der Naturhistorischen Gesellschaft zu Hannover* **2**, 1–66.

Lombi, E., Tearall, K. L., Howarth, J. R., Zhao, F. J., Hawkesford, M. J. and McGrath, S. P. (2002) Influence of iron status on cadmium and zinc uptake by different ecotypes of the hyperaccumulator *Thlaspi caerulescens*. *Plant Physiology* **128**, 1359–1367.

Lou, L. Q., Shen, Z. G. and Li, X. D. (2004) The copper tolerance mechanisms of *Elsholtzia haichowensis*, a plant from copper-enriched soils. *Environmental and Experimental Botany* **51**, 111–120.

Macklin, M. G. and Smith, R. S. (1990) Historic riparian vegetation development and alluvial metallophyte plant communities in the Tyne Basin, North-east England. In: *Vegetation and Erosion* (ed. J. B. Thornes), pp. 239–256. John Wiley & Sons, Chichester, UK.

Macnair, M. R. and Cumbes, Q. (1987) Evidence that arsenic tolerance in *Holcus lanatus* L. is caused by an altered phosphate uptake system. *New Phytologist* **107**, 387–394.

Macnair, M. R., Smith, S. E. and Cumbes, Q. J. (1993) The heritability and distribution of variation in degree of copper tolerance in *Mimulus guttatus* at Copperopolis, California. *Heredity* **71**, 445–455.

MAGS (1975) *Environmental problems caused by heavy metals in the region of Stolberg* (in German). Ministerium für Arbeits, Gesundheit und Soziales des Landes Nordrhein-Westfalen, Düsseldorf.

Malaisse, F. and Bamps, P. (2005) *Basanthe kisimbae* (Passifloracerae), a new species in Congo-Kinshasa. *Systematics and Geography of Plants* **75**, 263–265.

Malaisse, F., Baker, A. J. M. and Ruelle, S. (1999) Diversity of plant communities and leaf heavy metal content at Luiswishi copper/cobalt mineralization, Upper Katanga,

Dem. Rep. Congo. *Biotechnologie, Agronomie, Société et Environnement* **3**, 104–114.

Malaisse, F., Colonval-Elenkov, E. and Brooks, R. R. (1983) Studies on copper and cobalt tolerance in three closely-related taxa within the genus *Silene* L. (Caryophyllaceae) from Zaïre. *Plant Systematics and Evolution* **142**, 207–221.

Mengoni, A., Gonnelli, C., Hakvoort, H. W. J. *et al.* (2003) Evolution of copper-tolerance and increased expression of a 2b-type metallothionein gene in *Silene paradoxa* L. populations. *Plant and Soil* **257**, 451–457.

Myers, N. R. A., Mittermeier, R. A., Mittermeier, C. G., da Fonseca, G. A. B. and Kent, J. (2000) Biodiversity hotspots for conservation priorities. *Nature* **403**, 853–858.

Nordal, I., Haraldson, K. B., Ergon, A. and Eriksen, A. (1999) Copper resistance and genetic diversity in *Lychnis alpina* (Caryophyllaceae) populations on mining sites. *Folia Geobotanica* **34**, 471–481.

Noret, N., Meerts, P., Vanhaelen, M., Dos Santos, A. and Escarré, J. (2007) Do metal-rich plants deter herbivores? A field test of the defence hypothesis. *Oikos* **152**, 92–100.

Olivieri, I. and Vitalis, R. (2001) Biology of extinctions (in French). *Médecines Science* **17**, 63–69.

Pardey, A. (2002) Nature conservation of heavy metal sites. A survey of the present situation in Germany, Belgium and the Netherlands (in German). *Naturschutz und Landschaftsplanung* **34**, 145–151.

Pardey, A., Kalkkuhl, R., Heibel, E. and Haese, U. (1999) Concept for the conservation of heavy metal vegetation (in German). Landesanstalt für Ökologie, Bodenordnung und Forsten/Landesamt fur Agrarordnung Nordrhein-Westfalen. *LÖBF Schriftenreihe* **16**, 1–272.

Pawlowska, T. E., Blaszkowski, J. and Rühling, A. (1996) The mycorrhizal status of plants colonizing a calamine spoil mound in southern Poland. *Mycorrhiza* **6**, 499–505.

Pollard, A. J. and Baker, A. J. M. (1997) Deterrence of herbivory by zinc hyperaccumulation

in *Thlaspi caerulescens* (Brassicaceae). *New Phytologist* **135**, 655–658.

Prat, S. (1934) The heredity of copper resistance (in German). *Berichte der Deutschen Botanischen Gesellschaft* **52**, 65–67.

Punz, W. and Mucina, L. (1997) Vegetation on anthropogenic metalliferous soils in the Eastern Alps. *Folia Geobotanica* **32**, 283–295.

Raskin, R. (2003) Can heavy metal vegetation be restored? (in German). *Mitteilungen der Landesanstalt für Ökologie, Bodenordnung und Forsten* **3**, 18–21.

Reeves, R. D. and Baker, A. J. M. (2000) Metal-accumulating plants. In: *Phytoremediation of Toxic Metals: Using Plants to Clean Up the Environment* (eds. I. Raskin and B. D. Ensley), pp. 193–229. John Wiley & Sons, New York.

Regvar, M., Vogel, K., Irgel, N. *et al.* (2003) Colonisation of pennycress (*Thlaspi* sp.) of the Brassicaceae by arbuscular mycorrhizal fungi. *Journal of Plant Physiology* **160**, 615–626.

Robyns, W. (1932) Plant growth and flora on the copper-enriched soils in Upper Katanga (in Dutch). *Natuurwetenschappelijk Tijdschrift* **14**, 101–107.

Robyns, A. (1995) Passifloraceae. *Flora of Central Africa* (Zaïre, Rwanda, Burundi) *Spermatophyta* (in French). Jardin Botanique national de Belgique, Meise, Belgium. 75 pp.

Rodwell, J. S., Morgan, V., Jefferies, R. G. and Moss, D. (2007) *The European Context of the British Lowland Grassland*. JNCC No. 349, Chapter 7. *Metallophyte Vegetation*. Joint Nature Conservation Committee, Peterborough, UK.

Ruelle, S. (1995) *A study of metal pollution at the Paposo site (II Region, Chile)* (in French). Travail de Fin d'Etudes 'Ingénieur Agronome, Faculté Universitaire des Sciences Agronomiques de Gembloux, Belgium.

Schat, H., Sharma, S. S. and Vooijs, R. (1997) Heavy metal-induced accumulation of free proline in a metal-tolerant and a nontolerant ecotype of *Silene vulgaris*. *Physiologia Plantarum* **101**, 477–482.

Schat, H., Vooijs, R. and Kuiper, E. (1996) Identical major gene loci for heavy metal tolerances that have independently evolved in different local populations and subspecies of *Silene vulgaris*. *Evolution* **50**, 1888–1895.

Schubert, R. (1953) The heavy metal plant communities in the eastern Harz foreland (in German). *Wisssenschaftliche Zeitschrift der Martin-Luther-Unversität Halle-Wittenberg, Mathematisch-Naturwissenschaftliche Reihe* **3**, 51–70.

Schubert, R. (1954) The heavy metal vegetation of the Bottendorf hills (in German). *Wisssenschaftliche Zeitschrift der Martin-Luther-Unversität Halle-Wittenberg, Mathematisch-Naturwissenschaftliche Reihe* **4**, 99–120.

Schulz, A. (1912) On the phanerogams growing on heavy metal enriched soils in Germany (in German). *Jahresberichte des Westfälischen Provincialvereins für Wissenschaft und Kunst* **40**, 209–227.

Schwickerath, M. (1931) *Violetum calaminariae* of the zinc soils in the vicinity of Aachen. A plant phytosociological study (in German). *Beiträge zur Denkmalspflege* **14**, 463–503.

Sebald, O. (1988) The genus *Becium* Lindley (Lamiaceae) in Africa and on the Arabian Peninsula (Part 1). *Stuttgarter Beiträge zur Naturkunde, Serie A (Biologie)* **419**, 1–74.

SERNAGEOMIN (1989) *Survey of tailings storage facilities in Chile, Stage A, Regions V and XIII* (in Spanish). Servicio Nacional de Geología y Minería, Santiago, Chile.

SERNAGEOMIN (1990) *Survey of tailings storage facilities in Chile, Stage B, Regions IV, V, and VII* (in Spanish). Servicio Nacional de Geología y Minería, Santiago, Chile.

Shewry, P. R., Woolhouse, H. W. and Thompson, K. (1979) Relationships of vegetation to copper and cobalt in the copper clearings of Haut-Shaba, Zaire. *Botanical Journal of the Linnean Society* **79**, 1–35.

Simon, E. (1978) Heavy metals in soils, vegetation development and heavy metal tolerance in plant populations from metalliferous areas. *New Phytologist* **81**, 175–188.

Smith, R. F. (1979) The occurrence and need for conservation of metallophytes on mine wastes in Europe. *Minerals and the Environment* **1**, 131–147.

Sotiaux, A., de Zuttere, P h., Schumacker, R., Pierrot, R. B. and Ulrich, C. (1987) *Scopelophila cataractae* (Mitt.) Broth. (Pottiaceae, Musci), new for continental Europe in France, Belgium, The Netherlands, and Germany (in French). *Cryptogamie, Bryologie et Lichénologie* **8**, 95–108.

Thalius, J. (1588) *Flora of the Harz Mountains, or an Enumeration of Indigenous Plant Species in the Mountains and Their Surroundings* (in Latin). Frankfurt a.M.

Tonin, C., Vandenkoornhuyse, P., Joner, J., Straczek, E. J. and Leyval, C. (2001) Assessment of arbuscular mycorrhizal fungi diversity in the rhizosphere of *Viola calaminaria* and the effect of these fungi on heavy metal uptake by clover. *Mycorrhiza* **10**, 161–168.

Tuomainen, M. H., Nunan, N., Lesranta, S. J. *et al.* (2006) Multivariate analysis of protein profiles of metal hyperaccumulator *Thlaspi caerulescens* accessions. *Proteomics* **6**, 3695–3706.

Turnau, K. and Mesjasz-Przybylowicz, J. (2003) Arbuscular mycorrhiza of *Berkheya codii* and other Ni hyperaccumulating members of the Asteraceae from ultramafic soils in South Africa. *Mycorrhiza* **13**, 185–190.

Van der Ent, A. (2007) Possibilities for restoration of the zinc flora in the upper Geul valley (in Dutch). *De Levende Natuur* **108**, 14–19.

Van der Ent, A. (2008) Possibilities for restoration of heavy metal vegetation in the Geul valley (in Dutch). Radboud Universiteit Nijmegen. 126 pp.

Van de Riet, B. P., Bobbink, R, Willems, J. H., Lucassen, E. C. H. E. T. and Roelofs, J. G. M. (2005) *Advice on Heavy Metal Vegetation* (in Dutch). Directie Kennis, Ministerie van LNV, The Hague.

Van Ginneken, L., Meers, E., Guisson, R. *et al.* (2007) Phytoremediation of heavy metal-contaminated soils combined with bioenergy production. *Journal of Environmental Engineering and Landscape Management* **15**, 227–236.

Van Hoof, N. A. L. M., Hassinen, V. H., Hakvoort, H. W. J. *et al.* (2001) Enhanced copper tolerance in *Silene vulgaris* (Moench) Garcke populations from copper mines is associated with increased transcript levels of a 2b-type metallothionein gene. *Plant Physiology* **126**, 1519–1526.

Viladevall, M., Santibáñez, R., Ponce, J. *et al.* (1994) Vegetation analysis of 'tholas' as an exploration method for antimony-gold mineralization in the high Andes of Bolivia (in Spanish). Actas *7° Congreso Geológico Chileno*, Volumen II. pp. 1264–1267.

Villagrán, C. and Hinojosa, L. F. (1997) The story of forest in South America II: phytogeography (in Spanish). *Revista Chilena de Historia Natural* **70**, 241–267.

Vogel-Mikuš, K., Pongrac, P., Kump, P. *et al.* (2007) Localisation and quantification of elements within seeds of Cd/Zn hyperaccumulator *Thlaspi praecox* by micro-PIXE. *Environmental Pollution* **147**, 50–59.

Von Hodenberg, A. and Finck, A. (1975) Investigations on the toxic growth damage of cereals and beet in the Harz area (in German). *Landwirtschaftliche Forschung* **28**, 322–332.

Weber, M., Harada, E., Vess, C., van Roepenack-Lahaye, E. and Clemens, S. (2006) Comparative transcriptome analysis of toxic metal responses in *Arabidopsis thaliana* and the Cd^{2+}-hypertolerant facultative metallophyte *Arabidopsis halleri*. *Plant, Cell and Environment* **29**, 950–963.

Weeda, E., Schaminée, J. H. J. and van Duuren, L. (2002) *Atlas of the Plant Communities in the Netherlands. Grassland, Fringes, and Dry Heathlands* (in Dutch). KNNV Uitgeverij, Utrecht. pp. 88–89.

Whitfield, L., Richards, A. and Rimmer, D. (2004) Effects of mycorrhizal colonisation of *Thymus polytrichus* from heavy-metal-contaminated sites in northern England. *Mycorrhiza* **14**, 47–54.

Whiting, S. N., Reeves, R. D., Richards, D. *et al.* (2004) Research priorities for the conservation of metallophyte biodiversity and their potential for restoration and site remediation. *Restoration Ecology* **12**, 106–116.

Wieshammer, G., Unterbrunner, R., Garcia, T. B. *et al.* (2007) Phytoextraction of Cd and Zn from agricultural soils by *Salix* sp. and intercropping of *Salix caprea* and *Arabidopsis halleri*. *Plant and Soil* **298**, 255–264.

Wild, H. (1968) Geochemical anomalies in Rhodesia. 1. The vegetation of copper bearing soils. *Kirkia* **7**, 1–71.

Wild, H. (1970) Geobotanical anomalies in Rhodesia. 3. The vegetation of nickel bearing soils. *Kirkia* **7** (Suppl), 1–72.

Wilson, J. B. (1988) The cost of heavy-metal tolerance: an example. *Evolution* **42**, 408–413.

Wu, L., Bradshaw, A. D. and Thurman, D. A. (1975) The rapid evolution of heavy metal tolerance in plants. III. The rapid evolution of copper tolerance in *Agrostis stolonifera*. *Heredity* **34**, 165–167.

WWF International and IUCN (1999) Metals from the forests. Mining and forest degradation. Special issue of the newsletter *Arborvitae*. January 1999, pp. 1–40.

Xing, J. P., Jiang, R. F., Ueno, D. *et al.* (2008) Variation in root-to-shoot translocation of cadmium and zinc among different accessions of the hyperaccumulators *Thlaspi caerulescens* and *Thlaspi praecox*. *New Phytologist* **178**, 315–325.

Xiong, Z. T., Wang, T., Liu, K. *et al.* (2008) Differential invertase activity and root growth between Cu-tolerant and non-tolerant populations in *Kummerowia stipulacea* under Cu stress and nutrient deficiency. *Environmental and Experimental Botany* **62**, 17–27.

Zhao, F. J., Hamon, R. E., Lombi, E., McLaughlin, M. J. and McGrath, S. P. (2002) Characteristics of cadmium uptake in two contrasting ecotypes of the hyperaccumulator *Thlaspi caerulescens*. *Journal of Experimental Botany* **53**, 535–543.

CHAPTER THREE

Lichens and industrial pollution

OLE WILLIAM PURVIS

Introduction

Lichens are composite organisms including at least one fungus (mycobiont) and an alga or cyanobacterium (photobiont) living in a mutualistic symbiosis (Hawksworth & Honegger 1994). The lichen symbiosis may involve multiple and different bionts, especially photobionts, at various stages of its life history (Hawksworth 1988; Jahns 1988). Lichenised fungi are ecologically obligate biotrophs acquiring carbon from their photobionts (Honegger 1997). Lichens colonize bark, rocks, soil and various other substrata, and occur in all terrestrial ecosystems, covering more than 6% of the Earth's land surface. They are dominant in Arctic and Antarctic tundra regions where they form the key component of ecosystem processes, as a part of global biogeochemical cycles and also the food chain. Arctic and sub-Arctic lichen heaths are readily visible from space using remote sensing techniques and the effects of 'point source' smelters in creating 'industrial barrens' and emission reductions leading to recovery of *Cladonia*-rich heaths are well-documented (Tømmervik et al. 1995, 1998, 2003).

Lichens play a major role in plant ecology and the cycling of elements, such as C, N, P, heavy metals and radionuclides (Knops et al. 1991; Nash 1996b) and are extremely tolerant of ionising radiation (Brodo 1964). Lichens contribute to soil formation and stabilization (Jones 1988). Lichenized fungi may well be ancestral to fruit-body forming ascomycetes (Eriksson 2005). There are around 13 500 known species and 18 000–20 000 estimated worldwide (Sipman & Aptroot 2001). The lichen habit is dispersed through many different ascomycete and a few basidiomycete orders, and there is much convergent evolution in thallus form (Grube & Hawksworth 2007). Lichens play a significant role in global processes, ecosystem function and the maintenance of biodiversity (Hawksworth 1988, 1991, 2006; Nash 1996a; Gorbushina 2006). Growth rates tend to be slow varying from less than 1 mm to a maximum of a few centimetres a year. Some are rapid colonisers of bare ground, including metal-contaminated

Ecology of Industrial Pollution, eds. Lesley C. Batty and Kevin B. Hallberg. Published by Cambridge University Press. © British Ecological Society 2010.

sites (Gilbert 1980a, 1990). Lichen colonisation, growth and succession are poorly understood. In extreme Antarctic conditions, cryptoendolithic lichens occur in a yeast-like, cellular form buried within rocks, only forming recognisable thalli when environmental conditions improve (Friedmann 1982). Lichens contain over 800, in many cases unique, chemical products, present up to 50% of the dry weight of the thallus (Elix 1996; Huneck & Yoshimura 1996; Huneck 1999; Stadler & Keller 2008). The main function of these compounds, which are formed by the fungal partner, is most likely as antifungals, antimicrobials and anti-herbivory. It has been speculated that they might be stress-induced (Lange 1992). Lacking a protective cuticle and roots, lichens absorb nutrients and trace elements, including metals from dry and wet atmospheric deposition (De Bruin & Hackenitz 1986; Bargagli 1998; Ceburnis & Steinnes 2000; Loppi & Pirintsos 2003).

Links between lichen abundance and human activities were made in 1797 (Brightman 1982) and 1807 (Laundon & Waterfield 2007), long before their symbiotic nature was established. Mynydd Parys Cu-Pb-Zn mines dominated copper production in the early industrial revolution (Jenkins *et al.* 2000). Mr Arthur Aikin, whilst touring North Wales, visited the copper works on Parys Mountain on 13 August 1796, over 30 years since 'the fortunate discovery of copper took place…' thus converting a piece of ground, originally of little value, into one of the most profitable estates in the kingdom (Aikin 1797). 'The nearer we approached the scene', he wrote, the 'more penetrating was the fume of sulphur'. He tells us that the copper ores contained up to 25% sulphur and 25% copper (p. 368) and mentions 'We had no difficulty in distinguishing this celebrated mountain, for it is perfectly barren from the summit to the plain below; not a single shrub, and hardly a blade of grass, being able to live in the sulphurous atmosphere' (p. 367). He indirectly refers to lichens through a quote from Erasmus Darwin's poem on the subject (the scene described refers elsewhere):

> 'No grassy mantle hides the sable hills,
> No flowery chaplet crowns the trickling rills,
> Nor tufted moss nor leathery lichen creeps
> In russet tapestry o'er the crumbling steeps'

It was not until forests and other vegetation began to die out in the vicinity of smelters in Central Europe and Britain in the late 1800s that biologists looked toward emissions from smelters that were often central to the area impacted (Bell & Treshow 2002). Sulphur dioxide was identified by P. Sorauer in 1886 as being the specific pathogen (Bell & Treshow 2002). Nowadays, Parys Mountain's volcanic-associated massive sulphide (VMS) ore deposits create special conditions favouring a specialised microbiology and metallophyte lichens (Jenkins *et al.* 2000). Peter James and the author selected eight sites on this mountain (Fig. 3.1) for designation as Sites of Special Scientific Interest (SSSI) for lichens in 1987.

Figure 3.1. Peter James examining sulphide-rich boulders (SSSI) at Parys Mountain colonised by *Acarospora sinopica*, 15 March 2003.

Forty years ago Oliver Gilbert's pioneering doctoral study on Biological Indicators of Air Pollution (involving lichens, bryophytes, fungi, terrestrial algae, phanerogams and invertebrates) carried out in Newcastle-upon-Tyne, in an area famous for its collieries and industrial heritage, was published and presented at the first (and only) European Congress on the Influence of Air Pollution on Plants and Animals held in Wageningen (Gilbert 1969). Air SO_2 concentrations were correlated with the sulphur contents of *P. saxatilis* thalli (Gilbert 1965, 1968, 1969) and in *Hypogymnia physodes* (Griffith 1966; Hawksworth 1973a). Gilbert's detailed mapping studies of lichens and bryophytes led to the first zone scales as they came under increasing pollution stress. These correlated with SO_2 levels, a product of fuel combustion and smelting (Gilbert 1969, 1970). Gilbert and other pioneers established that shrubby beard-like lichens were most sensitive to SO_2 and crustose lichens least sensitive. In 1972, 15 000 school children across the UK used a simplified version of Gilbert's scale which involved observing lichens on trees, stone and concrete. They found a link between pollution dominated by sulphur dioxide and the presence or absence of selected species (Gilbert 1974; Mabey 1974). The simple scale compared favourably with the more sophisticated scales based on correlations between epiphytic lichen diversity and winter mean average SO_2 concentrations in lowland England and Wales (Hawksworth & Rose 1970). Around the same time in London, Jack Laundon produced maps showing *Xanthoria parietina* limits in London related to sulphur dioxide levels (Laundon 1967, 1970). These were probably the first such maps ever with isopleths and dots (David Hawksworth, personal communication). Mark Seaward studied distribution patterns of *Lecanora muralis* on roofing tiles (asbestos cement tile and sheet, concrete and mortar tiles and siliceous wall capstone) in relation to distance from Leeds city centre, substrate mean daily SO_2 concentrations and rainwater pH. His research indicated 'substrate switches' indicative of environmental change and the existence of an urban ecotype (Seaward 1976). At the end of the 1960s, a quantitative recording method, the Index of Atmospheric Purity (IAP) was developed by De Sloover and LeBlanc for quantifying environmental conditions using lichens as bioindicators (Kricke & Loppi 2002). The IAP considers the number of species at each monitoring site and their sensitivity towards environmental stressors, primarily air pollution, and was applied particularly in Switzerland and Germany.

Interest in problems associated with air pollution, lichens and the First International Mycological Congress held at Exeter in 1971 provided the impetus for publication of the first book devoted to air pollution and lichens (Ferry *et al.* 1973). At that time, it was recognised that air pollutants which influence lichens include heavy metals, radionuclides, dusts, fertilisers, fungicidal sprays and selective weed killers in addition to HF and SO_2 (James 1973). The concept of excess levels of air pollutant concentrations leading to adverse effects on organisms was considered early in vascular plants (Nash 1973), and standards for sulphur dioxide air quality in Europe were later produced (Farmer 1995). Ozone was recognized in the late 1950s as the principal phytotoxic component of photochemical smog and by the 1970s considered by many to be one of the most important air pollutants in the world (Nash 1973). Widely regarded as the probable cause of lichen decline in southern California (Sigal & Nash 1983) where O_3 also caused considerable damage to agricultural crops and pine (Ashmore 2002). Recent experimental research calls the influence of O_3 on lichens into question (Riddell *et al.* 2008).

Critical levels of SO_2 for the especially sensitive cyanobacterial lichens ($10\,\mu g\,m^{-3}\,yr^{-1}$), were established (UNECE 2004). Much literature followed, totalling over 1500 papers (Nash & Wirth 1988; Nimis *et al.* 2002) and books

(Ferry *et al.* 1973; Nash & Wirth 1988; Richardson 1992; Nimis *et al.* 2002); those published up to 2000 were included in a series 'Literature on air pollution and lichens' in *The Lichenologist* from 1974.

Over the past 40 years, significant environmental change has occurred globally as well as in science, society and public policy. Changes in lichen diversity and abundance in the UK, Europe and worldwide have occurred over space and time. This has stimulated research in lichens from the cellular to community levels, elemental analysis of lichen tissues and the development of monitoring techniques. Lichen distribution (and of other organisms) depends on many ecological variables other than air pollution including climate, substrate, light, dust etc. As SO_2 levels have dramatically fallen, the Index of Atmospheric Purity (IAP) was modified by researchers in Italy to reflect not only air pollution but an estimation of the degree of alteration from 'normal conditions,' i.e., of baseline conditions (Kricke & Loppi 2002). A meeting held in Rome in 2000 sponsored by the Italian National Agency for the Environment (ANPA) led to further modifications concerning objectivity in the sampling process and enabling a phytosociological approach, as largely followed in France (Van Haluwyn & Lerond 1988; Asta *et al.* 2002).

Spatial and temporal changes in the elemental composition of lichens have also occurred in many regions of the world with certain element concentrations increasing even in moderately remote regions, e.g., Isle Royale National Park, Michigan (Bennett 1995). This suggested the onset of chronic air pollution from a number of sources. Elevated element concentrations in four lichen species in Voyageurs National Park, Minnesota, sampled up to 24 km downwind of a large source of sulphur compounds and heavy metals suggested a high probability of physiological impairment of lichens (Bennett & Wetmore 1997). However, as tissue concentrations for the period 1990–1992 were declining, this impairment would probably cease should element concentrations continue to fall. Following the phasing out of leaded gasoline, many studies have reported declining concentrations of lead, testifying to the success of measures to reduce levels of this element of environmental concern (Bargagli 1998; Garty 2001; Purvis *et al.* 2008).

Recovery in lichen diversity has been recorded in many areas following SO_2 emission reductions, including near formerly major point polluting sources (e.g., Gunn *et al.* 1995). Establishing factors responsible for these changes and how lichens respond to environmental change are nevertheless challenging questions which help to drive research today. This chapter focuses on selected 'case studies' from the UK and implications for public policy and education.

London's lichens

Herbarium samples provide temporal and spatial information on lichen (and other organisms) distributions and help to interpret environmental change.

London and other cities have provided valuable refuges for certain lichen species, including during conditions of high pollution (Laundon 1967, 1970, 1973; Gilbert 1990). New lichen species to science have been found in London and designated as 'type material' (i.e., the original specimen used to describe the species). In London, All Saint's churchyard in Fulham is the type locality of *Lepraria eburnea*, and Blackheath is the type locality for *Vezdaea leprosa* (Waterfield 2002). The latter species is considered a stress-tolerant 'ruderal' (Gilbert 1990, 1992). Stress-tolerant ruderals occur in habitats in which the character of the vegetation is determined by the coincidence of moderate intensities of stress and disturbance (Grime 1979). London and other cities also provide habitats for rare lichen species. At Kew Gardens, beautiful yellow patches of the rare, Red Data Book species *Cyphelium notarisii* were found colonising garden benches near the Palm house in 1998 and were still present in 2005 (James & Wolseley 2005).

Air pollution was reported from cities as early as 1273, when the burning of coal in London was prohibited as being 'prejudicial to health' (Laundon 1973). Laundon suggests it is likely that air pollution was already having some effect on lichens at that time. Needless to say, medieval legislation was ineffective. By the seventeenth century, ever increasing coal combustion in London led to a serious and growing problem and the first records of damage to vegetation were fascinatingly portrayed by the English diarist John Evelyn in 1661 in 'Fumifugium: or the inconvenience of the aer and smoake of London dissipated' (Bell & Treshow 2002). The first known lichen records to be published for London were those of James Petiver who recorded *Cladonia coccifera* (a red-fruited 'pixie cup' lichen) from Putney Heath before 1695. This species was noted by Laundon as 'gratifyingly being present nearly 300 years later' (Laundon 1967). *Cladonia gracilis*, collected from Hampstead Heath on 7 March 1696/7 (probably by Reverend Adam Buddle), is now in the Sloane Herbarium of the British Museum (Natural History) (Hawksworth & Seaward 1977). The Heath is the type locality of *Cladonia peziziformis*, a Biodiversity Action Plan species no longer found there. The type specimen collected by Dillenius is in the Sherard Herbarium at Oxford (Waterfield 2002). *Cladonia* species occur mainly on acidic, humus-rich substrata (Purvis *et al.* 1992) and are considered as 'acidophytes' (species which prefer acidic habitats) (Van Herk 1999; Sutton *et al.* 2008). They thrive near point pollution sources such as smelters emitting SO_2 and metal particulates (Mikhailova & Vorobeichik 1995; Williamson *et al.* 1996; Purvis *et al.* 2004; Purvis & Pawlik-Skowrońska 2008). Evidence suggests these species in Britain are generally declining in recent years although quantitative data are lacking (Coppins *et al.* 2002).

The most intensively studied woodland lichen assemblage in relation to atmospheric conditions and ecological continuity is Epping Forest, lying northeast of London (Hawksworth 1973a; Laundon 1973; Hawksworth & McManus 1992;

James & Davies 2003). Edward Forster (1765–1849) was the first lichenologist to examine the area, and he made collections and notes mainly between 1784 and 1796 (Laundon 1967) which are preserved in the Natural History Museum (BM) herbarium and library (Hawksworth 1973a). Collections made over a 300-year period show that numbers of lichen species recorded have varied tremendously over this period. Lichen species declined when SO_2 concentrations peaked, with the almost total extinction of lichens sensitive to air pollution and woodland management (Rose 1976; Rose & Coppins 2000). However, the species returning today are not all the same ones which are present in the BM herbarium. More species were found growing closer to London than in more remote regions (James & Davies 2003; Davies 2005).

In the winters of the late nineteenth and twentieth centuries when stationary high pressure systems settled over Western Europe, wind speeds fell and temperature inversions formed. Pollutant concentrations increased and fog became widespread in Britain, with London severely affected by these conditions (Brimblecombe 1987/88; NSCA 2002). The famous smog which began on 5 December 1952 led to the abandonment of La Traviata at Sadler's Wells theatre and death of cattle at Smithfield market, and within 12 hours large numbers of people showed respiratory illnesses. This led to the Clean Air Act of 1956 and Smoke Control Areas. Partly through legislation, social, economic and technological changes, concentrations of sulphur dioxide have fallen from an annual high mean of $350\,\mu g\,m^{-3}$ in the 1970s to an annual average across London in 2001 of $3\,\mu g\,m^{-3}$ with no exceedances of the EU objectives for health or vegetation expected (Bell *et al.* 2004). During peak concentrations, the city of London was so polluted that it was termed a 'lichen desert' ('lavöknen'). This term was originally coined by Sernander (Sernander 1926) where the trunks in city centres around gas-manufacturing works, railway stations (steam trains were in service) were devoid of macrolichens (Laundon 1973). Laundon recorded only nine lichens growing on trees within 16 km from Charing Cross in central London and a single species in Central London (Laundon 1967). During the high SO_2 pollution climate prevailing in the 1960s and 1970s, two-thirds of the lichen flora occurred in churchyards and cemeteries, which is attributed to the pH buffering capacity of limestones and other calcareous stones. At Mitcham churchyard in London, *Caloplaca flavovirescens* occurred on over 80% of limestone memorials erected in the eighteenth century. The absence of thalli on modern stones suggested pollution limited recolonisation, i.e., representing a 'relict flora' (Laundon 1967). Oliver Gilbert also suggested that colonies of *Parmelia saxatilis* near Blyth, Northumberland, were of a relict nature and perhaps represented a separate ecotype (Gilbert 1971).

A significant improvement was recorded in the 1980s including species that had not been seen in London for over 200 years (Rose & Hawksworth 1981; Gilbert 1986; Purvis 1987; Hawksworth & McManus 1989). Oliver Gilbert discovered *Hypogymnia physodes* with the 'pollution lichen' at Chelsea Physic Garden in 1986, at that time the innermost record of a foliose (leaf-like) lichen in central London (Gilbert 1986). The specimen is now in the Natural History Museum (BM) lichen herbarium. However, the order of species return did not follow the order of loss ('zone skipping'), reflecting availability of colonising sources and biological factors (Hawksworth & McManus 1989). Clearly, previously defined indicator species, frequency data and zone scales no longer reflected air quality changes. There were 38 species recorded in Regent's Park in central London in 2001, including the new colonist, *Flavoparmelia soredians* (James *et al.* 2002) on young oak trees, a species with an essentially Mediterranean distribution (Rose 1995). Ornamental cherry (*Prunus* spp.) trees were colonised by healthy *Lecanora conizaeoides*. A SO_2 sensitive cyanobacterial species, *Peltigera neckeri*, was found by an 11-year-old (and who identified it as a *Peltigera* species) during a British Lichen Society field visit on 7 January 2001 to Brompton Cemetery, 5.3 km from Charing Cross (the accepted centre of London). It was probably imported with chippings (Hawksworth *et al.* 2001). Subsequently, two quantitative studies (using different recording methods) were carried out by Linda Davies (on ash–*Fraxinus*) using fine ADMS dispersion modelled pollution data and René Larsen (on oak–*Quercus*) using monthly average pollution concentrations and trees georeferenced with GPS. Trees were selected in both studies away from roads where NH_3 and NO_x concentrations would be lower than those at roadsides (Cape *et al.* 2004). Davies recorded 74 lichen species (and 14 moss, 7 fungal and 3 algal species) on ash (Davies *et al.* 2007). Larsen recorded 64 lichen species (and 4 bryophyte species) on oak (Larsen *et al.* 2007). In areas of highest NO_x, species recorded belong almost exclusively to the families *Candelariaceae*, *Physciaceae* and *Teloschistaceae* (Davies *et al.* 2007). Both studies established correlations between frequency, NO_x and bark pH, which suggested that transport-related pollution and bark pH influence lichen and bryophyte diversity today. *Hypogymnia physodes*, the first macrolichen to be recorded in central London (Gilbert 1986), was virtually restricted to outer London, apart from some of the larger parks, as at Wimbledon Common and Hyde Park. It is no longer present in Chelsea Physic Garden. Spatial distribution patterns in both Larsen's and Davies' studies suggest phytotoxic effects and critical level exceedances for NO_x as shown previously for SO_2. Although there has been a dramatic decrease in SO_2 concentrations, emissions of oxides of nitrogen (NO_x) in London have changed very little over the period since Laundon's pioneering study (Laundon 1967, 1970). However, the sources and emission heights are now quite different (GLA 2002). Historically, NO_x was emitted at chimney height and was a product of domestic and industrial coal burning. Nowadays, NO_x in London is mainly emitted from vehicles at ground level where dispersion is often impeded (Davies *et al.* 2007). Climatic factors, although not yet quantified in London in relation to lichen diversity, are important. Biomonitoring, therefore, has a practical role to assess London's air quality under changing atmospheric conditions (Davies *et al.* 2007; Larsen *et al.* 2007).

The Natural History Museum wildlife garden is located at a very busy road junction in central London. Located within the garden is an automated urban roadside monitoring station, 4 m from the roadside where sulphur dioxide, nitric oxide, nitrogen dioxide, carbon monoxide and PM_{10} are measured. Lichen diversity has changed significantly with many species recorded (and some imported) when the garden was established now dying out (as in the case of the 'Pollution Lichen' and new species recolonising (Honey *et al.* 1998; Leigh & Ware 2003; Bell *et al.* 2004). Lichen transplants were introduced into the garden alongside temporary gauges in order to evaluate the impact of urban air pollution on different lichen species. Digital photography and chlorophyll fluorescence confirmed that the healthiest lichens were those transplanted adjacent to the road where peak NO, NO_2, NO_x and VOC air concentrations were recorded (DEFRA 2002; Bell *et al.* 2004). Further experiments are required under different climatic conditions and pollution gradients. Trees are now developing a flora characterised by increasing numbers of nitrogen and particle-tolerant species, including those not formerly recorded in urban areas. If, as predicted, air quality continues to improve given recent advances in 'clean vehicle technologies', we may expect an increase in lichen diversity here (Leigh & Ware 2003).

Early studies carried out under a high SO_2 pollution climate indicated that shrubby lichens were most sensitive to pollution and crustose species least. *Usnea* species were absent from an area 48 km south from the centre of London (Rose 1973). *Usnea* spp., at one time widespread and luxuriant, almost entirely disappeared from a major area of England and Wales covering at least $68\,000\,km^2$ and at least $6\,000\,km^2$ of lowland Scotland, mainly as a result of the increase in atmospheric pollution (Seaward 1998) (Fig. 3.2a). Linda Davies and the author visited Regent's Park, central London on 1 June 2007 to monitor *Usnea* sp. on *Tilia* sp. she discovered during her earlier study (Davies 2005). Whilst the return of *Usnea* to central London is encouraging, thalli are small (Fig. 3.2). Re-colonisation is progressing. The number of recorded 10×10 km grid squares it has returned to has increased by more than 55% within the past 10 years (Mark Seaward, personal communication) (Fig. 3.2b). However, a problem of recolonisation by selected genotypes remains. This is graphically illustrated by a pioneering study of 231 specimens of *Parmelia sulcata* which showed there were three rRNA genotypes, but only one was found to be recolonising London (Crespo *et al.* 1999).

The 'pollution' lichen *Lecanora conizaeoides*

Decline of *L. conizaeoides* (Fig. 3.3a) has occurred (Wirth 1999; Bates *et al.* 2001; Kirschbaum & Hanewald 2001). The species is generally recognised as being

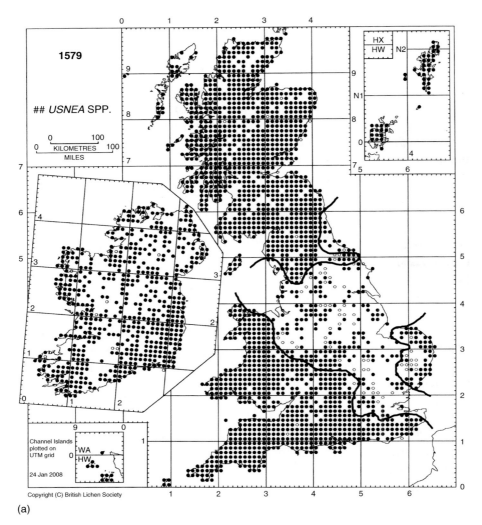

(a)

Figure 3.2. (a) Distribution of *Usnea* spp. In the British Isles (January 2008) to show its disappearance from an area of *c.* 68 000 km^2 during the period *c.* 1800–1970, and its subsequent re-establishment at numerous sites (in 233 recording units 10 km × 10 km in January 2008; Seaward pers. comm.) within that area; open dot = pre-1960 (usually 19th C) records, filled dot = 1960 onwards records. Modified from a map constructed (January 2008) from records stored on the database of the British Lichen Society's Mapping Scheme housed at the University of Bradford. (b) *Usnea* sp. growing on *Tilia* sp. Regent's Park. Linda Davies and William Purvis. 1 June 2007. Scale in mm.

(b)

Figure 3.2. (*cont.*)

the most SO$_2$ tolerant in Britain (Hawksworth & Rose 1970; Laundon 1973; Seaward & Hitch 1982). *L. conizaeoides* prefers acidic bark (Tønsberg 1992; Wirth 1999; Massara 2004). It was formerly widespread in lowland areas of Europe subjected to air pollution and reported from several eastern North American cities and near sulphur springs in Iceland (Purvis *et al.* 1992). Its occurrence on birch (*Betula*) in Iceland near sulphur springs has been called into question and is believed to reflect an unintentional introduction (Bailey 1968; Laundon 2003). Laundon (2003) considered the lichen to be exceptional for being unknown before 1860. He studied many eighteenth and early nineteenth century herbaria to locate early specimens without success. The earliest specimens he recorded came from two disjunct areas. *L. conizaeoides* was originally collected in England *c.* 1862 at Twycross, Leicestershire, by the Reverend A. Bloxam (BM) at a time when the county was probably beginning to experience the effects of background air pollution (Laundon 2003). A second collection was made in 1865 from larch (*Larix*) from Mittersill in the Salzach valley in the Austrian Alps (Laundon 2003). *L. conizaeoides* is probably native to southern Central Europe where it grows in woods with dwarf mountain pine (*Pinus mugo*), especially in boggy areas (Wirth 1985). The species is believed to have reached England by windborne propagules and was widespread by 1880 (Laundon 2003).

L. conizaeoides may have colonised the railway network first (Fig. 3.3B), especially track-side fences along the smoky, windy corridors created by steam

(a)

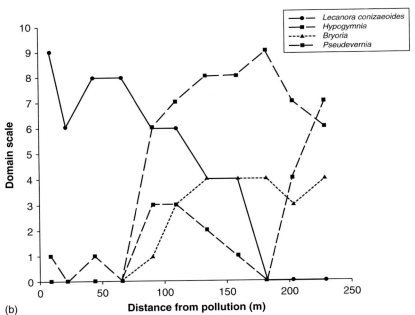

(b)

Figure 3.3. (a) Pollution lichen. *Lecanora conizaeoides* Nyl. Epping Forest. Essex. J. M. Crombie [19th century, undated]. (b) *L. conizaeoides* and other lichen species in

trains. Thus some of the first records from northeast England came from rural railway stations (Gilbert 1980b). Background air pollution then sustained its progress. Britain was the first country in the world to be industrialised, and it was the resulting air pollution here which created favourable conditions and enabled it to spread so rapidly (Laundon 1973). *L. conizaeoides* became ubiquitous in areas where winter mean atmospheric SO_2 concentrations ranged from 50 to 150 $\mu g\,m^{-3}$ (Seaward & Hitch 1982). In Scandinavia, *Lecanora conizaeoides* was recorded much later, in 1915 in Sweden (Gøteborg) and 1947 in Norway (Stavanger). The Norwegian distribution agrees with Ahti's (1965) that is essentially suboceanic in distribution. However, occurrence at Hamar indicates that it can withstand considerable cold temperatures in winter (Tønsberg 1992).

Imperial College London has a detailed record of the lichen flora on free-standing oak trees along a 70-km transect extending south-southwest from central London almost to the southeast coast of England (Bates *et al.* 1990, 2001; Bell *et al.* 2004). Over a 21-year period, declines were observed in *Lecanora conizaeoides*. A decline in cover was recorded over the period 1979–1999 at nearly all stations, apart from the innermost site (Kensington Gardens), the only station to retain appreciable SO_2 levels. When the transect was started in 1979, it was absent at Putney Heath, by the mid 1980s as SO_2 levels fell, it increased dramatically in cover. Then as SO_2 levels fell still further. the cover fell to zero by 1998.

In North America, *Lecanora conizaeoides* was first recorded at St John's Newfoundland by Teuvo Ahti in 1956 (Ahti 1965). It subsequently appeared in several North American cities and is believed to have been introduced recently into eastern Massachusetts via soredia attached to lumber from a ship (LaGreca & Stuzman 2006). A similar pattern of invasion was recorded here as in Europe. In eastern Massachusetts, *L. conizaeoides* was first recorded as colonising highly acidic (pH < 4) wood (less often bark) in rural swamps of Atlantic white cedar, subsequently spreading to other suburban sites which contain sufficiently acidic substrates. However, in this case, acidification of tree species at the suburban sites cannot be explained by high atmospheric sulphur dioxide

Figure 3.3. (*cont.*)
relation to decreasing air pollution from the railway at Aviemore, Highland, Scotland. Pollution arose from the continuous burning of coal at an engine shed built in 1896 which was in constant daily use for the shunting of steam locomotives until 1960. The abundance of each lichen species was recorded by percentage cover or presence/absence (Domain scale) using 11 80 × 20 cm quadrats on randomly selected mature vertical silver birch *Betula pendula* boles in open-canopy birch woodland along a transect running northeast downwind from the shed alongside the old bridle-road between Aviemore and Dalfaber in 1868 (Laundon 2003).

concentrations because levels are much lower than in Europe during its expansion (LaGreca & Stuzman 2006). In Germany, an association between stemflow concentrations of sulphur and the cover of *Lecanora conizaeoides* was identified (Hauck *et al.* 2001). A role for lichen substances through influencing surface hydrophobicity was also identified (Hauck *et al.* 2008). No single factor can explain the present day distribution of *Lecanora conizaeoides*. *Buellia pulverea* is associated with *Lecanora conizaeoides*, and tolerant of both sulphur dioxide and fluoride pollution. The provenance of these and several other lichens in anthropogenic habitats is equally unclear. *Lecanora vinetorum*, a very rare Central European species known from Switzerland and northern Italy, is a saxicolous lichen remarkable for having colonized trees and wood sprayed with fungicides containing copper (Laundon 2003). It grows on walls, cherry trees (*Prunus avium*) and the worked wood of vineyard frames used for the cultivation of the grape-vine *Vitis vinifera* (Poelt & Huneck 1969; Laundon 2003). *Lecanora conizaeoides* is most likely to survive on acidic substrates and where local acidification persists. It is currently growing on acid gravestones in central London and Edinburgh and occurs in rural areas in similar habitats, as well as on wood and particular tree species, especially flowering cherry. 'Substrate switches', where lichens colonise different substrates under changing atmospheric conditions (usually in response to pH changes), have been recorded since the earliest days of lichen recording (Hawksworth & Rose 1976; Seaward 1976).

Nitrogen pollution

Over a relatively short time following a reduction in SO_2 emissions, changes in atmospheric pollutant concentrations across Europe have occurred. Nitrogen emissions have also changed during the same time period, but these are made up of several compounds produced from a variety of sources whose effects may vary with the compound and its source (Wolseley *et al.* 2006). The main sources of reactive nitrogen in the atmosphere are nitrogen oxides (NO_x) and ammonia (NH_3). Ammonia is largely a product of intensive farming whilst road traffic contributes a major part of the NO_x along with other high temperature combustion processes. Whilst there has been a substantial decline in NO_x emissions since 1990, NH_3 emissions have only decreased slightly ($<10\%$) and are still high especially in agricultural areas (NEGTAP 2001; Sutton *et al.* 2001). NH_3 levels have probably increased in urban areas due to catalytic converters (Cape *et al.* 2004). While nitrogen oxides and their reaction products may be carried over long distances in the atmosphere in a number of forms, ammonia is deposited as NH_3 along very local gradients of up to 300 m (Fowler *et al.* 1998; NEGTAP 2001; Wolseley *et al.* 2006).

Recorded changes in epiphytic lichen assemblages, initially in the Netherlands and subsequently across Europe, by independent researchers confirm ammonia effects on assemblage composition (Van Herk 1999; DEFRA 2002; Frati *et al.* 2007;

Branquino 2008; Sutton *et al.* 2008). Further research in the Netherlands also demonstrated that an increase in nitrophytes was correlated with increasing bark pH. Long distance nitrogen deposition was affecting sensitive lichen communities across Europe (Van Herk 2001, 2003). Pioneering research in Pembrokeshire confirms the value of new growth of lichens on twigs as sensitive indicators of environmental change and that indicator species can be used to detect trends towards acidification or eutrophication (Wolseley & Pryor 1999; Larsen-Vilsholm *et al.* 2009). The frequency of lichens on both twigs and trunks of standard oak trees was tested across known gradients in NH_3 concentrations and in different climatic regions in 'continental' Norfolk and 'oceanic' Devon. At mean NH_3 concentrations $>2\,\mu g\,m^{-3}$, trunks continued to carry relict lichen communities due to either a legacy of previous acidification or ecological continuity. A loss of 'acidophytes' occurred prior to the establishment of 'nitrophytes', indicating the importance of establishing levels of ammonia at which sensitive communities are at risk (DEFRA 2002; Wolseley *et al.* 2006; Larsen-Vilsholm *et al.* 2009). Field measurements at the farm, landscape and national scales in the vicinity of the UK National Ammonia Monitoring Network confirm the value of distinguishing particular lichens either as 'acidophyte' and 'nitrophyte' (Sutton *et al.* 2004; Leith *et al.* 2005; Sutton *et al.* 2009). Data indicate that twig lichens are even more sensitive to ammonia than lichens on trunks, have a faster turnover and may respond more rapidly to pollution changes. As suggested by Van Herk (1999), 'acidophytes' avoid a high supply of reactive nitrogen and 'nitrophytes' prefer it (Sutton *et al.* 2008).

A new method focusing on more-easy-to-identify macrolichens on trunks and twigs has been developed (Sutton *et al.* 2009; Wolseley *et al.* 2009). An overall nitrogen index (LAN) is calculated: LAN = LA − LN, where LA = acidophyte index and LN = nitrophyte index.

A positive value indicates a site with lower NH_3 concentrations dominated by acidophyte lichen species, and a negative value indicates a site with higher NH_3 concentrations dominated by nitrophytes. LAN correlates with NH_3 concentrations across the UK, thus providing a new method to determine NH_3 in regions where monitoring stations are absent (Wolseley *et al.* 2009).

Using the data, a critical level was derived, as the NH_3 concentration at which the observed LAN value is significantly lower than the LAN value for background NH_3 concentrations (Sutton *et al.* 2008). Expressed graphically (Fig. 3.4) as the point at which the least-squares regression best-fit line shows a value of LAN which is smaller than the minimum confidence interval for LAN at the cleanest study location. Ammonia concentrations are highly variable, and regional background values far from sources are of the order of 0.03–$0.3\,\mu g\,m^{-3}$ according to the time of the year. In the Netherlands, 'background' concentrations are typically larger than $3\,\mu g\,m^{-3}$, and thus this approach would not be applicable in this and other regions with high

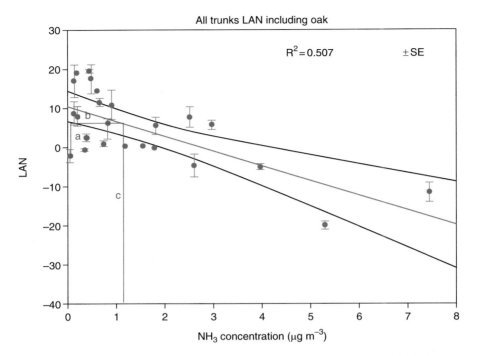

Figure 3.4. Relationship between the abundance of sensitive lichens (L_{AN} index) on tree trunks and NH_3 concentrations across the UK. The 95% confidence limits of the regression line are shown (Sutton *et al.* 2009). Figure 3.4 was originally published as Fig. 6.5. in Sutton, M. A., Wolseley, P. A., Leith, I. D., Van Dijk, N., Tang, Y. S., James, P. W., Theobald, M. R. & Whitfield, C. (2009) 'Estimation of the ammonia critical level for epiphytic lichens based on observations at farm, landscape and national scales' in Sutton, M., Reis, S. & Baker, M. H. (eds.), *Atmospheric Ammonia – Detecting Emission Changes and Environmental Impacts*, p. 80. Netherlands, Springer. © 2009 Springer Science + Business Media B. V. Reproduced with kind permission of Springer Science and Business Media.

background concentrations. Through this research, a revised long-term critical level of $1\,\mu g\,m^{-3}$ NH_3 has been proposed at an Expert Workshop held under the UNECE Convention on Long-range Transboundary Air Pollution (WG1 2006) to protect epiphytic lichen communities and bryophytes. It has now been adopted, and the new text will soon be inserted into 'the Mapping Manual', the standard reference (UNECE 2004). However, whilst it is believed the critical level would be protective over a period of several years, the long-term consequences (20–30 years) are far less certain (Sutton *et al.* 2009).

Nitric acid (HNO_3) and ozone (O_3), secondary products of photochemical reactions of nitrogen oxides (NO_x) and volatile organic compounds, are important pollutants in regions with large outputs from petrol combustion (Riddell *et al.* 2008). In the Los Angeles air basin, nitrogen dry deposition rates in forests downwind of the urban areas can reach 35–40 kg ha^{-1} year^{-1}), roughly equivalent to the amount of N used to fertilise agricultural fields. Laboratory fumigations and field transplant experiments of *Ramalina menziesii* suggest that gaseous HNO$_3$, not O$_3$, may be responsible for the impoverishment of lichen communities in these semi-arid regions, where it is a major component of air pollution during ambient summer conditions in the Los Angeles air basin (Riddell *et al.* 2008).

Lichen biodiversity in metal-enriched environments

Waste dump and metalliferous habitats contribute to the world's industrial and mineralogical heritage, e.g., the World Heritage Site, Falun Mine, Sweden (Lindeström 2002). They also provide important refuges for biodiversity, including rare 'metallophyte' lichens restricted to particular chemical environments. The preservation of biodiversity was highlighted in the Convention on Biodiversity (CBD) held in Rio de Janeiro in 1993. The Robert Brooks Memorial Workshop on Metal-tolerant Plants supported by Rio Tinto and the Royal Botanic Gardens, Kew, held in July 2001 identified geobotanical exploration/ field surveys and the establishment of collections from mineralised areas prior to mining activities as key priorities (Whiting *et al.* 2004, 2005). Metal-tolerant plants stabilise surfaces and are particularly useful for land restoration, e.g., following mine closure. Vascular plant species which 'hyperaccumulate' metals to concentrations in excess of those found in many ores are increasingly used in 'phytomining' (particularly nickel) and in remediation (Callahan *et al.* 2006). But lichens grow too slowly and are too challenging to grow under these conditions to have direct application.

There has been a long tradition of studying lichens found on metalliferous spoil heaps and naturally mineralised rocks especially in Central Europe (Hilitzer 1923; Schade 1933; Poelt 1955; Lange & Ziegler 1963; Noeske *et al.* 1970; Wirth 1972; James *et al.* 1977; Purvis 1985; Purvis & James 1985; Purvis & Halls 1996; Huneck 2006). Poelt and Ullrich (1964) introduced the term 'chalcophile' (greek chalcos = ore) to describe lichens more-or-less restricted to metalliferous rocks and slags and ores. *Acarospora sinopica*, a rust-coloured species occurs on highly mineralised weathered rocks containing iron and copper sulphide minerals (Fig. 3.1). Volkmar Wirth was the first to suggest that it was the low pH rather than the high concentrations of iron and other metals that was responsible for their development (Wirth 1972). Sulphide minerals are the principal acid-forming constituents of mine spoils which liberate dilute sulphuric acid as a result of bacterially assisted oxidative weathering and which leads to 'acid mine drainage', a major source of metal contamination (Jenkins *et al.* 2000). A role for lichen substances has also been suggested (Hauck *et al.* 2007).

Crust-like 'metallophyte' lichens are species which colonise metalliferous habitats that support few, if any, higher plants. Many grow directly on minerals, including copper and uranium secondary phosphates and arsenates from which they may derive nutrition (McLean *et al.* 1998; Purvis *et al.* 2004; Haas & Purvis 2006). Distinctive lichen assemblages occur in upland regions of the UK and Europe on rocks in environments where copper sulphides predominate with associated basic secondary copper compounds often with significant calcium carbonate ($CaCO_3$) creating a higher pH environment. A new community characterised by 'Copper Lichen' (*Lecidea inops*) was described from Cu-rich environments in Scandinavia and the UK (Purvis & Halls 1996). *L. inops* was originally described in 1874 from a Swedish copper mine. In the UK, this rare species is now included on Schedule 8 of the UK Wildlife and Countryside Act (1981). At Coniston Copper Mines in the English Lake District, it occurs on malachite ($Cu_2CO_3(OH)_2$), azurite ($Cu_3(CO_3)_2(OH)_2$) and with smaller amounts of chrysocolla ($Cu_2H_2Si_2O_5(OH)_4$), covelline (CuS) and bornite (Cu_5FeS_4) cementing fractured quartz (Stanley 1979; Purvis & James 1985). *Psilolechia leprosa* occurs within the same community and was described from material sampled from Poldice Mine, Redruth, Cornwall where it grows on north-facing granitic walls of mine buildings, associated with mortar containing copper minerals (Coppins & Purvis 1987). It was subsequently found by Ken Sandell and the author on the first church they examined (in Oxfordshire) on mortar beneath copper grilles. *Psilolechia leprosa* is most frequent on church walls beneath copper lightning conductors (Purvis 1996). Churchyards provide havens for different metallophyte species associated with various metal structures and their weathering products. *Psilolechia leprosa* was later found by P. W. James and one author on volcanic rocks and relict cloud forests in madeira and the Azores where it was not obviously associated with copper mineralisation (Natural History Museum lichen herbarium).

Lecidea inops contains the only known British record of the mineral moolooite ($CuC_2O_4 . nH_2O$ (n $\sim$ 0.4–0.7) (Chisholm *et al.* 1987)) in which it occurs as vivid blue crystalline inclusions (< 0.5 mm) (Young 1987; Ryback & Tandy 1992). The species is most frequent in mountainous regions in Scandinavia and has been found on rocks and wood subject to irrigation by Cu-bearing ground waters in Norway and Slovakia (Purvis & James 1985). Christian Sommerfelt described the greenish lichen *Lecidea theiodes* as a species new to science in 1826 from Norway. The green colour was considered to be due to a lichen acid (the green rhizocarpic acid), like the 'map lichen' often used in dating studies. Analytical studies confirmed over 150 years later the green layer contained Cu, possibly as a Cu-norstictic acid complex (Purvis *et al.* 1987). They supported field observations emphasising the importance of deposition and the reaction between cations and coordinating ligands in biomass (Purvis 2000; Haas & Purvis 2006).

One of the most obvious effects of mineralization on lichens is the strong rust colour which occurs in many different and unrelated lichenised groups, one of which is *Acarospora* (Acharius 1798; Weber 1962; Schwab 1986). Field observations suggest the rust colour in these lichens reflects rock chemistry and is usually considered to be of no taxonomic significance (Hawksworth 1973b). In copper-rich environments, these populations often have a distinct green colour. This unusual colouration has resulted in copper-accumulating populations being described as distinct species at least three times (Sampaio 1920; Clauzade *et al.* 1982; Alstrup 1982). Molecular phylogenetic methods are helping to unravel species concepts and understand evolution in *Acarospora smaragdula* sens. lat, one of the most enigmatic in the European lichen floras (Crewe *et al.* 2006; Wedin *et al.* 2009). Modern SEM's and microanalytical (EPMA, XRD, FTIR) methods were employed to investigate mineralisation in the lichen cortex of the rust red *Acarospora sinopica* and paler yellow-orange *A. sinopica* f. *subochracea* sampled from Swedish mines. Results confirmed the distinctive colours are not simply due to hydrated iron oxides, 'rust', as previously believed. Analysis suggests mixed sulphide and oxide phases with little crystallinity, as well as other elements arising from clay minerals are present (Purvis *et al.* 2008a). Photobiont influences are also to be expected. Species found in these specialised acidic habitats were found to contain a single photobiont species, whilst several photobiont species occurred in 'nitrophyte' species (Douglas *et al.* 1995; Beck 1999).

Lecanora cascadensis (previously misidentified and now called *Lecanora sierrae* (Ryan & Nash 1993) was found associated with rocks containing bornite (Cu_5FeS_4) and other copper minerals at Lights Creek District, Plumas County, California (Czehura 1977). Colour differences in thalli were found to correspond well with copper mineralisation in soil. The intensity and size of the lichen's green perimeter increased under the influence of copper mineralisation enabling 1000 ppm and greater than 2000 ppm anomalies to be distinguished in soil. The ash of anomalous green '*L. cascadensis*' contained 4% Cu, a concentration approximately three times that of associated species collected at the same site. This evidence suggests that *L. cascadensis* (i.e., *Lecanora sierrae*) could be a valuable geobotanical prospecting tool (Czehura 1977).

Conclusions

As highlighted in this chapter, mapping studies, long-term monitoring and herbarium samples identify rapid changes in lichen diversity in Britain and Europe in response to the success of measures to reduce pollutants and socio-economic factors. This provides further evidence (Purvis *et al.* 2007a, b) that lichen biomonitoring is a fitting tribute to the 25th anniversary of the Convention on Long-range Transboundary Air Pollution in 1979, stimulated by the link between sulphur emissions in Europe and acidification of lakes in Scandinavia, first suspected back in the 1960s. Lichens continue to be highly effective biological indicators of air pollution effects, though sensitivity to newly emerging pollutant issues must be established. Long-term influences of low levels of eutrophication, gaseous pollutants (particularly globally rising background ozone concentrations) (Bell 2002) on lichen communities and succession under changing climatic conditions (Aptroot & Van Herk 2007) are unknown. Other factors, including soil-plant relationships and a pollution legacy must also be considered (Gilbert 1976; Blake *et al.* 1999; Hauck & Runge 1999; Blake & Goulding 2002; Purvis & Pawlik-Skowrońska 2008; Purvis *et al.* 2008). Integrated biomonitoring approaches are clearly important in this challenging, but highly rewarding, field of research (Purvis *et al.* 2007a, b). There are further exciting outreach opportunities in schools, for instance, as part of the Big Lottery funded Open Air Laboratories (OPAL) project, aimed at getting the public involved with using selected organisms to monitor environmental health, across England (OPAL 2007).

Lichen-mineral interactions continue to be an exciting field and to stimulate research across diverse disciplines since the pioneering studies of Lounamaa in Finland 50 years ago (Lounamaa 1956; Haas & Purvis 2006; Purvis & Pawlik-Skowrońska 2008). Understanding metal uptake and retention by lichens is important for environmental monitoring (including the survey of heavy-metal concentration in European mosses within the UNECE International Cooperative Programme on Effects of Air Pollution on Natural Vegetation and Crops (http://icpvegetation.ceh.ac.uk)), understanding global geochemical cycles and learning how organisms tolerate potentially toxic elements. Study of lichens in novel mineralogical and geochemical environments will yield new species and data concerning speciation, adaptation, tolerance and the fate of potential environmental contaminants. The possibility of controlling the properties of materials by tailoring their substructure at the nanometre scale is a current topic of great interest (Soare *et al.* 2006; Purvis *et al.* 2008b). New mineral species and phases are likely to be found in lichens in other mineralised habitats. 'Hyperaccumulating' plants and other organisms fix extraordinary quantities of metal ions without being poisoned and may provide 'natural clean-up solutions' (McLean *et al.* 1998; Whiting *et al.* 2004, 2005; Callahan *et al.* 2006). This research could even shed new evidence to support the existence of life on other planets through evidence of 'biosignatures' (Blackhurst *et al.* 2004; Souza-Egipsy *et al.* 2006).

Acknowledgements

Dedicated to Peter James on the occasion of his 80th birthday (28th April). I am very grateful to Lesley Batty (University of Birmingham) for kindly inviting me to participate in the BES 'Ecology of Industrial Pollution meeting'. I thank my family, friends and colleagues too numerous to mention for helping stimulate my interest and support my studies since commencing the special honours botany course at Sheffield University 30 years ago. I thank staff in image resources (NHM) for photography (Fig. 3.3a), Jack Laundon for permission to adapt Fig. 3.3b, Mark Seaward (British Lichen Society Mapping Recorder) for providing the map used to construct Fig. 3.2a and Neil Cape (Centre for Ecology and Hydrology) for Fig. 3.4 and Jim Bennett (University of Wisconsin) for his extremely helpful comments, discussions in California and for drawing literature to my attention.

References

Acharius, E. (1798) *Lichenographiae svecicae Prodromus*. Björn, Linköping.

Ahti, T. (1965) Notes on the distribution of *Lecanora conizaeoides*. *Lichenologist* **3**, 91–92.

Aikin, A. (1797) Article 7. An account of the great copper works in the Isle of Anglesey. *A Journal of Natural Philosophy, Chemistry, and the Arts* **1**, 367–372.

Alstrup, V. (1982) A comparative study of *Acarospora isortoquensis* and *A. undata*. *Lichenologist* **16**, 205–206.

Aptroot, A. and Van Herk, C. M. (2007) Further evidence of the effects of global warming on lichens, particularly those with Trentepohlia phycobionts. *Environmental Pollution* **146**, 293–298.

Ashmore, M. R. (2002) Effects of oxidants at the whole plant and community level. In: *Air Pollution and Plant Life* (ed. J. N. B. Bell and M. Treshow M), pp. 88–118. John Wiley and Sons, Chichester, UK.

Asta, J., Erhardt, W., Ferretti, M. *et al.* (2002) Mapping lichen diversity as an indicator of environmental quality. In: *Monitoring with Lichens – Monitoring Lichens 7* (eds. P. L. C. Scheidegger and P. A. Wolseley), pp. 273–279. Kluwer Academic, Dordrecht.

Bailey, E. H. (1968) *Lecanora conizaeoides* in Iceland. *Lichenologist* **4**, 73.

Bargagli, R. (1998) *Trace Elements in Terrestrial Plants: an Ecophysiological Approach to Biomonitoring and Biorecovery*. Berlin, Springer.

Bates, J. W., Bell, J. N. B. and Farmer, A. M. (1990) Epiphyte recolonization of oaks along a gradient of air-pollution in South-East England, 1979–1990. *Environmental Pollution* **68**, 81–99.

Bates, J. W., Bell, J. N. B. and Massara, A. C. (2001) Loss of *Lecanora conizaeoides* and other fluctuations of epiphytes on oak in SE England over 21 years with declining SO_2 concentrations. *Atmospheric Environment* **35**, 2557–2568.

Beck, A. (1999) Photobiont inventory of a lichen community growing on heavy-metal-rich rock. *Lichenologist* **31**, 501–510.

Bell, J. N. B. (2002) Conclusions and future directions. In: *Air Pollution and Plant Life*. 2nd edn (eds. J. N. B. Bell and M. Treshow), pp. 455–458. John Wiley and Sons, Chichester, UK.

Bell, J. N. B. and Treshow, M. (2002) Historical perspectives. In: *Air Pollution and Plant Life* (eds. J. N. B. Bell and M. Treshow), pp. 5–21. John Wiley and Sons, Chichester, UK.

Bell, J. N. B., Davies, L. and Honour, S. (2004) Air pollution research in London. In: *Urban Air Pollution, Bioindication and Environmental Awareness* (eds. A. Klumpp, W. Ansel and G. Klumpp), pp. 3–16. Cuvillier, Göttingen.

Bennett, J. P. (1995) Abnormal chemical element concentrations in lichens of Isle Royale National Park. *Environmental and Experimental Botany* **35**, 259–277.

Bennett, J. P. and Wetmore, C. M. (1997) Chemical element concentrations in four lichens on a transect entering Voyageurs-National Park. *Environmental and Experimental Botany* **37**, 173–185.

Blackhurst, R. L., Jarvis, K. and Grady, M. (2004) Biologically–induced elemental variations in Antarctic sandstones: a potential test for Martian micro-organisms. *International Journal of Astrobiology* **3**, 97–106.

Blake, L. and Goulding, K. W. T. (2002) Effects of atmospheric deposition, soil pH and acidification on heavy metal contents in soils and vegetation of semi-natural ecosystems at Rothamsted Experimental Station, UK. *Plant and Soil* **240**, 235–251.

Blake, L., Goulding, K. W. T., Mott, C. J. B. and Johnston, A. E. (1999) Changes in soil chemistry accompanying acidification over more than 100 years under woodland and grass at Rothamsted Experimental Station, UK. *European Journal of Soil Science* **50**, 401–412.

Branquino, C. (2008) *UNECE Ammonia Workshop*, Edinburgh, December 2006.

Brightman, F. H. (1982) Erasmus Darwin claims the earliest mention of lichens and air pollution (1790). *Bulletin of the British Lichen Society* **51**, 18.

Brimblecombe, P. (1987/88) *The Big Smoke*. Methuen, London.

Brodo, I. M. (1964) Field studies of the effects of ionizing radiation on lichens. *The Bryologist* **67**, 76–87.

Callahan, D. L., Baker, A. J. M., Kolev, S. D. and Wedd, A. G. (2006) Metal ion ligands in hyperaccumulating plants. *Journal of Biological Inorganic Chemistry* **11**, 2–12.

Cape, J. N., Tang, Y. S., Van Dijk, N., Love, L., Sutton, M. A. and Palmer, S. C. F. (2004) Concentrations of ammonia and nitrogen dioxide at roadside verges, and their contribution to nitrogen deposition. *Environmental Pollution* **132**, 469–478.

Ceburnis, D. and Steinnes, E. (2000) Conifer needles as biomonitors of atmospheric heavy metal deposition: comparison with mosses and precipitation, role of the canopy. *Atmospheric Environment* **34**, 4265–4271.

Chisholm, J. E., Jones, G. C. and Purvis, O. W. (1987) Hydrated copper oxalate, moolooite, in lichens. *Mineralogical Magazine* **51**, 715–718.

Clauzade, G., Roux, C. and Wirth, V. (1982) *Acarospora undata* sp. nov. *Bulletin du Musée d'Histoire Naturelle de Marseille* **41**, 35–39.

Coppins, B. J. and Purvis, O. W. (1987) A review of Psilolechia. *Lichenologist* **19**, 29–42.

Coppins, B. J., Hawksworth, D. L. and Rose, F. (2002) Lichens. In: *The Changing Wildlife of Great Britain and Ireland. The Systematics Association Special Volume Series 62* (ed. D. L. Hawksworth), pp. 126–147. Taylor & Francis, London.

Crespo, A., Bridge, P. D., Hawksworth, D. L., Grube, M. and Cubero, O. F. (1999) Comparison of rRNA genotype frequencies of *Parmelia sulcata* from long-established and recolonizing sites following sulphur dioxide amelioration. *Plant Systematics and Evolution* **217**, 177–183.

Crewe, A. T., Purvis, O. W. and Wedin, M. (2006) Molecular phylogeny of Acarosporaceae (Ascomycota) with focus on the proposed genus *Polysporinopsis*. *Mycological Research* **110**, 521–526.

Czehura, S. J. (1977) A lichen indicator of copper mineralization, Lights Creek District, Plumas County, California. *Economic Geology* **72**, 796–803.

Davies, L. (2005) A study of the lichens of London under contemporary atmospheric conditions. Unpublished PhD thesis, Department of Environmental Science and Technology, Imperial College London, University of London, London.

Davies, L., Bates, J. W., Bell, J. N. B., James, P. W. and Purvis, O. W. (2007) Diversity and sensitivity of epiphytes to oxides of nitrogen in London. *Environmental Pollution* **146**, 299–310.

De Bruin, M. and Hackenitz, E. (1986) Trace element concentrations in epiphytic lichens and bark substrate. *Environmental Pollution* **11**, 153–160.

DEFRA (2002) Effects of NOx and NH3 on lichen communities and urban ecosystems. A pilot study. In: *Air Pollution Research in London (A.P.R.I.L.) Network*. Imperial College London and Natural History Museum, London.

Douglas, G. E., John, D. M. and Purvis, O. W. (1995) The identity of phycobionts from lichens in the *Acarosporion sinopicae* alliance. *The Phycologist* **40**, 33.

Elix, J. A. (1996) Biochemistry and secondary metabolites. In: *Lichen Biology* (ed. I. T. H.

Nash), pp. 154–180. Cambridge University Press, Cambridge, UK.

Eriksson, O. (2005) Ascomyceternas ursprung och evolution – Protolichenes-hypotesen. *Svensk Mykologisk Tidskrift* **26**, 22–33.

Farmer, A. (1995) *Reducing the Impact of Air Pollution on the Natural Environment.* Joint Nature Conservation Committee, Peterborough, UK.

Ferry, B. W., Baddeley, M. S. and Hawksworth, D. L. (1973) *Air Pollution and Lichens.* Athlone Press, London.

Fowler, D., Pitcairn, C. E. R., Sutton, M. A. *et al.* (1998) The mass budget of atmospheric ammonia within 1 km of livestock buildings. *Environmental Pollution (Nitrogen Conference Special Issue)* **102** (S1), 343–348.

Frati, L., Santoni, S., Nicolardi, V. *et al.* (2007) Lichen biomonitoring of ammonia emission and nitrogen deposition around a pig stockfarm. *Environmental Pollution* **146**, 311–316.

Friedmann, I. (1982) Endolithic microorganisms in the Antarctic cold desert. *Science* **215**, 1045–1053.

Garty, J. (2001) Biomonitoring atmospheric heavy metals with lichens: theory and application. *Critical Reviews in Plant Sciences* **20**, 309–371.

Gilbert, O. L. (1965) Lichens as indicators of air pollution in the Tyne Valley. In: *Ecology and the Industrial Society* (eds. G. Goodman, R. W. Edwards and J. M. Lambert), pp. 35–47. Blackwell Scientific Publications, Oxford, UK.

Gilbert, O. L. (1968) Biological estimation of air pollution. In: *Plant Pathologist's Pocketbook* (ed. C. M. Institute), pp. 206–207. Commonwealth Mycological Institute, Kew, UK.

Gilbert, O. L. (1969) The effect of SO₂ on lichens and bryophytes around Newcastle upon Tyne. *Air Pollution, Proceedings of the first European Congress on the Influence of Air Pollution on Plants and Animals*, April 22–27, 1968. Centre for Agricultural Publishings and Documentation, Wageningen, the Netherlands, pp. 223–233.

Gilbert, O. L. (1970) A biological scale for the estimation of sulphur dioxide pollution. *New Phytologist* **69**, 629–634.

Gilbert, O. L. (1971) Studies along the edge of a lichen desert. *Lichenologist* **5**, 11–17.

Gilbert, O. L. (1974) An air pollution survey by school children. *Environmental Pollution* **6**, 175–180.

Gilbert, O. L. (1976) An alkaline dust effect on epiphytic lichens. *Lichenologist* **8**, 173–178.

Gilbert, O. L. (1980a) Effect of land-use on terricolous lichens. *Lichenologist* **12**, 117–124.

Gilbert, O. L. (1980b) A lichen flora of Northumberland. *Lichenologist* **12**, 325–395.

Gilbert, O. L. (1986) Lichens return to central London. *British Lichen Society Bulletin* **58**, 28–29.

Gilbert, O. L. (1990) The lichen flora of urban wasteland. *Lichenologist* **22**, 87–101.

Gilbert, O. L. (1992) Lichen reinvasion with declining air pollution. In: *Bryophytes and Lichens in a Changing Environment* (eds. J. W. Bates and A. M. Farmer), pp. 159–177. Clarendon Press, Oxford, UK.

Greater London Authority (GLA) (2002) *Cleaning London's Air. The Mayor's Air Quality Strategy.* Greater London Authority, London.

Gorbushina, A. A. (2006) Fungal activities in subaereal rock-inhabiting microbial communities. In: *Fungi in Biogeochemical Cycles.* (ed. G. M. Gadd), pp. 344–376. Cambridge University Press, Cambridge, UK.

Griffith, J. L. (1966) Some aspects of the effect of atmospheric pollution on the lichen flora to the west of Consett, C. Durham. University of Durham, Durham, UK.

Grime, J. P. (1979) *Plant Strategies and Vegetation Processes.* John Wiley and Sons, Chichester, UK.

Grube, M. and Hawksworth, D. L. (2007) Trouble with lichen: the re-evaluation and re-interpretation of thallus form and fruit body types in the molecular era. *Mycological Research* **111**, 1116–1132.

Gunn, J., Keller, W., Negusanti, J., Potvin, R., Beckett, P. and Winterhalder, K. (1995) Ecosystem recovery after emission reductions: Sudbury, Canada. *Water Air and Soil Pollution* **85**, 1783–1788.

Haas, J. R. and Purvis, O. W. (2006) Lichen biogeochemistry. In: *Fungi in Biogeochemical Cycles* (ed. G. M. Gadd), pp. 344–376. Cambridge University Press, Cambridge, UK.

Hauck, M. and Runge, M. (1999) Occurrence of pollution-sensitive epiphytic lichens in woodlands affected by forest decline: a new hypothesis. *Flora* **194**, 159–168.

Hauck, M., Hesse, V., Jung, R., Zöller, T. and Runge, M. (2001) Long-distance transported sulphur as a limiting factor for the abundance of *Lecanora conizaeoides* in montane spruce forests. *Lichenologist* **33**, 267–269.

Hauck, M., Huneck, S., Elix, J. A. and Paul, A. (2007) Does secondary chemistry enable lichens to grow on iron-rich substrates? *Flora* **202**, 471–478.

Hauck, M., Jurgens, S. R., Brinkmann, M. and Herminghaus, S. (2008) Surface hydrophobicity causes SO_2 tolerance in lichens. *Annals of Botany* **101**, 531–539.

Hawksworth, D. L. (1973a) 3, Mapping Studies. In: *Air Pollution and Lichens* (eds. B. W. Ferry, M. S. Baddeley and D. L. Hawksworth), pp. 38–76. Athlone Press, London.

Hawksworth, D. L. (1973b) Ecological factors and species delimitation in the lichens. In: *Taxonomy and Ecology. Systematics Association Special Volume No. 5* (ed. V. H. Heywood), pp. 31–69. Academic Press, London.

Hawksworth, D. L. (1988) The variety of fungal-algal symbioses, their evolutionary significance, and the nature of lichens. *Botanical Journal of the Linnean Society* **96**, 3–20.

Hawksworth, D. L. (1991) The fungal dimension of biodiversity: magnitude, significance, and conservation. *Mycological Research* **95**, 641–656.

Hawksworth, D. L. (2006) Mycology and mycologists. In: *8th International Mycological Congress, Cairns (Australia), August 20–25, 2006* (eds. W. Meyer and C. Pierce), pp. 65–72. Medimond, Bolonga.

Hawksworth, D. L. and Honegger, R. (1994) The lichen thallus: a symbiotic phenotype of nutritionally specialized fungi and its response to gall producers. In: *Plant Galls: Organisms, Interactions, Populations* (ed. M. A. J. Williams), pp. 77–98. Clarendon Press, Oxford, UK.

Hawksworth, D. L. and McManus, M. (1989) Lichen recolonisation in London under conditions of rapidly falling sulphur dioxide levels, and the concept of zone skipping. *Botanical Journal of the Linnean Society* **100**, 99–109.

Hawksworth, D. L. and McManus, P. M. (1992) Changes in the lichen flora on trees in Epping Forest through periods of increasing and then ameliorating sulphur dioxide air pollution. *Epping Naturalist* **71**, 92–101.

Hawksworth, D. L. and Rose, F. (1970) Qualitative scale for estimating sulphur dioxide air pollution in England and Wales using epiphytic lichens. *Nature* **227**, 145–148.

Hawksworth, D. L. and Rose, F. (1976) *Lichens as Pollution Monitors*. Edward Arnold, London.

Hawksworth, D. L. and Seaward, M. R. D. (1977) *Lichenology in the British Isles 1568–1975. An Historical and Bibliographical Survey*. Richmond Publishing, Richmond, UK.

Hawksworth, D. L., Hitch, C. J. B. and Purvis, O. W. (2001) Lichens of West Brompton Cemetery. *British Lichen Society Bulletin* **88**, 19–22.

Hilitzer, A. (1923) Les lichens des rochers amphiboliques aux environs de Vseruby. *Casopis Narodniho Musea* **97**, 1–14.

Honegger, R. (1997) Metabolic interactions at the mycobiont-photobiont interface. In: *Plant Relationships* (eds. G. C. Carroll and P. Tudzynki), pp. 209–221. Springer-Verlag, Berlin.

Honey, M. R., Leigh, C. and Brooks, S. J. (1998) The fauna and flora of the newly created Wildlife Garden in the grounds of The Natural History Museum, London. *The London Naturalist* **77**, 17–47.

Huneck, S. (1999) The significance of lichens and their metabolites. *Naturwissenschaften* **86**, 559–570.

Huneck, S. (2006) Der Flechten der Kupferschieferhalden um Eislebn, Mansfield und Sangerhausen. *Mitteilungen zur floristischen Kartierung Sachsen-Anhalt* **4**, 1–62.

Huneck, S. and Yoshimura, I. (1996) *Identification of Lichen Substances*. Springer, Berlin.

Jahns, H. (1988) The establishment, individuality and growth of lichen thalli. *Botanical Journal of the Linnean Society* **96**, 21–29.

James, P. W. (1973) The effect of air pollutants other than hydrogen fluoride and sulphur dioxide on lichens. In: *Air Pollution and Lichen* (eds. B. W. Ferry, M. S. Baddeley and D. L. Hawksworth), pp. 143–175. Athlone Press, London.

James, P. W. and Davies, L. (2003) Conservation and management. Resurvey of the corticolous lichen flora of Epping Forest. *Essex Naturalist (New Series)* **20**, 67–82.

James, P. W. and Wolseley, P. (2005) January meetings. Field visit to Royal Botanic Gardens Kew. *British Lichen Society Bulletin* **96**, 17–22.

James, P. W., Hawksworth, D. L. and Rose, F. (1977) Lichen communities in the British Isles: a preliminary conspectus. In: *Lichen Ecology* (ed. M. R. D. Seaward), pp. 295–413. Academic Press, London.

James, P. W., Purvis, O. W. and Davies, L. (2002) Epiphytic lichens in London. *Bulletin of the British Lichen Society* **90**, 1–3.

Jenkins, D. A., Johnson, A. H. and Freeman, C. (2000) Mynydd Parys Cu-Pb-Zn Mines: mineralogy, microbiology and acid mine drainage. In: *Environmental Mineralogy: Microbial Interactions, Anthropogenic Influences, Contaminated Land and Waste Management* (eds. J. D. Cotter-Howells, L. S. Campbell, E. Valsami-Jones and M. Batchelder), pp. 161–179. Mineralogical Society of Great Britain and Ireland, London.

Jones, D. (1988) Lichens and pedogenesis. In: *Handbook of Lichenology. Vol. 3* (ed. M. Galun), pp. 109–124. CRC, Boca Raton.

Kirschbaum, U. H. and Hanewald, K. (2001) Changes in the lichen vegetation at the long time investigated areas of Melsungen and Limburg, Hesses, Germany between 1997 and 1999. *Journal of Applied Botany* **75**, 20–30.

Knops, J. M. H., Nash, T. H., Boucher, V. L. and Schlesinger, W. H. (1991) Mineral cycling and epiphytic lichens – implications at the ecosystem level. *Lichenologist* **23**, 309–321.

Kricke, R. and Loppi, S. (2002) Bioindication: the I.A.P. approach. In: *Monitoring with Lichens-Monitoring Lichens* (eds. P. L. Nimis, C. Scheidegger and P. A. Wolseley), pp. 21–37. Kluwer Academic, Dordrecht, the Netherlands.

Lagreca, S. and Stuzman, B. W. (2006) Distribution and ecology of *Lecanora conizaeoides* (Lecanoraceae) in eastern Massachusetts. *The Bryologist* **109**, 335–357.

Lange, O. L. (1992) Pflanzenleben unter Streß. Flechten als Pioniere der Vegetation am Extremstandorten der Erde. University of Würzburg, Würzburg.

Lange, O. L. and Ziegler, H. (1963) Der schwermetallgehalt von flechten aus dem Acarosporetum sinopicae auf erzschlackenhalden des Harzes. I. Eisen und kupfer. *Mitteilungen der Floristisch-soziologischen ArbeitsGemeinschaft* **10**, 156–183.

Larsen, R. S., Bell, J. N. B., James, P. W. *et al.* (2007) Lichen and bryophyte distribution on oak in London in relation to air pollution and bark acidity. *Environmental Pollution* **146**, 332–340.

Larsen-Vilsholm, R., Wolseley, P. A., Søchting, U. and Chimonides, P. J. (2009) Biomonitoring with lichens on twigs, *Lichenologist*, **41**, 189–202.

Laundon, J. R. (1967) A study of the lichen flora of London. *Lichenologist* **3**, 277–327.

Laundon, J. R. (1970) London's lichens. *London Naturalist* **49**, 20–69.

Laundon, J. R. (1973) Urban lichen studies. In: *Air Pollution and Lichens* (eds. B. W. Ferry, M. S. Baddeley and D. L. Hawksworth), pp. 109–123. Athlone Press, London.

Laundon, J. R. (2003) Six lichens of the *Lecanora varia* group. *Nova Hedwigia* **76**, 83–111.

Laundon, J. R. and Waterfield, A. (2007) William Borrer's lichens in the Supplement to the English Botany 1826–1866. *Botanical Journal of the Linnean Society* **154**, 381–392.

Leigh, C. and Ware, C. (2003) The development of the flora, fauna and environment of the Wildlife Garden at The Natural History Museum, London. *The London Naturalist* **82**, 75–134.

Leith, I. D., Van Dijk, N., Pitcairn, C. E. R., Wolseley, P. A., Whitfield, C. P. and Sutton, M. A. (2005) *Biomonitoring Methods for Assessing the Impacts of Nitrogen Pollution: Refinements and Testing.* JNCC Report. Joint Nature Conservancy Council, Peterborough, UK.

Lindeström, L. (2002) *The Environmental History of the Falun Mine*, Uppsala, Stiftelsen Stora Kopparberget & ÅF-Miljöforskargruppen AB.

Loppi, S. and Pirintsos, S. A. (2003) Epiphytic lichens as sentinels for heavy metal pollution at forest ecosystems (central Italy). *Environmental Pollution* **121**, 327–332.

Lounamaa, J. (1956) Trace elements in plants growing wild on different rocks in Finland. A semi-quantitative spectrographic study. *Annales Botanici Societatis Zoologicae Botanicae Fennicae 'Vananmo'* **29**, 1–196.

Mabey, R. (1974) *The Pollution Handbook: The ACE/ Sunday Times Clean Air and Water Surveys.*

Massara, A. C. (2004) The ecology and physiology of the pollution tolerant lichen, *Lecanora conizaeoides*. Department of Biology. Imperial College London, London.

McLean, J., Purvis, O. W., Williamson, B. J. and Bailey, E. H. (1998) Role for lichen melanins in uranium remediation. *Nature* **391**, 649–650.

Mikhailova, I. and Vorobeichik, E. L. (1995) Epiphytic lichenosynusia under conditions of chemical pollution: dose-effect dependencies. *Russian Journal of Ecology* **26**, 425–431.

Nash, T. H., III. (1973) The effect of air pollution on other plants, particularly vascular plants. In: *Air Pollution and Lichen* (eds. B. W. Ferry, M. S. Baddeley and D. L. Hawksworth), pp. 192–223. Athlone Press, London.

Nash, T. H., III (1996a) *Lichen Biology.* Cambridge University Press, Cambridge, UK.

Nash, T. H., III (1996b) Nutrients and mineral cycling. In: *Lichen Biology* (ed. T. H. Nash III), pp. 136–153. Cambridge University Press, Cambridge, UK.

Nash, T. H., III and Wirth, V. (1988) *Lichens. Bryophytes and Air Quality.* J. Cramer, Berlin.

NEGTAP. (2001) *Transboundary Air Pollution: Acidification, Eutrophication and Ground-level Ozone in the UK. NEGTAP.* Department for Environment, Food and Rural Affairs, London.

Nimis, P. L., Scheidegger, C. and Wolsley, P. A. (2002) *Monitoring with Lichens - Monitoring Lichens.* Kluwer Academic Publishers, Dordrecht.

Noeske, O., Läuchli, A., Lange, O. L. and Ziegler, H. (1970) Konzentration und lokalisierung von schwermetallen in flechten der erzschlackenhalden des Harzes. *Deutsche Botanische Gesellschaft Neue Folge* **4**, 67–79.

NSCA. (2002) The clean air revolution: 1952–2052. Marking 50 years since the great London smog. *Clean Air and Environmental Protection* **32**, 1–75.

OPAL. (2007) Big Lottery Fund open air laboratories. http://www3.imperial.ac.uk/ newsandeventspggrp/imperialcollege/ newssummary/news_20-8-2007-13-21-51? newsid=15994.

Poelt, J. (1955) Flechten der Schwarzen Wand in der Grossarl. *Verhandlung Zoologische-Botanische Gesellschaft. Wien* **95**, 107–113.

Poelt, J. and Huneck, S. (1969) *Lecanora vinetorum* nova spec., ihre Vergesellschaftung, ihre Ökologie und ihre Chemie. *Österreichische Botanische Zeitschrift* **115**, 411–422.

Poelt, J. and Ullrich, H. (1964) Über einige chalkophile Lecanora - Arten der Mittelueuropäischen Flora (Lichenes, Lecanoraceae). *Österreichische Botanische Zeitschrift* **111**, 257–268.

Purvis, O. W. (1985) The effect of mineralization on lichen communities with special reference to cupriferous substrata. Unpublished PhD Thesis, Imperial College, London and British Museum (Natural History), London.

Purvis, O. W. (1987) Another foliose lichen found in Inner London. *Bulletin of the British Lichen Society* **61**, 4.

Purvis, O. W. (1996) Interactions of lichens with metals. *Science Progress* **79**, 283–309.

Purvis, O. W. (2000) *Lichens*. Natural History Museum, London.

Purvis, O. W. and Halls, C. (1996) A review of lichens in metal-enriched environments. *Lichenologist* **28**, 571–601.

Purvis, O. W. and James, P. W. (1985) Lichens of the Coniston Copper Mines. *Lichenologist* **17**, 221–237.

Purvis, O. W. and Pawlik-Skowrońska, B. (2008) Lichens and metals. In: *Stress in Yeasts and Filamentous Fungi* (eds. S. V. Avery, M. Stratford and P. Van West), pp. 175–200. Elsevier, Amsterdam.

Purvis, O. W., Bailey, E. H., Mclean, J., Kasama, T. and Williamson, B. J. (2004) Uranium biosorption by the lichen *Trapelia involuta* at a uranium mine. *Geomicrobiology Journal* **21**, 159–167.

Purvis, O. W., Chimonides, P. J., Jones, G. C. *et al.* (2004) Lichen biomonitoring near Karabash Smelter Town, Ural Mountains, Russia, one of the most polluted areas in the world. *Proceedings of The Royal Society of London Series B-Biological Sciences* **271**, 221–226.

Purvis, O. W., Chimonides, P. D. J., Jeffries, T. E., Jones, G. C., Rusu, A. M. and Read, H. (2007a) Multi-element composition of historical lichen collections and bark samples, indicators of changing atmospheric conditions. *Atmospheric Environment* **41**, 72–80.

Purvis, O. W., Coppins, B. J., Hawksworth, D. L. H., James, P. W. and Moore, D. M. (1992) *The Lichen Flora of Great Britain and Ireland.* Natural History Museum Publications in association with the British Lichen Society, London.

Purvis, O. W., Crewe, A. T., Wedin, M., Kearsley, A. and Cressey, G. (2008a) Mineralization in rust-coloured Acarospora. *Geomicrobiology* **25**, 142–148.

Purvis, O. W., Dubbin, W., Chimonides, P. D., Jones, G. C. and Read, H. (2008) The multi-element content of the lichen *Parmelia sulcata*, soil, and oak bark in relation to acidification and climate. *The Science of the Total Environment* **390**, 558–568.

Purvis, O. W., Elix, J. A., Broomhead, J. A. and Jones, G. C. (1987) The occurrence of copper norstictic acid in lichens from cupriferous substrata. *Lichenologist* **19**, 193–203.

Purvis, O. W., Pawlik-Skowrońska, B., Cressey, G., Jones, G. C., Kearsley, A. and Spratt, J. (2008b) Mineral phases and element composition of the copper hyperaccumulator lichen *Lecanora polytropa*. *Mineralogical Magazine* **72**, 539–548.

Purvis, O. W., Seaward, M. R. D. and Loppi, S. (2007b) Lichens in a changing pollution environment: an introduction. *Environmental Pollution* **146**, 291–292.

Richardson, D. H. S. (1992) *Pollution Monitoring with Lichens*. Richmond Publishing, Slough, UK.

Riddell, J., Nash, T. H. and Padgett, P. (2008) The effect of HNO_3 gas on the lichen *Ramalina menziesii*. *Flora* **203**, 47–54.

Rose, F. (1973) Detailed mapping in South-East England. In: *Air Pollution and Lichen* (eds. B. W. Ferry, M. S. Baddeley and D. L. Hawksworth), pp. 77–88. Athlone Press, London.

Rose, F. (1976) Lichenological indicators of age and environmental continuity in woodlands. In: *Lichenology: Progress and Problems* (eds. D. H. Brown, D. L. Hawksworth and R. H. Bailey), 279–307. Academic Press, London.

Rose, F. (1995) 1018 *Parmelia soredians* Nyl. In: *Lichen Atlas of the British Isles* (ed. M. R. D. Seaward). British Lichen Society, London.

Rose, F. and Coppins, S. (2000) Site assessment of epiphytic habitats using lichen indices. *NATO Advanced Research Workshop on Lichen Monitoring; Monitoring with Lichens: Monitoring Lichens*. Orielton Field Centre. Kluwer Academic Publisher, London.

Rose, C. I. and Hawksworth, D. L. (1981) Lichen recolonization in London's cleaner air. *Nature* **289**, 289–292.

Ryan, B. D. and Nash, T. H., III (1993) *Lecanora* section *Placodium* (lichenized ascomycotina) in North-America: new taxa in the *L. garovaglii* group. *The Bryologist* **96**, 288–298.

Ryback, G. and Tandy, P. (1992) Eighth supplementary list of British Isles minerals (English). *Mineralogical Magazine* **56**, 261–275.

Sampaio, G. (1920) *Liquenes inéditos.* 5–6.

Schade, A. (1933) Das Acarosporetum sinopicae als Charaktermerkmal der Flechtenflora sächsischer Bergwerkshalden. *Sitzungsberichte und Abhandlungen der Naturwissenschaftlichen Gesellschaft Isis in Dresden 1932.* pp. 131–160.

Schwab, A. J. (1986) Rostfärbene Arten der Sammelgattung *Lecidea* (Leanorales) Revision der Arten mittel-und Nordeuropas. *Mitteilungen Botanishe Staatssamlung München* **22**, 221–476.

Seaward, M. R. D. (1976) Performance of *Lecanora muralis* in an urban environment. In: *Lichenology: Progress and Problems* (eds. D. H. Brown, D. L. Hawksworth and R. H. Bailey), pp. 323–357. Academic Press, London.

Seaward, M. R. D. (1998) Time-space analyses of the British lichen flora, with particular reference to air quality surveys. *Folia Cryptogamica Estonica* **32**, 85–96.

Seaward, M. R. D. and Hitch, C. J. B. (1982) *Atlas of the Lichens of the British Isles. Vol. 1.* Natural Environment Research Council, Cambridge, UK.

Sernander, R. (1926) *Stockholms Natur.* Almquist and Wiksells, Uppsala, Sweden.

Sigal, L. L. and Nash, T. H., III (1983) Lichen communities on conifers in southern California: an ecological survey relative to oxidant air pollution. *Ecology* **64**, 1343–1354.

Sipman, H. and Aptroot, A. (2001) Where are the missing lichens? *Mycological Research* **105**, 1433–1439.

Soare, L. C., Lemaitre, J., Bowen, P. and Hofmann, H. (2006) A thermodynamic model for the precipitation of nanostructured copper oxalates. *Journal of Crystal Growth* **289**, 278–285.

Souza-Egipsy, V., Ormo, J., Bowen, B. B., Chan, M. A. and Komatsu, G. (2006) Ultrastructural study of iron oxide precipitates: implications for the search for biosignatures in the Meridiani hematite concretions, Mars. *Astrobiology* **6**, 527–545.

Stadler, M. and Keller, N. P. (2008) Paradigm shifts in fungal secondary metabolite research. *Mycological Research* **112**, 127–130.

Stanley, C. J. (1979) *Mineralogical Studies of Copper, Lead, Zinc and Cobalt Mineralization in the English Lake District.* University of Aston, Birmingham, UK.

Sutton, M. A., Pitcairn, C. E. R. and Whitfield, C. P. (2004) *Bioindicator and Biomonitoring Methods for Assessing the Effects of Atmospheric Nitrogen on Statutory Nature Conservation Sites.* JNCC Report. Joint Nature Conservancy Council, Peterborough, UK.

Sutton, M. A., Tang, Y. S., Dragosits, U. *et al.* (2001) A spatial analysis of atmospheric ammonia and ammonium in the UK. In: *International Nitrogen Conference; Optimizing Nitrogen Management in Food and Energy Production and Environmental Protection* (ed. J. Galloway). Balkema, Potomac, MD.

Sutton, M. A., Wolseley, P. A., Leith, I. D. *et al.* (2009) Estimation of the ammonia critical level for epiphytic lichens based on observations at farm, landscape and national scales. In: *Atmospheric Ammonia – Detecting Emission Changes and Environmental Impacts – Results of an Expert Workshop under the Convention on Long-range Transboundary Air Pollution* (eds. M. Sutton, S. Reis and S. Baker), 71–86. Springer Publishers, New York.

Tømmervik, H., Hogda, K. A. and Solheim, L. (2003) Monitoring vegetation changes in Pasvik (Norway) and Pechenga in Kola Peninsula (Russia) using multitemporal Landsat MSS/TM data. *Remote Sensing of Environment* **85**, 370–388.

Tømmervik, H., Johansen, B. E. and Pedersen, J. P. (1995) Monitoring the effects of air-pollution on terrestrial ecosystems in Varanger (Norway) and Nikel-Pechenga (Russia) using remote-sensing. *Science of the Total Environment* **161**, 753–767.

Tømmervik, H., Johansen, M. E., Pedersen, J. P. and Guneriessen, T. (1998) Integration of remote sensed and in-situ data in an analysis of the air pollution effects on terrestrial ecosystems in the border areas between Norway and Russia. *Environmental Monitoring and Assessment* **49**, 51–85.

Tønsberg, T. (1992) The sorediate and isidiate, corticolous, crustose lichens in Norway. *Sommerfeltia* **14**, 1–331.

UNECE. (2004) *Manual on Methodologies and Criteria for Modelling and Mapping Critical Loads and Levels and Air Pollution Effects, Risks and Trends.* UNECE Convention on Long-range Transboundary Air Pollution.

Van Haluwyn, C. and Lerond, M. (1988) Lichénosociologie et qualité de l'air; protocole opératoire et limites. *Cryptogamie Bryologie et Lichénologie* **9**, 313–336.

Van Herk, C. M. (1999) Mapping of ammonia pollution with epiphytic lichens in the Netherlands. *Lichenologist* **31**, 9–20.

Van Herk, C. M. (2001) Bark pH and susceptibility to toxic air pollutants as independent causes of changes in epiphytic lichen composition in space and time. *Lichenologist* **33**, 419–442.

Van Herk, C. M. (2003) Long distance nitrogen air pollution effects on lichens in Europe. *Lichenologist* **35**, 413–415.

Waterfield, A. (2002) Herbarium records of London lichens. *The London Naturalist* **81**, 35–48.

Weber, W. A. (1962) Environmental modifications and the taxonomy of the crustose lichens. *Svensk Botanisk Tidskrift* **56**, 293–333.

Wedin, M., Westberg, M., Crewe, A. T., Tehler, A. and Purvis, O. W. (2009) Species delimitation and evolution of metal bioaccumulation in the lichenized *Acarospora smaragdula* (Ascomycota, Fungi) complex. *Cladistics* **25**, 1–12.

WG1. (2006) Workgroup 1: Critical levels for NH₃. Expert Workshop under the UNECE Convention on Long-range Transboundary Air Pollution. *Atmospheric Ammonia: Detecting Emission Changes and Environmental Impacts, 4–6 December 2006.* Edinburgh, Scotland.

Whiting S. N., Reeves, R. D., Richards, D. *et al.* (2004) Research priorities for conservation of metallophyte biodiversity and their potential for restoration and site remediation. *Restoration Ecology* **12**, 106–116.

Whiting, S. N., Reeves, R. D., Richards, D. *et al.* (2005) Use of plants to manage sites contaminated with metals. In: *Plant Nutritional Genomics* (eds. M. R. Broadley and P. J. White), pp. 106–116. Blackwell Publishing, London.

Williamson, B. J., Purvis, O. W., Bartók, K. *et al.* (1996) Chronic pollution from mineral processing in the town of Zlatna, Apuseni Mountains (Romania). *Studia Universitatis Babes-Bolyai, Geologia* **41**, 87–93.

Wirth, V. (1972) Die Silikatflechten. Gemeinshaften im ausseralpinen Zentraleuropa. *Dissertationes Botanicae Lehre* **17**, 1–306, 109.

Wirth, V. (1985) Zur Ausbreitung, Herkunft und Ökologie anthropogen geförderter Rinden- und Holzflechten. *Tuexenia II* **5**, 523–535.

Wirth, V. (1999) Beiträg zur kenntnis der dynamik epiphytischer flechtenbestände. *Stuttgarter Beiträge zur Naturkunde* **A595**, 1–17.

Wolseley, P. A. and Pryor, K. V. (1999) The potential of epiphytic twig communities on *Quercus petraea* in a Welsh woodland site (Tycanol) for evaluating environmental changes. *Lichenologist* **31**, 41–61.

Wolseley, P. A., James, P. W., Theobald, M. R. and Sutton, M. A. (2006) Detecting changes in epiphytic lichen communities at sites affected by atmospheric ammonia from agricultural sources. *Lichenologist* **38**, 161–176.

Young, B. (1987) *Glossary of the Minerals of the English Lake District and Adjoining Areas.* British Geological Survey, Newcastle upon Tyne, UK.

CHAPTER FOUR

The impacts of metalliferous drainage on aquatic communities in streams and rivers

LESLEY C. BATTY, MONTSERRAT AULADELL
AND JON SADLER

Introduction

Metals are naturally occurring elements of the Earth's crust that can be released into the aquatic environment through the processes of weathering and erosion, where they are present in trace amounts and do not normally constitute an environmental problem. However, the activity of humans has increased the release of many metals to the environment. It is difficult to assess what the natural background levels would have been in affected areas, particularly where the influence has been prolonged, but it has been reported that in mining areas concentrations of metals in waters and associated sediments can be 3–4 orders of magnitude higher following mineral ore extraction (Runnells *et al.* 1992; Helgen & Moore 1996). Although some of the metals released by human activities, such as Fe and Zn, are essential elements for the successful growth and functioning of biota, the presence of these substances in elevated concentrations or in other chemical forms can be potentially toxic. In addition, many metals, such as Pb and Cd, have no known role in biological functioning and can be toxic to organisms at very low concentrations. Therefore, the release of metals into the environment poses a significant threat to the fauna and flora of receiving water courses.

The release of metals to the environment has a strong historical dimension and can be traced back many hundreds, if not thousands, of years; the extraction of metal ores, for example, is reported to date back to the Bronze Age in the UK (e.g., Ixer & Budd 1998), and evidence of environmental impacts is found to occur as far back as pre-Roman times (Rothwell 2007). However, it was not until the major increase in industrial activity in the form of mining, smelting, and processing of ores and coal within the eighteenth and nineteenth centuries that significant environmental and human health impacts were noted (Goudie 1993). The metals were either released directly into the river systems through surface runoff (including spoil heap discharge) and direct drainage from adits

Ecology of Industrial Pollution, eds. Lesley C. Batty and Kevin B. Hallberg. Published by Cambridge University Press. © British Ecological Society 2010.

and processing works or could be deposited from aerial sources via smelting. The proliferation of manufacturing industries in the twentieth century increased the sources, types and forms of metals within the environment and led to large-scale impacts within affected ecosystems (e.g., see Chapter 13). Agriculture has also been a significant source of metals due to their presence in many pesticides, herbicides and fertilisers (e.g., Nicholson *et al.* 2003; Nziguheba & Smolders 2008). During the latter half of the twentieth century increased environmental legislation together with a decline in manufacturing within many countries led to a decrease in the occurrence of metalliferous discharges (see, for example, Chapters 13 and 14); however, the legacy of these previous activities continues to act as both diffuse and point sources to the aquatic environment via drainage from mine workings and mobilisation of contaminated floodplain sediments (e.g., Ullrich *et al.* 2007; Hutchinson & Rothwell 2008; Tiefenthaler *et al.* 2008). Increasing urbanisation has also led to the creation of different sources of metals including urban runoff due to wear and tear of motor vehicles (e.g., Sansalone & Buchberger 1997), street dust and tanker effluents (Brown and Peake 2006). More recently metals such as silver and tin in the form of nanoparticles have been identified as a potential source of contamination (see Chapter 5). These sources of metal contamination are more difficult to control and pose a potential threat to ecosystems. In addition, in areas of the world where legislation is less stringent the 'traditional' sources of metals still act as significant contributors of pollutants to the environment. For example, mercury has been shown to be a significant contaminant in areas such as Latin America, Asia and Sub-Saharan Africa particularly from gold mining (Ullrich *et al.* 2007; Marrugo-Negrete *et al.* 2008), and lead exposure is also a major problem in developing countries with significant sources including metal smelters (Meyer *et al.* 2008).

In this chapter, we review the impacts of metal pollution on freshwater macroinvertebrates, a key basal trophic component of the aquatic ecosystem. This is a group of organisms often used in the monitoring of rivers and is a major component of many national monitoring schemes such as that of the Environment Agency in the UK (see Chapter 6 and Metcalf-Smith 1994). Macroinvertebrates are also key organisms in the transfer of energy through trophic levels and fluxes of matter within river systems, and therefore any effects of metals are likely to be transferred and/or biomagnified through the food web (Petersen *et al.* 1989; Maltby 1996; Barata *et al.* 2005). Within this chapter we therefore aim to provide a review of the key chemical variables that affect invertebrate communities within rivers receiving metalliferous discharges, the importance of the source of the contaminant and the potential risks to the food chain. We highlight the key research areas that need to be addressed, particularly in light of recent developments within the relevant legislation, in order to effectively manage these affected ecosystems in the future.

Characteristics of metalliferous discharges

Metalliferous releases to the environment occur in two main forms: net-acidic and net-alkaline. Many of the sources of metals from industry and urban runoff tend to be net-alkaline, but some, such as mining activities result in the release of sulphuric acid and hydrogen ions due to the weathering of sulphide minerals (predominantly pyrite) resulting in a net-acidic discharge (e.g., Younger et al. 2002). The speciation of metals within aquatic systems is controlled by many environmental factors of which pH and Eh (redox potential) are probably the most important, although the presence of organics, temperature and salinity can also be significant controlling factors (Stumm & Morgan 1995). Where discharges are acidic in nature, the speciation of metals is such that they become bioavailable, although the point at which this occurs varies with the metal in question. In contrast, within net-alkaline discharges metals are more likely to be sorbed to particulate matter or present as precipitates and unavailable for biological uptake.

Historically, many metalliferous discharges into water courses would have been from a point source, such as the output from a processing plant. However, today the majority of inputs are probably chronic and diffuse and derived from urban road runoff, contaminated floodplain sediments and mine workings. The nature of these releases means that the transfer of pollutants is strongly influenced by local weather conditions, particularly rainfall events, and therefore varies temporally as well as spatially. Although traditionally the release of metals following contact with waters (either storm waters or flooding of mine workings) are thought the follow the 'first flush' model (e.g., Gray 1998; Lee et al. 2002; Younger et al. 2002), departures from this classic model have been reported. For example, concentrations of metals were found to show the weakest first flush patterns within urbanised areas, particularly in larger catchments (Tiefenthaler et al. 2008). In addition Lawler et al. (2006) found that fine sediment (one of the main ways in which metals may be transported) influxes into a river arrived much later due to the presence of distal sources such as roads and mine workings. There is, however, little information on the relative importance of acute and chronic toxicity and the associated patterns of metal inputs into rivers on controlling macroinvertebrate community structure which should be an important consideration both in management and recovery of rivers following remediation efforts.

The potential impact of metals on biota therefore can vary greatly from site to site, depending upon specific environmental conditions and the sources of metals. However, there are some overall effects that have been reported in response to industrial metalliferous discharges, although it should be noted that the majority of research has focused upon the effects of mining activities.

General responses of macroinvertebrates to metal pollution

On the whole, macroinvertebrate communities tend to show a reduction in biodiversity and abundance in metal-polluted sites (Schultheis *et al.* 1997; García Criado *et al.* 1999; Gray 1998; Watanabe *et al.* 2008), with an increase in the numbers of tolerant taxa (Kelly 1988). In one example in the UK, rivers affected by mining and industry were found to contain severely reduced invertebrate communities with only 12 taxonomic families, dominated by pollution-tolerant species (Amisah & Cowx 2000). Ephemeroptera (mayflies), Trichoptera (caddisflies) and Plecoptera (stoneflies) are often reduced in terms of abundance and richness in metal-contaminated streams (Rutherford & Mellow 1994; Gray 1996a; Beltman *et al.* 1999; Malmqvist & Hoffstein 1999; Courtney & Clements 2002; DeNicol & Stapleton 2002; Solà *et al.* 2004; Woodcock & Huryn 2005; Tripole *et al.* 2006; Doi *et al.* 2007; Merricks *et al.* 2007). This observation is supported by experimental studies in artificial streams where mayflies and stoneflies were shown to be most sensitive to Cu and Zn and caddisflies and chironomids were least sensitive (Hickey & Golding 2002). Although there is often a decrease in species richness, abundance or density is not always affected (Beltman *et al.* 1999). This is because a reduction in diversity can allow an increase in the numbers of tolerant taxa such as chironomids. Chironomids are known for their resistance to high concentrations of metals (e.g., Rousch *et al.* 1997).

Studies of urban runoff identified metal-tolerant (Zn and Ni) families as including Hydrophilidae, Asellidae, Ephemerellidae, Philoptamidae and Chloroperlidae with metal-sensitive families being Leptophlebiidae, Ephemeridae, Leuctridae and Hydrobiidae (Beasley & Kneale 2003). Chloroperlidae, however, were reported to be absent in the presence of Acid Mine Drainage (AMD) in rivers in the Iberian Peninsula (García Criado *et al.* 1999), although Leptophlebiidae were also absent. This variation in response is sometimes noted in the literature and may be due to added effects of different metals and acidity (see later section on metal toxicity). In addition, many authors have identified the high degree of variation in the response of families and genera to metals and acidity (Gower *et al.* 1994; Hickey & Golding 2002; Hirst *et al.* 2002; Beasley & Kneale 2003; Gray & Delaney 2008), which makes it extremely important to identify to species level when monitoring responses of macroinvertebrate communities to industrial discharges.

In some isolated cases, no response has been measured in community composition and ecological function to the presence of industrial discharges such as coal ash (Reash 2004), which may be due to circum-neutral pH and the low bioavailability of metals (the importance of acidity is discussed later). Downstream of discharges, there may be some evidence of recovery as dilution and removal of metals through precipitation improves water quality, although in many cases the entire community does not fully recover (Gray 1998) (see later section on recovery).

In the majority of studies, the authors have assessed impacts in terms of metrics such as species richness and abundance with some summarisation achieved through the use of biological indices. National monitoring of riverine systems also utilise these assessment techniques, the most commonly used being the Biological Monitoring Working Party (BMWP) score (see Chapter 6). However, this system was based upon organic pollution rather than acidity and metals contamination and authors have questioned the use of this for industrial pollution (e.g., Metcalfe-Smith 1994). Grey and Delaney (2008) compared scoring systems in the assessment of acid mine drainage and found that the BMWP score was reasonably well correlated with abiotic factors such as pH, sulphate and Zn. However, a similar study carried out in Spain found that BMWP score was not entirely appropriate because some families with the highest BMWP scores were found at both polluted and unpolluted sites in rivers impacted by mine drainage (García-Criado et al. 1999). Alternative biological indices such as the Acid Water Indicator Community (AWIC) index has been proposed, which can distinguish between acid and neutral sites (e.g., Wade et al. 1989; Davy-Bowker et al. 2005; Riipinen et al. 2008), but these have not been proven to be appropriate for discharges that involve significant concentrations of metals. Given the previously highlighted need to identify organisms to species level combined with the complicated interaction between metals and pH, simple biological indices may not be appropriate for assessment of the ecological impact of metalliferous contamination.

Direct toxicity of metals

Toxicity within organisms only occurs when the rate of metal uptake exceeds the combined rate of excretion and detoxification of the metabolically available form of the metal (Rainbow 2002). The effects of metals on organisms can range from those which are undetectable, through sub-lethal to lethal. Mechanisms of metal toxicity are mainly linked to biochemical reactions involving (a) competitive blockage of a functional group or macromolecule at the cell membrane, this can disrupt transport and membrane stability (Gerhardt 1993); (b) displacement of essential cations by toxic metals, zinc can cause cross-linking of DNA-molecule inhibiting transcription process, whereas copper can depress the nervous system (Gerhardt 1993); and (c) conformational change in proteins, copper can bind to certain enzymes and inhibit their action (Flemming & Trevors 1989). Heavy-metal physiological effects are manifested mainly as hypoxia (deficiency of oxygen reaching the body tissues), caused by a reduction in gas exchange due to coagulation and precipitation of mucus or cytological damage (Koryak et al. 1972; Sridhar et al. 2001; Niyogi et al. 2002).

This information has mainly been obtained through the use of laboratory based toxicity tests. The main problem with toxicity tests is that they are rather simplistic given that the majority use single species and/or single contaminants and are environmentally unrealistic (see Buikema & Voshell 1993 for a compre-hensive review). However, many of, and in fact the majority of, metalliferous discharges within the environment have a number of different metals within them, and in fact can contain other contaminants such as Polyaromatic Hydro-carbons (PAHs) (e.g., Tiefenthaler *et al.* 2008). This can have a significant effect upon the bioavailability and toxicity of the elements to organisms. There are three basic ways in which they can interact with each other (Gerhardt 1993):

1. Additivity. This occurs where the action of the combined pollutants is equal to the sum of the individual effects.
2. Synergism. The action of the combined pollutants is greater than the sum of the individual effects.
3. Antagonism. The action of the combined pollutants is less than the sum of the individual effects.

There have been limited investigations into the interactions between two or more metals and their effects on biota within the same environment. An antagonistic effect was found to occur between Zn and Cd in *Daphnia magna* (Bat *et al.* 1998; Barata *et al.* 2002; Komjarova & Blust 2008). Both additivity and antagonism have been reported in previous studies of AMD toxicity in fish (Finlayson & Verrue 1982). Studies in a mining catchment in West Virginia showed that the impacts on biota were a result of a suite of chemical constitu-ents (Al, Ni and Mn) rather than a single critical toxin (Freund & Petty 2007). The individual concentrations of toxins were at levels lower than the critical threshold value for biological impairment, and the toxic effects were possibly due to the interactions between multiple stressors. Amphipods were also reported to show greater responses to a mixture of metals, rather than a single contaminant (Besser *et al.* 2009), and toxicity has also been found to increase in *Daphnia magna* (Negiliski *et al.* 1981; Biesinger *et al.* 1986). However, suppression of metal uptake has also been reported to occur in amphipods with uptake of Cu, Ni and Zn being suppressed by Cu (Komjarova & Blust 2008). The reported metal interactions were found to predominantly occur at low concentrations where binding sites are not completely occupied. Interactions in field condi-tions are, therefore, likely to be dependent upon the concentrations of metals within the water column together with the mixture of metal species. The potential importance of the interaction between organic and inorganic pollu-tants has not been adequately researched to date.

The diversity and abundance of macroinvertebrates has been found to be reduced in the presence of metals such as Cu and Zn, independent of pH (Soucek *et al.* 2000; Hickey & Golding 2002; Hirst *et al.* 2002). More specifically, mayflies have been reported to be significantly affected by the presence of metals (Cu, Zn, As) with a loss in numbers (Mori *et al.* 1999; Hickey & Golding 2002). This agrees with overall responses to metalliferous discharges. However, some mayfly species have been shown to be particularly tolerant, for example the taxa *Deleatidium* was one of the few found where iron concentrations were above 3 mg L^{-1} and pH was <4 (Winterbourn *et al.* 2000). Although oligochaetes are often considered to be tolerant of pollution, they have been found to be absent in the presence of some metals (Mori *et al.* 1999). The reduction in numbers of organisms is probably due to a reduction in survival rate together with emerging and hatching success as demonstrated in the chironomid *Chironomus tentans* (Hruska & Dubé 2004).

The possible causes of sub-lethal or toxic effects are many although most investigations have been laboratory based (see Buikema & Voshell 1993). Enzymatic studies on populations of *Hydropsyche exocellata* within rivers receiving metals showed signs of oxidative stress in the organisms. However, it is possible that this was also caused by additional pollutants within the system such as PAHs (Barata *et al.* 2005).

Organisms with thin cuticles such as chironomids have been shown to be more susceptible to metals such as Cu than damselflies, caddisflies and stoneflies (Kosalwat & Knight 1987). This could be due to increased uptake of metals into tissues as invertebrates with sclerotised exoskeletons have been shown to have reduced uptake of metals (Elangovan *et al.* 1999; Winterbourn *et al.* 2000). This is supported by some field observations: an absence of floating chironomid pupal exuviae was found to be the result of metal contamination in ponds as a result of smelting activities (Brooks *et al.* 2005). In addition, there were some changes in chironomid assemblages in response to changes in Cd inputs with time. However, this does not correspond to the majority of responses recorded in communities within rivers where chironomids are often the most tolerant species (e.g., Gower *et al.* 1994; Gray 1998; Groenendijk *et al.* 2002).

Much attention has been given to the effects of metals on chironomid larvae due to the potential for these organisms to be used as a bioindicator of pollution (Warwick 1990). Deformities in chironomid larvae have been reported to occur in response to heavy-metal contamination with the main changes occurring in the mouthparts (Wiederholm 1984; Kosalwat & Knight 1987; Michailova & Belcheva 1990; Moore *et al.* 1994; Dickman and Rygiel 1996; Hudson & Ciborowski 1996; Bird 1997). The deformities can include fused teeth, split median tooth, missing teeth, extra teeth, Kohn gap, deformed mandible and abnormally shaped median tooth (de Bisthoven *et al.* 2001; Martinez *et al.* 2004). The occurrence of high deformity rates at moderate concentrations of metals has been reported and suggested to be due to their tolerance of metals (Martinez *et al.* 2002). At higher concentrations, susceptible individuals were killed, and therefore, deformity rates decreased (Martinez *et al.* 2004). Acute waterborne toxicity to metals including Cu, Cd, Ni and Zn can occur in chironomids, and this tends to be more prevalent in the first larval instar, which is more sensitive (e.g., Gauss *et al.* 1985; Pascoe *et al.* 1989; Béchard *et al.* 2008).

Therefore, the use of the *Chironomus* genera as an indicator of metal pollution may be limited by the extent of contamination. However, the alternative genera *Procladius* may be a viable alternative as it is more tolerant to industrial contamination and tends to show morphological response to contaminant levels where *Chironomus* is highly affected or has been eliminated (Diggins & Stewart 1993). Fluctuating asymmetry, which is a disturbance in the form of a bilaterally symmetrical organism (such as *Chironomus*), has also been reported to occur as a result of metal contamination, although this is seen as a manifestation of stress at a genetic level (Groenendijk *et al.* 1998; Ilyashuk *et al.* 2003).

The presence of some macroinvertebrate species in areas with high concentrations of metals is thought to be related to inherited metal tolerance. Some species that have been found in acidic metal-contaminated streams are not normally known for their tolerance to metals and/or acidity, and it has been proposed that this is the result of evolutionary exposure to naturally acidic streams in the area (Winterbourn *et al.* 2000). Adaptation to metals has been implicated in the lack of differences observed between uncontaminated and contaminated samples within moderately polluted rivers (De Jonge *et al.* 2008). Metal adapted genes have been found in populations of *Chironomus riparius* exposed to Zn and Cu (Groenindjik *et al.* 2002). Consistent patterns in adaptation in field conditions have, however, not been found, and this is thought to be due to the influx of non-adapted genes from local non-tolerant populations via invertebrate drift from unpolluted sites (Groenindjik *et al.* 2002). The mechanism of tolerance to metals is not fully understood, although some tolerant species such as *Chironomus riparius* have been shown to lose metals such as Cd and Zn during metamorphosis (Groenendijk *et al.* 1999). In addition, adapted midge populations have been shown to have an increased storage capability in the guts and increased excretion efficiency (Postma *et al.* 1996). An increase in gill ventilation in response to acid mine drainage could suggest that organisms use this as a way of removing toxic substances from this sensitive region (Gerhardt *et al.* 2005). As noted earlier the stage of life cycle of the organism is important with earlier stages tending to be more sensitive than later stages as observed in *Chironomus tetans* and *Corbicula manitensis* (Wang 1987).

Resistance to metals by some species and a decrease in interspecific competition because of low variety of benthic organisms in polluted zones could explain the proliferation of certain species within impacted communities (Mori *et al.* 1999).

Importance of pH

The speciation and bioavailability of metals is strongly dependent upon pH. Metals respond in different ways to pH, but in general, as pH decreases, the availability of metals increases. In many cases the effect of metals at circum-neutral pH has been shown to be minimal; for example, studies in southwest

UK streams showed that there was a lack of significant change in community structure in relation to metals, but the acid-base status of the river was much more important (Hirst *et al.* 2002). However, under acidic conditions the toxicity of metalliferous drainage can be enhanced. A study by Gerhardt *et al.* (2005) showed that the effects of acid mine drainage were found to be more noticeable when pH values were less than 5.

In response to low pH, there is generally a reduction in taxonomic diversity and a change in community structure (Winterbourn *et al.* 2000; Liu *et al.* 2003). Ephemeroptera are particularly affected by pH with reports of many taxa within this group being removed at pH < 6.3 (Guérold *et al.* 2000). Crustaceans and molluscs are also strongly affected by low pH (Albers & Camardese 1993). It is evident that many community responses to both metals and acidity are similar, and therefore, it is difficult to separate them out in rivers receiving acidic metalliferous discharges.

There is limited information on the physiological responses of macroinvertebrates to acidity but pH is thought to affect organisms directly due to an increase in cell membrane permeability at low pH (Camargo 1995; Havas & Advokaat 1995), which can cause problems with osmoregulation (Cameron & Iwana 1989). Harvas and Advokaat (1995) noted that, in acidic environments, insects with less permeable cell membranes, such as *Corixa punctata* (Hemiptera, water bug), were more tolerant. In fish, it is also reported that Ca levels can be dramatically reduced within tissues which interfere with their reproductive cycle (EPA 1980). Necrosis of the gill epithelium of fish has also been reported, although this may be related to increased aluminium availability rather than a direct effect of the acidity (Cronan & Schofield 1979).

Although the bioavailability of metals is thought to increase at low pH, uptake of metals may actually be reduced in acidic conditions. Gerhardt *et al.* (2005), for example, reported that metal body burdens in mayflies were shown to decrease at low pH. There have been a number of explanations proposed for this pattern but the most viable are:

1. The presence of hydrogen ions at membrane receptors could effectively compete with metal ions, therefore reducing uptake (e.g., Campbell & Stokes 1985; Yan *et al.* 1990; Gerhardt 1993).
2. Metals that are present at the cellular surface of organisms could be desorbed at lower pH (Krantzberg & Stokes 1988).

Some organisms have been shown to be adapted to low pH environments and mayflies have been reported to occur at pH as low as 4.5 (Kelly 1991). Acidophilic organisms often maintain their cytoplasm at the same pH as neutrophilic relatives through the use of mechanisms such as proton uptake, cell wall permeability and internal buffering (e.g., Pick 1999). Experiments using the mayfly *Choroterpes picteti* have shown that sub-lethal behavioural effects can be observed at low pH (3.3) with decreased locomotion and increased diurnal activity (Gerhardt et al. 2005). It should be noted that, when metals were also present, this caused more significant effects in terms of locomotion and ventilation than acidity alone (Gerhardt et al. 2005).

The presence of adapted populations demonstrates the importance of the history of contaminated environments as it is fluctuations away from the 'norm' that cause impacts upon communities (Courtney & Clements 1998), and this can be significant even in very short episodes (Felten & Guérold 2006). However, as noted earlier there is little information on the importance of short episodes of acidification and/or metal contamination on organisms in the environment, particularly in the process of recovery.

The effects of acidity and metals are extremely difficult to separate out, and within the literature there are no satisfactory reports that effectively do this. If we are to fully appreciate and understand the effects of different types of metalliferous discharge then this should be a priority within the ecotoxicological community. Recent advances in metabolomic studies provide an opportunity to separate out these two stresses using experimental approaches. Some reports have shown that there may be potential biomarkers that can be used to monitor responses to metals (see Morgan et al. 2007 for a comprehensive review of this area), but further work is needed before metals and acid toxicity are fully understood.

Iron hydroxides

Iron is an essential element to both faunal and floral growth and therefore is not often examined in toxicological studies, despite often being present at high concentrations within industrial discharges (Nordstrom et al. 2000; Younger et al. 2002). In the majority of waters, iron is transported in the particulate form (e.g., Förstner & Wittman 1979); however, under acidic conditions iron can be present as Fe(II), which is the most toxic form (Gerhardt 1992). There is little information on the direct toxicity of Fe to aquatic invertebrates, but it is possible that mechanisms could include its role in DNA and membrane damage as observed in vertebrates (Vuori 1995). However, although direct toxicity is unlikely, except in very extreme environments iron poses another problem to aquatic communities. Iron can react with oxygen within receiving waters to produce iron hydroxides which are present as precipitates within the water column and quickly settle out onto the riverbed producing the characteristic red/orange colouration of receiving water courses (e.g., Younger et al. 2002) (Fig. 4.1). The chemical reactions involved in the production of iron hydroxides also generate acidity within the environment (equations 1 and 2)

$$Fe^{2+} + 1/4O_{2(aq)} + 2^1/_2H_2O \rightarrow Fe(OH)_{3(s)} + 2H^+ \qquad (1)$$
$$Fe^{3+} + 3H_2O \leftrightarrow Fe(OH)_{3(s)} + 3H^+ \qquad (2)$$

Figure 4.1. Acid mine drainage from Parys Mountain, Anglesey, UK, showing the characteristic orange colour of iron hydroxide deposits coating the riverbed. Photograph taken 3 km downstream of the main discharge. See colour plate section.

Investigations in a catchment receiving mine effluents revealed that all of the rivers were impacted in terms of macroinvertebrate assemblages when Fe was present at concentrations $>0.44\,mg\,L^{-1}$ (Freund & Petty 2007). This is because of the highly reactive nature of iron under oxidising conditions. The water quality limit in the United States is set at $0.5\,mg\,L^{-1}$, because above this concentration iron hydroxide precipitates can coat stream beds. Within the EU, there is currently no specific water quality standard for iron, although the drinking water standard is set at $200\,\mu g\,L^{-1}$. However, the Water Framework Directive will require good ecological and chemical status for water bodies, and therefore the potential impact of low concentrations of iron on biota through the precipitation of hydroxide minerals may increase the importance of iron release to the environment. Metal precipitates forming in rivers receiving metalliferous discharges are not only iron hydroxides but can also be manganese oxides and aluminium hydroxides (e.g., Younger *et al.* 2002; Balistrieri *et al.* 2003). Al, Fe and Mn hydroxide precipitates have been observed on invertebrate species (Fig. 4.2) and on the sediments within affected streams (e.g., McKnight & Feder 1984; Gerhardt 1992), with crevices being filled by flocculants (Schmidt *et al.* 2002).

All these metal precipitates act in a similar way to any other suspended solids within rivers and can impact macroinvertebrates in a number of ways.

Figure 4.2. *Habrophlebia fusca,* showing evidence of iron hydroxide deposits on its appendages. See colour plate section.

1. Clogging of gills and feeding structures (e.g., Cordone & Kelley 1961; Hynes 1970; Gerhardt 1992). Physical damage was shown to cause inhibition of sediment inhabitation of *Hydropsyche orientalis* and *Cheumatopsyche brevilineata* (Sasaki *et al.* 2005) and to lead to chronic lethal effects in hydropsychid caddisflies (DeNicol & Stapleton 2002).
2. Limitation of food resources. Suspended solids can reduce the growth of mosses and other plants which act as attachment sites for periphyton and also accumulate fine debris which can act as a food supply for collector-gatherers (Suren & Winterbourn 1992). Grazers can also be affected in terms of reduced uptake of sediment contaminated periphyton by the presence of suspended solids (Newcombe & MacDonald 1991; Broekhuizen *et al.* 2001). Ochre may cover leaves preventing access for shredders (Nelson 2000).
3. Lowering primary production. Suspended solids reduce the amount of light penetrating the water column and therefore reduce the growth of algae (Bilotta & Brazier 2008) potentially disrupting the food web.
4. Loss of habitat. The precipitation of significant quantities of hydroxides on the beds of rivers which is characteristic of iron contamination (Younger *et al.* 2002) can destroy potential habitat for a wide variety of benthic organisms (e.g., Gray 1996a, b).
5. Damage to organisms. Langer (1980) has also shown that suspended solids can scour and abrade benthic organisms damaging respiratory organs.

Iron precipitates may also be taken up by aquatic organisms in their food and the presence of iron-enriched food can actually reduce the absorption of Fe via the gut membrane by causing the formation of Fe hydroxide precipitates on the gut wall (Gerhardt 1995).

As well as affecting individual macroinvertebrates, iron and aluminium contamination and the resulting suspended solids were also found to be an important factor in affecting community structure (Gower *et al.* 1994; Auladell 2009). McKnight and Feder (1984) and Sasaki *et al.* (2005) reported that macro-invertebrates were inhibited when the riverbed was covered with metal pre-cipitates. Grazers feeding on periphyton and other biofilms have been reported to be the first invertebrates eliminated by increasing Fe concentrations (Rasmussen & Lindegaard 1988). Increased drift of invertebrates has also been reported to be caused by an increase in suspended solids (Gammon 1970; Ryder 1989), and this has also been linked to changes in drift in response to acid mine drainage (Elliott *et al.* 1988).

The proliferation of certain groups in communities impacted by iron-rich discharges could reflect the adaptation of certain taxa to living in deposited sediments such as oligochaetes and some chironomids (Wiederholm 1984; Mackay 1992; Smith 2003). This correlates with the changes in communities reported downstream of mine drainage (see earlier).

Iron hydroxides are also extremely reactive substances and other metals can be sorbed to their surface (e.g., He *et al.* 1997), posing a toxicological risk to inverte-brates within the system in terms of direct uptake through food (see section below).

This suggests that, where metalliferous discharges are dominated by iron (or aluminium and to a lesser extent manganese), the processes by which organ-isms are impacted are very different from those for non-ferruginous waters. This has implications not only in the assessment of impacts using biological indices (the differential response by groups of organisms and individual species suggests that a simple biological index such as BMWP would not be sufficient) but also on the recovery following remediation activities.

Uptake and bioaccumulation of metals

The presence of adapted organisms within metal-contaminated environments may pose a risk to the food chain. This will depend upon whether metals are taken up within the receiving organisms and are bioaccumulated or biomag-nified up the food chain. Within aquatic environments, organisms can take up metals directly through the water column (Rhea *et al.* 2006) but they may also be taken up via food or when burrowing in sediments.

The majority of studies have shown that macroinvertebrates are able to take up metals from sediments with direct relationships between tissue and sedi-ment concentrations of Cu, Cd, Zn and Pb being demonstrated for chironomids

(Martinez *et al.* 2001), Hydropsyche (Solà *et al.* 2004) and a range of macroinvertebrates (Goodyear & McNeill 1999). However, a correlation between sediment metal concentrations and invertebrate body burdens has not always been found (Bervoets *et al.* 1997, 1998), which may be due to specific responses in different species or groups. Macroinvertebrates source their food from a number of niches within the freshwater environment and are generally grouped into collector-gatherers, scraper-grazers, predators, shredders and filterers (e.g., Begon *et al.* 1990), so-called functional feeding groups. Research has shown that the uptake of metals into organisms is strongly affected by the feeding niche (Smock 1983; Timmermans *et al.* 1989; Kiffney & Clements 1993). Predatory insects in particular have been shown to obtain the majority of bioaccumulated metal from food (Hare *et al.* 2003), although food borne aluminium was found not to be important in the removal of predators downstream of mine drainage and the cause was thought to be related to direct toxicity via gills (Soucek *et al.* 2002). Van Hattum *et al.* (1991) found that, for invertebrates of different feeding habits, Zn concentrations decreased in the order predator > filter feeder > deposit feeder. In other studies, however, no difference between herbivore-detritivores and predators was found in terms of their body burdens of metals (Winterbourn *et al.* 2000). Within metal-contaminated environments metals are often at highest concentrations within biofilms and sediments (Farag *et al.* 1998). Biofilms are located at the bottom of the food web, supporting, in conjunction with detritus, most of the aquatic ecosystem, and often forming the interface between water toxicants and the food web (Sabater *et al.* 2007). This niche is utilised preferentially by shredder-scrapers, and this group has been shown in many investigations to take up the largest concentrations of metals such as Cd, Pb and Zn (Farag *et al.* 1998; Rhea *et al.* 2006). Investigations of rivers in Idaho that receive mine drainage found that the primary route of exposure of invertebrates to Cu was via periphyton (Beltman *et al.* 1999) and concentrations of metals in biofilms were a good predictor of metals in invertebrates (Rhea *et al.* 2006). In contrast, Borgmann *et al.* (2007) showed that concentrations in epiphytic and detrital food did not affect uptake of metal into *Hyallela azteca*. The prevalence of contradictory information within the literature suggests that the sources and patterns of uptake into macroinvertebrates may be both specific to the species and metal involved. Although the majority of studies have focused upon sediment and food concentrations it has also been suggested that invertebrates could collect metals from the sediment pore water and the boundary layer between the surface of the sediments and the overlying water column (Bervoets *et al.* 1998) adding another potential variable. Leaf litter deposited in mine waters also accumulates large amounts of metals such as Fe, Mn and Zn (Nelson 2000; Batty & Younger 2007), and this may be particularly associated with fine particulate organic matter (FPOM) due its larger surface area in comparison to coarse particulate organic matter (CPOM) (Watanabe *et al.* 2008), potentially constituting a risk to shredders. Particulate matter has also been found to be an important source of metals in *Hydropsyche exocellata* (Barata *et al.* 2005).

In terms of biomagnification within the food web, studies have been limited and those that have been reviewed have shown that there is no evidence for magnification of the metals Zn and Cu between feeding guilds, although there may be some potential for magnification of Pb in certain circumstances (Goodyear & McNeill 1999).

The uptake of high concentrations of metals by macroinvertebrates is likely to result in direct toxicity, and therefore some groups may be preferentially removed in impacted streams. For example, collector-browsers and collector-filterers showed a significant reduction in response to elevated levels of Cu and Zn (Hickey & Golding 2002). Survivorship of juvenile amphipods was also found to be significantly reduced when fed on copper contaminated diets, and reduced feeding was also observed in adults (Roberts *et al.* 2006). In contrast, some authors have suggested that there is no specific evidence for toxicity even where metals are accumulated within invertebrates (Borgmann *et al.* 2007). This is because heavy metals can be stored in organisms as non-toxic species or bound to metallothionein (Gerhardt 1993). De Schamplherlaere *et al.* (2004) demonstrated that, although dietary copper was effectively taken up by *Daphnia magna*, this caused an increase in reproduction and growth which was considered to be due to the storage of Cu in a non-toxic form.

The pathways by which metals are taken up by aquatic organisms within metal-polluted environments, therefore, remain rather unclear with contradictory evidence and, therefore, require further investigation. It is likely that the mixture of metals present, the acidity of the environment and the presence of precipitate material all influence the relative importance of metal sources within the environment on uptake into organisms.

Ecosystem function

As illustrated in previous sections some major groups are more affected by metals and acidity than others. This has the potential to change the structure of the community and has been reported in response to Mn and Zn (Nelson 2000); Al, Mn and Fe (Doi *et al.* 2007); and acidification (Guérold *et al.* 2000). Although direct toxicity has been demonstrated, indirect effects on community structure may also be caused by metals and acidity. Communities and food webs within acidified streams tend to be structurally simpler due to the absence of herbivores, which is thought to be caused by the absence of algae as a food source due to acidity or associated metal toxicity rather than direct toxicity to the invertebrates (Morris *et al.* 1989; Nelson 2000; Sasaki *et al.* 2005). Ledger and Hildrew (2000a) reported the absence of specialist herbivorous species in conditions of low pH and high metal concentrations.

However, a change in community structure does not necessarily have an impact upon function within the community. When we refer to function, we are referring to the way in which an ecosystem can contribute to fundamental processes within the landscape and may include functions such as oxygen consumption, and production, carbon mineralisation, organic matter production and sedimentation. One of the most common methods of assessing impacts of pollutants on ecosystem function is through the measurement of leaf litter processing as this can be used to monitor the effects of changes in bacterial, fungal and invertebrate communities. To the authors' knowledge, this is the only ecological function that has been directly assessed in rivers receiving metalliferous discharges to date.

It is widely reported that leaf litter decomposition is reduced in waters impacted by metals and acidity (e.g., Giesy 1978; Carpenter *et al.* 1983; Burton *et al.* 1985; Leland & Carter 1985; Maltby & Booth 1991; Carlisle & Clements 2005; Batty & Younger 2007; Auladell 2009). The causes of this remain unclear, although some authors suggest that the functionally dominant taxa was the most sensitive to contamination (Carlisle & Clements 2005) or that lethal and sub-lethal effects affected both fungal and/or invertebrate assemblages responsible for processing of leaf material (Newman *et al.* 1987; Cheng *et al.* 1993; Maltby *et al.* 1995). This has been directly linked to the abiotic environment (metal and nutrient concentrations, pH and conductivity) (Woodcock & Huryn 2005). However, other studies have shown that, although community structure is affected by the presence of metals in impacted rivers, no corresponding reduction in leaf litter processing was reported (Nelson 2000). This is thought to be due to redundancy within the system where metal-sensitive shredder species are replaced by tolerant equivalents that perform the same function (Schindler 1987; Nelson 2000; Woodcock & Huryn 2005). In freshwaters, a number of organisms are involved in plant litter decomposition including tubificid worms, chironomid larvae, isopods and amphipods (Mason 1976), and therefore, it appears that there is potential significant capacity for redundancy in the system. This has also been demonstrated in acidic streams for organisms involved in utilising biofilm and loose (FPOM) resources where acid-tolerant species demonstrate generalist feeding patterns maintaining the link between invertebrates and algae through the presence of shredder detritivores (Ledger & Hildrew 2000a, b, 2005). Therefore, ecosystem function appears to be maintained in some cases, even in polluted conditions. However, further research is required to determine the effects on other functions within the ecosystem, and the limitations of redundancy.

Importance of sediments

Most studies of metal pollution have examined the response to waterborne pollutants. This has led to a predominance of water quality legislation such as the Water Framework Directive (EU/2000/60/E). However, although the presence of metals within river waters has been shown to be important in responses of macroinvertebrate communities, in many cases concentrations of metals are actually below toxicity thresholds or water quality objectives. This does not necessarily result in a 'clean' environment, because high concentrations of metals may be present in sediments and particularly those that are fine grained (e.g., Petersen *et al.* 1998; Brumbaugh *et al.* 2007). For example, sediments within the Britannia Creek in Vancouver were found to be significantly enriched with copper, although active mining ceased in 1974 (Levings *et al.* 2004). Structural and functional changes have been reported in rivers that have received metals historically but no longer have a direct metals source due to the presence of metals within sediments (Besser *et al.* 2009). The extent to which this may affect communities is unclear as there is little information on the ranges and degree of contamination in sediments, particularly in relation to urban pollutants (Beasley & Kneale 2003).

Metal sludge deposition was reported to cause significant toxicity to amphipods related to Cu, Zn and Cd. Zn in sediment was found to be particularly important in toxicity to cladocerans (Finlayson *et al.* 2000). Zn accumulation is found to be in direct proportion to sediment levels, and therefore, the aqueous and particulate sources are important in controlling the concentrations of Zn found in macroinvertebrates (Goodyear & McNeill 1999). Sediment Fe and Mn concentrations were found to be closely related to the growth and reproduction of amphipods (Besser *et al.* 2007); however, pore water concentrations of Ni, Zn, Cd and Pb were also found to be important in amphipod survival. The process by which organisms are affected by sediment metal concentrations is not clear (e.g., Milani *et al.* 2003), although it is possible that direct toxicity may be due to ingestion (see earlier section). Where Fe is present then it is more likely that invertebrate response is affected by the presence of iron hydroxide precipitates (see earlier section).

It is thought that continued remobilisation of metals within the hyporheic zone may be the cause of changes in the hyporheic community, which has been reported to occur even 7 years after the source of metals was removed (Nelson & Roline 1999). This was attributed to changes in the chemical characteristics in the hyporheic zone from influxes of water at lower pH which would alter the form in which the metal was present. In floodplain sediments, changes in the water table can cause fluctuations in the oxygen status of the deposits and, therefore, release metals from their insoluble form, but this is less likely to occur within river channel sediments (Hudson-Edwards *et al.* 1998). When metals are present in the more mobile fractions and, therefore, have higher bioavailable concentrations, then they can cause toxic responses in sediment associated organisms (Riba *et al.* 2006). Scouring and dispersal of river sediments may also occur during periods of high flow, potentially releasing metals, but there are no reports of the impact this has on metal concentrations or the ecology within the literature.

Experiments conducted in field situations revealed that once iron-coated rocks were placed in a clean environment desorption of sulphate occurred from the precipitates, and this resulted in destabilisation of oxyhydroxides enough to release them from the rock surface (DeNicol & Stapleton 2002). The release of fine flocculated precipitate was thought to have a much greater impact upon the fauna than the adhering precipitate, which was not directly toxic. The dispersal and deposition of sediments, therefore, can affect metal toxicity temporally (Besser *et al.* 2009). This has important implications in remediation as once the source of contamination is removed the accumulated iron precipitates on the sediments will no longer be in equilibrium with the aqueous environment and, thus, could be a potential source of metals to the environment (DeNicol & Stapleton 2002).

Not all reports support the importance of sediment chemistry. In other studies, water chemistry was found to be more important than sediments in the toxicity to benthic communities (Van Damme *et al.* 2008). This was also true when comparing surface water to pore water quality, which is closely related to sediment chemistry (Battaglia *et al.* 2005). Al and Fe contaminated sediments were also found not to be directly toxic and that it was the pore waters and water column that were the cause of toxicity in invertebrates (Schmidt *et al.* 2002). Again, it is likely that the toxicity is related to specific environmental characteristics and the metals present.

If we return to the question of the Water Framework Directive, this legislation requires that surface waters should be of 'good ecological and chemical status' by the year 2015. Although this does not directly refer to sediment quality, it is clear from the reports above that sediments may act as a source of metals to the environment and affect the ecological status of the system. It is, therefore, imperative that sediment quality is also considered in the management of river catchments. Macklin *et al.* (2003) noted that there are no specific European standards for heavy metals in sediments and this has not changed to date. However, a recent report by the Environment Agency in the UK (Hudson-Edwards *et al.* 2008) has highlighted the potential importance of mining sediments as a risk to ecosystem health and reported on the development of sediment quality guidelines for the assessment of harm from such deposits. The development of guidelines is a major step forward in tackling this potential environmental threat, although it is recognised that more toxicological data are required for sediments.

Recovery of macroinvertebrate communities

There have been significant efforts to prevent the release of industrial discharges to rivers over recent years following changes in environmental legislation. This has been largely successful in terms of point sources, although diffuse sources and the residual sediment pollution remain a major challenge. There have been a number of studies into the recovery of riverine systems following disturbance from channel modification, flooding, drought, acidification and contamination with pesticides (see Milner 1994; Ormerod & Durance 2009). However, studies into recovery following pollution by metals and organics is less common, possibly due to the difficulty in addressing the diffuse sources of these pollutants. However, over recent years, there have been some efforts to remove point sources of metals, particularly where mining activities have ceased and/or remediation strategies applied. For example, within the UK, constructed wetlands have been used to remove metals (mostly iron) from coal mine effluent in a number of sites (Coal Authority 2009). The removal of point sources of metals may lead to the recovery of stream communities, and there has been some attempt to quantify rates and success of this recovery.

It is often difficult to assess recovery, due to an absence of information on the required endpoint. This is particularly true for mining effluents where the pollution may have been affecting the stream community for many hundreds or thousands of years (Runnells *et al.* 1992). In this case, it is clearly impossible to choose an endpoint that is the original community due to a lack of data. Alternatively, a reference site may be chosen where the community can be used as a model for the recovering community (Milner 1994); however, unexploited areas that are geologically similar to mined regions are extremely rare, and therefore, the endpoint may not be an accurate reflection of the true 'natural' community. Recovery in urbanised areas is often even more of a challenge to assess quantitatively, largely due to the presence of multiple sources of pollutants. Langford *et al.* (Chapter 13) provide details of the recovery of an urbanised catchment. Function and production endpoints have also been suggested as possible aims for remediation activities (Milner 1994), but in most examples within the literature, it is a simple measure of density and/or species richness which is used to assess recovery following remediation of metalliferous discharges.

A study in the coal fields of southwest Virginia showed that following remediation action total taxa richness showed consistent improvement over time corresponding to an increase in pH from 4.70 to 6.97 and a decrease in Al concentration (Simon *et al.* 2006). This response was also recorded following the closure of metal mines in China (Watanabe *et al.* 2000) and Idaho, USA (Holland *et al.* 1994). However, in none of these examples was recovery found to be complete. More specifically Ephemeroptera did not recover, most likely due to the longer amount of time taken for recolonisation of this group (Simon *et al.* 2006), but even when recovery is monitored over a long period of time, complete recovery is not achieved. This may be due to a lack of suitable habitat (Holland *et al.* 1994; Armitage *et al.* 2007), continued input of contaminants from diffuse sources (Holland *et al.* 1994; Prat *et al.* 1999; Simon *et al.* 2006) or insufficient remediation of the pollutants (Watanabe *et al.* 2000). It is, therefore, evident that the success and the rate of recovery will depend on a number of factors such as the type of disturbance, the duration, habitat conditions and availability and the location of potential colonisers as previously reported (Williams & Hynes 1977; Niemi *et al.* 1990; Robinson *et al.* 1990; Wallace *et al.* 1991). A source of organisms is absolutely vital to recovery. Battaglia *et al.* (2005) noted that the lack of colonisation of clean sediment was due to an absence of colonising organisms. The community structure that does result will depend upon the clean communities that act as a source (Beltman *et al.* 1999). It is not as simple as this, however, as macroinvertebrates also require suitable biotic habitats such as macrophytes and algae as well as clean sediments (Laasonen *et al.* 1998).

Although previously it has been suggested that, in terms of recovery, the amount of time will be longer after physical disturbance than chemical disturbance (Smith 2003), it is clear that the time involved may be significant due to the difficulty in remediating all diffuse sources of metals within any given catchment. In addition, in order for recovery to be successful, the source of contamination must be completely removed as even short periods of contamination such as low pH can significantly affect the success of recovery (Bradley & Ormerod 2002; Felten & Guérold 2006). Even where sites have been remediated, pollutants that remain in sediments or floodplains can still affect the river course (Hutchinson & Rothwell 2008) (see the section on Importance of sediments).

Conclusions

Historically, metalliferous discharges to the aquatic environment have had significant impacts upon the associated communities, and although there have been advances in the treatment and prevention of these contaminant sources, diffuse pollution and new metal contaminants still pose a significant threat to the environment. It is clear that metals do directly and indirectly affect organisms within the environment, but where this is mixed with acidity, it is difficult to separate out their specific effects. Recent advances in biochemical analyses provide an opportunity to elucidate the specific stress responses in both flora and fauna to individual and mixed pollutants which may aid in prioritising treatment of specific compounds or discharges.

The impact of metals on organisms within the aquatic environment has focused upon either specific organisms or communities; however, it is clear that ecosystem function is a potentially important consideration when assessing the impacts of pollutants. Ecosystem function response to metals has only been assessed in any detail in terms of organic matter decomposition, where a certain amount of redundancy within the system has been reported. However, there are many other types of function within systems that should be investigated together with the identification of thresholds of redundancy.

Although the presence of metals in their dissolved form within waters has traditionally been seen as the most significant threat to organisms (particularly macroinvertebrates), it is becoming clear that the presence of metals associated with sediments constitute an equivalent (if not more important) threat. The longevity of such deposits and their reactivity to changing environmental conditions makes them a particularly important source of metal contamination to be considered for the future. It is vital, therefore, that the pathways by which the metals are transferred to organisms and the responses of sediment chemistry to changes in flow conditions and flooding amongst other environmental variables are identified in order to provide data for sediment quality guidelines for environmental organisations and enforcers.

References

Albers, P. H. and Camardese, M. B. (1993) Effects of acidification on metal accumulation by aquatic plants and invertebrates 1. Constructed wetlands. *Environmental Toxicology and Chemistry* **12**, 959–967.

Amisah, S. and Cowx, I. G. (2000) Impacts of abandoned mine and industrial discharges on fish abundance and macroinvertebrate diversity of the upper River Don in South Yorkshire, UK. *Journal of Freshwater Ecology* **15**, 237–250.

Armitage, P. D., Bowes, M. J. and Vincent, H. M. (2007) Long-term changes in macro-invertebrate communities of a heavy metal polluted stream: the River Nent (Cumbria, UK) after 28 years. *River Research and Applications* **23**, 997–1015.

Auladell, M. (2009) Ecological impact of mine drainage and its treatment on aquatic communities. Unpublished PhD Thesis, University of Birmingham, UK.

Balistrieri, L. S., Box, S. E. and Tonkin, J. W. (2003) Modeling precipitation and sorption of elements during mixing of river water and porewater in the Coeur d'Alene river basin. *Environmental Science and Technology* **37**, 4694–4701.

Barata, C., Lekumberri, I., Vila-Escalé, M., Prat, N. and Porte, C. (2005) Trace metal concentrations, antioxidant enzyme activities and susceptibility to oxidative stress in the trichoptera larvae *Hydropsyche exocellata* from the Llobregat river basin (NE Spain). *Aquatic Toxicology* **74**, 3–19.

Barata, C., Markich, S. J., Baird, D. J., Taylor, G. and Soares, A. M. (2002) Genetic variability in sublethal tolerance to mixtures of cadmium and zinc in clones of *Daphnia magna* Straus. *Aquatic Toxicology* **60**, 85–99.

Bat, L., Rafaelli, D. and Marr, J. L. (1998) The accumulation of copper, zinc and cadmium by the amphipod *Corophium volutator* (Pallas). *Journal of Experimental Marine Biology and Ecology* **223**, 167–184.

Battaglia, M., Hose, G. C., Turak, E. and Warden, B. (2005) Depauperate macroinvertebrates in a mine affected stream: clean water may be the key to recovery. *Environmental Pollution* **138**, 132–141.

Batty, L. C. and Younger, P. L. (2007) The effect of pH on plant litter decomposition and metal cycling in wetland mesocosms supplied with mine drainage. *Chemosphere* **60**, 158–164.

Beasley, G. and Kneale, P. E. (2003) Investigating the influence of heavy metals on macroinvertebrate assemblages using Partial Canonical Correspondence Analysis (pCCA). *Hydrology and Earth System Sciences* **7**, 221–233.

Béchard, K. M., Gillis, P. L. and Wood, C. M. (2008) Acute toxicity of waterborne Cd, Cu, Pb, Ni and Zn to first-instar *Chironomus riparius* larvae. *Archives of Environmental Toxicology and Chemistry* **54**, 454–459.

Begon, M., Harper, J. L. and Townsend, C. R. (1990) *Ecology*. 2nd edn. Blackwell Scientific Publications, London.

Beltman, D. J., Clements, W. H., Lipton, J. and Cacela, D. (1999) Benthic invertebrate metals exposure, accumulation, and community-level effects downstream from a hard-rock mine site. *Environmental Toxicology and Chemistry* **18**, 299–307.

Bervoets, L., Blust, R., de Wit, M. and Verheyen, R. (1997) Relationships between river sediment characteristics and trace metal concentrations in tubificid worms and chironomid larvae. *Environmental Pollution* **95**, 345–356.

Bervoets, L., Solis, D., Romero, A. M., van Damme, P. A. and Ollevier, F. (1998) Trace metal levels in chironomid larvae and sediments from a Bolivian river: impact of mining activities. *Ecotoxicology and Environmental Safety* **41**, 275–283.

Besser, J. M., Brumbaugh, W. G., Allert, A. L., Poulton, B. C., Schmitt, C. J. and Ingersoll, C. G. (2009) Ecological impacts of lead mining on Ozark streams: toxicity of sediment and pore water. *Ecotoxicology and Environmental Safety* **72**, 516–526.

Besser, J. M., Brumbaugh, W. G., May, T. W. and Schmitt, C. J. (2007) Biomonitoring of lead, zinc, and cadmium in streams draining lead-mining and non-mining areas, Southeast Missouri, USA. *Environmental Monitoring and Assessment* **129**, 227–241.

Biesinger, K. E., Christenen, G. M. and Fiandt, J. T. (1986) Effects of salt mixtures on *Daphnia magna* reproduction. *Ecotoxicology and Environmental Safety* **11**, 9–14.

Bilotta, G. S. and Brazier, R. E. (2008) Understanding the influence of suspended solids on water quality and biota. *Water Research* **42**, 2849–2861.

Bird, G. A. (1997) Deformities in cultured *Chironomus tentans* larvae and the influence of substrate on growth, survival and mentum wear. *Environmental Monitoring and Assessment* **45**, 273–283.

Borgmann, U., Couillard, Y. and Grapetine, L. C. (2007) Relative contribution of food and water to 27 metals and metalloids accumulated by caged *Hyalella azteca* in two rivers affected by metal mining. *Environmental Pollution* **145**, 753–765.

Bradley, D. C. and Ormerod, S. J. (2002) Long-term effects of catchment liming on invertebrates in upland streams. *Freshwater Biology* **47**, 161–171.

Broekhuizen, N., Parkyn, S. and Miller, D. (2001) Fine sediment effects on feeding and growth in the invertebrate grazers *Potamopyrgus antipodarum* (Gastropoda, Hydrobiidae) and *Deleatidium* sp. (Ephemeroptera, Leptophlebiidae). *Hydrobiologia* **457**, 125–132.

Brooks, S. J., Udachin, V. and Williamson, B. J. (2005) Impact of copper smelting on lakes in the southern Ural Mountains, Russia, inferred from chironomids. *Journal of Palaeolimnology* **33**, 229–241.

Brown, J. N. and Peake, B. M. (2006) Sources of heavy metals and polycyclic aromatic hydrocarbons in urban stormwater runoff. *Science of the Total Environment* **359**, 145–155.

Brumbaugh, W. G., May, T. W., Besser, J. M., Allert, A. L. and Schmitt, C. J. (2007) *Assessment of Elemental Concentrations in Streams of the New Lead Belt in southeastern Missouri, 2002–2005*. US Geological Survey Scientific Investigations Report 2007–5057.

Buikema, A. L., Jr and Voshell, J. R., Jr (1993) Toxicity studies using freshwater benthic macroinvertebrates. In: *Freshwater Biomonitoring and Benthic Macroinvertebrates* (eds. M. Rosenberg and V. H. Resh), pp. 344–398. Chapman and Hall, New York.

Burton, T. M., Stanford, R. M. and Allan, J. W. (1985) Acidification effects on stream biota and organic matter processing. *Canadian Journal of Fisheries and Aquatic Science* **42**, 669–675.

Camargo, J. A. (1995) Effect of body-size on the intraspecific tolerance of aquatic insects to low pH – a laboratory study. *Bulletin of Environmental Contamination and Toxicology* **54**, 403–408.

Cameron, J. N. and Iwama, W. T. W. (1989) Compromises between ionic regulation and acid-base regulation in aquatic animals. *Canadian Journal of Zoology* **67**, 3078–3084.

Campbell, P. G. C. and Stokes, P. M. (1985) Acidification and toxicity of metals to aquatic biota. *Canadian Journal of Fisheries and Aquatic Sciences* **42**, 2034–2049.

Carlisle, D. M. and Clements, W. H. (2005) Leaf litter breakdown and shredder production in metal-polluted streams. *Freshwater Biology* **50**, 380–390.

Carpenter, J., Odum, W. E. and Mills, A. (1983) Leaf litter decomposition in a reservoir affected by acid-mine drainage. *Oikos* **41**, 165–172.

Cheng, K., Wallace, J. B. and Grubaugh, J. W. (1993) The impact of insecticide treatment on abundance, biomass, and production of litterbag fauna in a headwater stream, a study of pre-treatment, treatment and recovery. *Limnologica* **23**, 93–106.

Coal Authority. (2009) http://www.coal.gov.uk/environmental/ minewatertreat/england/minewaterschemesengland.cfm (accessed 21 May 2009).

Cordone, A. J. and Kelley, D. W. (1961) The influences of inorganic sediment on the aquatic life of streams. *California Fish and Game* **47**, 189–228.

Courtney, L. A. and Clements, W. H. (1998) Effects of acidic pH on benthic macroinvertebrate communities in stream microcosms. *Hydrobiologia* **379**, 135–145.

Courtney, L. A. and Clements, W. H. (2002) Assessing the influence of water and substratum quality on benthic macroinvertebrate communities in a metal-polluted stream: an experimental approach. *Freshwater Biology* **47**, 1766–1778.

Cronan, C. S. and Schofield, C. S. (1979) Aluminium leaching response to acid precipitation: effects on high-elevation watersheds in the northeast. *Science* **204**, 304–305.

Davy-Bowker, J., Murphy, J. F., Rutt, G. R., Steel, J. E. C. and Furse, M. T. (2005) The development and testing of a macroinvertebrate biotic index for detecting the impact of acidity on streams. *Archiv für Hydrobiologie* **163**, 383–403.

De Bisthoven, L. J., Postma, J., Vermeulen, A., Goemans, G. and Ollevier, F. (2001) Morphological deformities in *Chironomus riparius* meigen larvae after exposure to cadmium over several generations. *Water Air and Soil Pollution* **129**, 167–179.

De Jonge, M., Van der Vijver, B., Blust, R. and Verbotes, L. (2008) Responses of aquatic organisms to metal pollution in a lowland river in Flanders: a comparison of diatoms and macroinvertebrates. *Science of the Total Environment* **407**, 615–629.

DeNicol, D. M. and Stapleton, M. G. (2002) Impact of acid mine drainage on benthic communities in streams: the relative roles of substratum vs. aqueous effects. *Environmental Pollution* **119**, 303–315.

De Schampherlaere, K. A. C. and Janssen, C. R. (2004) Effects of chronic dietary copper exposure on growth and reproduction of *Daphnia magna*. *Environmental Toxicology and Chemistry* **23**, 2038–2047.

De Schamplherlaere, K. A. C., Heijeric, D. G. and Janssen, C. R. (2004) Comparison of the effect of different pH buffering techniques on the toxicity of copper and zinc to *Daphnia magna* and *Pseudokirchneriella subcapitata*. *Ecotoxicology* **13**, 697–705.

Dickman, M. and Rygiel, G. (1996) Chironomid larval deformity frequencies, mortality, and diversity in heavy-metal contaminated sediments of a Canadian riverine wetland. *Environment International* **22**, 693–703.

Diggins, T. P. and Stewart, K. M. (1993)
Deformities of aquatic larval midges
(Chironomidae, Diptera) in the sediments
of the Buffalo River, New-York. *Journal of
Great Lakes Research* **19**, 648–659.

Doi, H., Takagi, A. and Kikuchi, E. (2007) Stream
macroinvertebrate community affected by
point-source metal pollution. *International
Reviews of Hydrobiologia* **92**, 258–266.

Elangovan, R., Balance, S., White, K. N.,
McCrohan, C. R. and Powell, J. J. (1999)
Accumulation of aluminium by the freshwater
crustacean *Asellus aquaticus* in neutral water.
Environmental Pollution **106**, 257–263.

Elliott, J. M., Humpesch, U. H. and Macon, T. T.
(1988). *Larvae of the British Ephemeroptera:
A Key With Ecological Notes*. 49. Freshwater
Biological Association, Ambleside, UK.

EPA (1980) *Acid Rain*. Environmental Protection
Agency, Washington, DC. 36 pp.

Farag, A. M., Woodward, D. F., Goldstein, J. N.,
Brumbaugh, W. and Meyer, J. S. (1998)
Concentrations of metals associated with
mining waste sediments, biofilm, benthic
macroinvertebrates, and fish from the
Coeur d'Alene River Basin, Idaho. *Archives of
Environmental Contamination and Toxicology* **34**,
119–127.

Felten, V. and Guérold, F. (2006) Short-term
physiological responses to a severe acid
stress in three macroinvertebrate species: a
comparative study. *Chemosphere* **63**, 1427–
1435.

Finlayson, B. and Verrue, K. (1982) Toxicity of
copper, zinc and cadmium mixtures to
juvenile Chinook salmon. *Transactions of the
American Fisheries Society* **111**, 645–650.

Finlayson, B., Fujimura, R. and Huang, Z. Z.
(2000) Toxicity of metal-contaminated
sediments from Keswick Reservoir,
California, USA. *Environmental Toxicology and
Chemistry* **19**, 485–494.

Flemming, C. A. and Trevors, J. T. (1989) Copper
toxicity in fresh-water sediment and
aeromonas-hydrophila cell-suspensions
measured using an O_2 electrode. *Toxicity
Assessment* **4**, 473–485.

Förstner, U. and Wittman, G. T. W. (1979) *Metal
Pollution in the Aquatic Environment*. Springer
Verlag, Berlin.

Freund, J. G. and Petty, J. T. (2007) Response of
fish and macroinvertebrate bioassessment
indices to water chemistry in a mined
Appalachian watershed. *Environmental
Management* **39**, 707–720.

Gammon, J. R. (1970) *The Effect of Inorganic
Sediment on Stream Biota*. USEPA Water
Pollution Cont. Res. Series, *18050 DWC 12/70*
USGPO, Washington, DC.

García-Criado, F., Tomé, A., Vega, F. J. and
Antolín, C. (1999) Performance of some
diversity and biotic indices in rivers
affected by coal mining in northwestern
Spain. *Hydrobiologia* **394**, 209–217.

Gauss, J. D., Woods, P. E., Winner, R. W. and
Skillings, J. H. (1985) Acute toxicity of
copper to three life stages of *Chironomus
tentans* as affected by water hardness-
alkalinity. *Environmental Pollution A* **37**,
149–157.

Gerhardt, A. (1992) Subacute effects of iron (Fe)
on *Leptophlebia marginata* (L.) (Insecta:
Ephemeroptera). *Freshwater Biology* **27**,
79–84.

Gerhardt, A. (1993) Review of impact of heavy
metals on stream invertebrates with special
emphasis on acid conditions. *Water, Air and
Soil Pollution* **66**, 289–314.

Gerhardt, A. (1995) Joint and single toxicity of
Cd and Fe related to uptake to the mayfly
Leptophlebia marginata (L.) (Insecta).
Hydrobiologia **306**, 229–240.

Gerhardt, A., de Bisthoven, L. J. and Soares, A. M.
(2005) Effects of acid mine drainage and
acidity on the activity of *Choroterpes picteti*
(Ephemeroptera: Leptophlebiidae). *Archives
of Environmental Contamination and Toxicology*
48, 450–458.

Giesy, J. P., Jr (1978) Cadmium inhibition of leaf
decomposition in an aquatic microcosm.
Chemosphere **7**, 467–476.

Goodyear, K. L. and McNeill, S. (1999)
Bioaccumulation of heavy metals by
aquatic macro-invertebrates of different

feeding guilds: a review. *The Science of the Total Environment* **229**, 1–19.

Goudie, A. (1993) *The Human Impact on the Natural Environment*. Blackwell Publishers, Oxford, UK.

Gower, A. M., Myers, G., Kent, M. and Foulkes, M. E. (1994) Relationships between macroinvertebrate communities and environmental variables in metal-contaminated streams in South-West England. *Freshwater Biology* **32**, 199–221.

Gray, N. F. (1996a) The use of an objective index for the assessment of the contamination of surface water and groundwater by acid mine drainage. *Journal of the Chartered Institution of Water and Environmental Management* **10**, 332–340.

Gray, N. F. (1996b) Field assessment of acid mine drainage contamination in surface and ground water. *Environmental Geology* **27**, 358–361.

Gray, N. F. (1998) Acid mine drainage composition and the implications for its impact on lotic systems. *Water Research* **13**, 2122–2134.

Gray, N. F. and Delaney, E. (2008) Comparison of benthic macroinvertebrate indices for the assessment of the impact of acid mine drainage on an Irish river below an abandoned Cu-S mine. *Environmental Pollution* **155**, 31–40.

Groenendijk, D., Kraak, M. H. S. and Admiraal, W. (1999) Efficient shedding of accumulated metals during metamorphosis in metal-adapted populations of the midge *Chironomus riparius*. *Environmental Toxicology and Chemistry* **18**, 1225–1231.

Groenendijk, D., Lücker, S. M. G., Plans, M., Kraak, M. H. S., Admiraal, W. (2002) Dynamics of metal adaptation in riverine chironomids. *Environmental Pollution* **117**, 101–109.

Groenendijk, D., Zenstra, L. W. M. and Postma, J. F. (1998) Fluctuating asymmetry and mentum gaps in populations of the midge *Chironomus riparius* (Diptera: Chironomidae) from a metal-contaminated river.

Environmental Toxicology and Chemistry **17**, 1999–2005.

Guérold, F., Boudot, J. P., Jacquemin, G., Vein, D., Merlet, D. and Rouiller, J. (2000) Macroinvertebrate loss as a result of headwater stream acidification in the Vosges Mountains (N-E France). *Biodiversity and Conservation* **9**, 767–783.

Hare, L., Tessier, A. and Borgmann, U. (2003) Metal sources for freshwater invertebrates: pertinence for risk assessment. *Human and Ecological Risk Assessment* **9**, 779–793.

Havas, M. and Advokaat, E. (1995) Can sodium regulation be used to predict the relative acid-sensitivity of various life-stages and different species of aquatic fauna. *Water, Air and Soil Pollution* **85**, 865–870.

He, M., Wang, Z. and Tang, H. (1997) Spatial and temporal patterns of acidity and heavy metals in predicting the potential for ecological impact on the Le An river polluted by acid mine drainage. *The Science of the Total Environment* **206**, 67–77.

Helgen, S. O. and Moore, J. N. (1996) Natural background determination and impact quantification in trace metal-contaminated river sediments. *Environmental Science and Technology* **30**, 129–135.

Hickey, C. W. and Golding, L. A. (2002) Response of macroinvertebrates to copper and zinc in a stream mesocosm. *Environmental Toxicology and Chemistry* **21**, 1854–1863.

Hirst, H., Jüttner, I. and Ormerod, S. J. (2002) Comparing the responses of diatoms and macroinvertebrates to metals in upland streams of Wales and Cornwall. *Freshwater Biology* **47**, 1752–1765.

Holland, W. K., Rabe, F. W. and Biggam, R. C. (1994) Recovery of macroinvertebrate communities from metal pollution in the South Fork and mainstem of the Coeur d'Alene River, Idaho. *Water Environment Research* **66**, 84–88.

Hruska, K. A. and Dubé, M. G. (2004) Using artificial streams to assess the effects of metal-mining effluent on the life cycle of the freshwater midge (*Chironomus tentans*)

in situ. *Environmental Toxicology and Chemistry*
23, 2709–2718.

Hudson, L. A. and Ciborowski, J. J. H. (1996)
Spatial and taxonomic variation in
incidence of mouthpart deformities in
midge larvae (Diptera: Chironomidae:
Chironomini). *Canadian Journal of Fisheries
and Aquatic Sciences* **53**, 297–304.

Hudson-Edwards, K. A., Macklin, M. G., Brewer,
P. A. and Dennis, I. A. (2008) *Assessment of
Metal Mining-contaminated River Sediments in
England and Wales*. Environment Agency
Science Report SC030136/SR4, Environment
Agency, Bristol, UK.

Hudson-Edwards, K. A., Macklin, M. G., Curtis,
C. D. and Vaughan, D. J. (1998) Chemical
remobilisation of contaminant metals
within floodplain sediments in an incising
river system: implications for dating and
chemostratigraphy. *Earth Surface Processes
and Landforms* **23**, 671–684.

Hutchinson, S. M. and Rothwell, J. J. (2008)
Mobilisation of sediment-associated metals
from historical Pb working sites on the
River Sheaf, Sheffield, UK. *Environmental
Pollution* **155**, 61–71.

Hynes, H. B. N. (1970) *The Ecology of Running
Waters*. Liverpool University Press,
Liverpool, UK.

Ilyashuk, B., Ilyashuk, E. and Dauvalter, V. (2003)
Chironomid responses to long-term metal
contamination: a paleolimnological study
in two bays of Lake Imandra, Kola
Peninsula, northern Russia. *Journal of
Paleolimnology* **30**, 217–230.

Ixer, R. A. and Budd, P. (1998) The mineralogy of
Bronze Age copper ores from the British
Isles: implications for the composition of
early metalwork. *Oxford Journal of Archaeology*
17, 15–41.

Kelly, M. (1988) *Mining and the Freshwater
Environment*. Elsevier Applied Science,
London.

Kelly, M. (1991) *Mining and the Freshwater
Environment*. Elsevier, London. 231 pp.

Kiffney, P. M. and Clements, W. H. (1993)
Bioaccumulation of heavy metals by
benthic invertebrates at the Arkansas River,
Colorado. *Environmental Toxicology and
Chemistry* **12**, 1507–1517.

Komjarova, I. and Blust, R. (2008) Multi-metal
interactions between Cd, Cu, Ni, Pb and Zn
in water flea *Daphnia magna*, a stable
isotope experiment. *Aquatic Toxicology*
90, 138–144.

Koryak, M., Sykora, J. L. and Shapiro, M. A. (1972)
Riffle Zoobenthos in streams receiving acid
mine drainage. *Water Research* **6**, 1239.

Kosalwat, P. and Knight, A. W. (1987) Chronic
toxicity of copper to a partial life-cycle of
the midge, *Chironomus decorus*. *Archives of
Environmental Contamination and Toxicology* **16**,
283–290.

Krantzberg, G. and Stokes, P. M. (1988) The
importance of surface adsorption and pH in
metal accumulation by chironomids.
Environmental Toxicology and Chemistry **7**,
653–670.

Laasonen, P., Muotka, T. and Kivijärvi, I. (1998)
Recovery of macroinvertebrate
communities from stream habitat
restoration. *Aquatic Conservation: Marine
Freshwater Ecosystems* **8**, 101–113.

Langer, O. E. (1980) Effects of sedimentation on
salmonid stream life. In: *Report on the
Technical Workshop on Suspended Solids and the
Aquatic Environment* (ed. L. Weagle).
Whitehorse, Yukon Territory.

Lawler, D. M., Petts, G. E., Foster, I. D. L. and
Harper, S. (2006) Turbidity dynamics during
spring storm events in an urban headwater
river system: The Upper Tame, West
Midlands, UK. *The Science of the Total
Environment* **360**, 109–126.

Lee, J. H., Bang, K. W., Ketchum, L. H., Choe, J. S.
and Yu, M. J. (2002) First flush analysis of
urban storm runoff. *The Science of the Total
Environment* **293**, 163–175.

Ledger, M. E. and Hildrew, A. G. (2000a)
Herbivory in an acid stream. *Freshwater
Biology* **43**, 545–556.

Ledger, M. E. and Hildrew, A. G. (2000b) Resource
depression by a trophic generalist in an
acid stream. *Oikos* **90**, 271–278.

Ledger, M. E. and Hildrew, A. G. (2005) The ecology of acidification and recovery: changes in herbivore-algal food web linkages across a stream pH gradient. *Environmental Pollution* **137**, 103–118.

Leland, H. V. and Carter, J. L. (1985) Effects of copper on production of periphyton, nitrogen fixation and processing of leaf litter in a Sierra Nevada, California stream. *Freshwater Biology* **15**, 155–173.

Levings, C. D., Barry, K. L., Grout, J. A. *et al.* (2004) Effects of acid mine drainage on the estuarine food web, Britannia Beach, Howe Sound, British Columbia, Canada. *Hydrobiologia* **525**, 185–202.

Liu, W. X., Coveney, R. M. and Chen, J. L. (2003) Environmental quality assessment on a river system polluted by mining activities. *Applied Geochemistry* **18**, 749–764.

Mackay, R. J. (1992) Colonisation of lotic macroinvertebrates: a review of processes and patterns. *Canadian Journal of Fisheries and Aquatic Sciences* **49**, 617–628.

Macklin, M. G., Brewer, P. A., Balteanu, D. *et al.* (2003) The long term fate and environmental significance of contaminant metals released by the January and March 2000 mining tailings dam failures in Maramures County, upper Tisa Basin, Romania. *Applied Geochemistry* **18**, 241–257.

Malmqvist, B. and Hoffstein, P. O. (1999) Influence of drainage from old mine deposits on benthic macroinvertebrate communities in central Swedish streams. *Water Research* **33**, 2415–2423.

Maltby, L. (1996) Detritus processing. In: *River Biota: Diversity and Dynamics* (eds. G. Petts and P. Calow), pp. 145–167. Blackwell Science, Oxford, UK.

Maltby, L. and Booth, R. (1991) The effect of coal-mine effluent on fungal assemblages and leaf breakdown. *Water Research* **25**, 247–250.

Maltby, L., Forrow, D. M., Boxall, A. B. A., Calow, P. and Betton, C. I. (1995) The effect of motorway runoff on freshwater ecosystems: 1. Field study. *Environmental Toxicology and Chemistry* **14**, 1079–1092.

Marrugo-Negrete, J., Benitez, L. N. and Olivero-Verbel, J. (2008) Distribution of mercury in several environmental compartments in an aquatic ecosystem impacted by gold mining in northern Columbia. *Archives of Environmental Contamination and Toxicology* **55**, 305–316.

Martinez, E. A., Moore, B. C., Schaumloffel, J. and Dasgupta, N. (2001) Induction of morphological deformities in *Chironomus tentans* exposed to zinc and lead spiked sediments. *Environmental Toxicology and Chemistry* **20**, 2475–2481.

Martinez, E. A., Moore, B. C., Schaumloffel, J. and Dasgupta, N. (2002) The potential association between menta deformities and trace elements in *Chironomidae* (Diptera) taken from a heavy metal contaminated river. *Archives of Environmental Contamination and Toxicology* **42**, 286–291.

Martinez, E. A., Moore, B. C., Schaumloffel, J. and Dasgupta, N. (2004) Effects of exposure to a combination of zinc- and lead-spiked sediments on mouthpart development and growth in *Chironomus tentans*. *Environmental Toxicology and Chemistry* **23**, 662–667.

Mason, C. F. (1976) *Decomposition*. Edward Arnold, Southampton, UK.

McKnight, D. M. and Feder, G. L. (1984) The ecological effect of acid conditions and precipitation of hydrous metal oxides in a Rocky Mountain stream. *Hydrobiologia* **339**, 73–84.

Merricks, T. C., Cherry, D. S., Zipper, C. E., Currie, R. J. and Valenti, T. W. (2007) Coal-mine hollow fill and settling pond influences on headwater streams in Southern West Virginia, USA. *Environmental Monitoring and Assessment* **129**, 359–378.

Metcalfe-Smith, J. L. (1994) Biological water quality assessment of rivers: use of macroinvertebrate communities. In: *The Rivers Handbook: Hydrological and Ecological Principles* Vol. 2 (eds. P. Calow and G. E. Petts), pp. 144–170. Blackwell Scientific Publications, Oxford, UK.

Meyer, P. A., Brown, M. J. and Falk, H. (2008) Global approach to reducing lead exposure and poisoning. *Reviews in Mutation Research* **659**, 166–175.

Michailova, P. and Belcheva, R. (1990) Different effects of lead on external morphology and polytene chromosomes of *Glyptotendipes barbipes* (Staeger) (Diptera: Chironomidae). *Folia Biol (Krakow)* **38**, 83–88.

Milani, A., Reynoldson, T. B., Borgmann, U. and Kolasa, J. (2003) The relative sensitivity of four benthic invertebrates to metals in spiked-sediment exposures and application to contaminated field sediment. *Environmental Toxicology and Chemistry* **22**, 845–854.

Milner, A. M. (1994) System recovery. In: *The Rivers Handbook: Hydrological and Ecological Principles* Vol. 2 (eds. P. Calow and P. E. Petts), pp. 76–98. Blackwell Scientific Publishers, Oxford, UK.

Moore, B. C., Thornberg, A., Schaumloffel, J., Creighton, J., Filby, R. H. and Funk, W. H. (1994) Mouthpart deformities in chironomids from metal contaminated sediments of the Couer d'Alene River. *Lake and Reservoir Management* **9**, 100.

Morgan, A. J., Kille, P. and Stürzenbaum, S. R. (2007) Microevolution and ecotoxicology of metals in invertebrates. *Environmental Science and Technology* **41**, 1085–1096.

Mori, C., Orsini, A. and Migon, C. (1999.) Impact of arsenic and antimony contamination on benthic invertebrates in a minor Corsican river. *Hydrobiologia* **393**, 73–80.

Morris, R., Raylor, E. W., Brown, D. J. A. and Brown, J. A. (1989) *Acid Toxicity and Aquatic Animals*. Cambridge University Press, Cambridge, UK.

Negiliski, D. S., Ahsanullah, M. and Mobley, M. C. (1981) Toxicity of zinc, cadmium and copper to the shrimp *Callianassa australiensis* II. Effects of paired and triad combinations on metals. *Marine Biology* **64**, 305–309.

Nelson, S. M. (2000) Leaf pack breakdown and macroinvertebrate colonization: bioassessment tools for a high-altitude regulated system? *Environmental Pollution* **110**, 321–329.

Nelson, S. M. and Roline, R. A. (1999) Relationships between metals and hyporheic invertebrate community structure in a river recovering from metals contamination. *Hydrobiologia* **397**, 211–226.

Newman, R. M., Perry, J. A., Tam, E. and Crawford, R. L. (1987) Effects of chronic chlorine exposure on litter processing in outdoor experimental streams. *Freshwater Biology* **18**, 415–428.

Newcombe, C. P. and McDonald, D. D. (1991) Effects of suspended sediments on aquatic ecosystems. *North American Journal of Fisheries Management* **11**, 72–82.

Nicholson, F. A., Smith, S. R., Alloway, B. J., Carlton-Smith, C. and Chambers, B. J. (2003) An inventory of heavy metals inputs to agri-soils in England and Wales. *The Science of the Total Environment* **311**, 205–219.

Niemi, G. J., DeVore, P., Detenbeck, N. *et al.* (1990) Overview of case studies on recovery of aquatic systems from disturbance. *Environmental Management* **14**, 571–587.

Niyogi, D. K., Lewis, W. M. and McKnight, D. M. (2002) Effects of stress from mine drainage on diversity, biomass, and function of primary producers in mountain streams. *Ecosystems* **5**, 554–567.

Nordstrom, D. K., Alpers, C. N., Ptacek, C. J. and Blowes, D. W. (2000) Negative pH and extremely acidic mine waters from Iron Mountain, California. *Environmental Science and Technology* **34**, 254–258.

Nziguheba, G. and Smolders, E. (2008) Inputs of trace elements in agricultural soils via phosphate fertilisers in European countries. *The Science of the Total Environment* **390**, 53–57.

Ormerod, S. J. and Durance, I. (2009) Restoration and recovery from acidification in upland Welsh streams over 25 years. *Journal of Applied Ecology* **46**, 164–174.

Pascoe, D., Willias, K. A. and Green, D. W. J. (1989) Chronic toxicity of cadmium to *Chironomus riparius* meigen-effects upon larval development and adult emergence. *Hydrobiologia* **175**, 109–115.

Petersen, J. C., Adamski, J. C., Bell, R. W., David, J. V., Femmer, S. R. and Freiwald, D. A. (1998) *Water Quality in the Ozark Plateaus, Arkansas, Kansas, Missouri and Oklahoma, 1992–5.* Circular 1158, US Geological Survey, Denver, Colorado.

Petersen, R. C., Cummins, K. W. and Ward, G. M. (1989) Microbial and animal processing in detritus woodland stream. *Ecological Monographs* **59**, 21–39.

Pick, U. (1999) *Dunaliella acidophilia*: a most extreme acidophilic alga. In: *Enigmatic Microorganisms and Life in Extreme Environments* (ed. J. Seckbach), pp. 465–478. Kluwer Academic Press, Dordrecht.

Postma, J. F., Van Nugteren, P. and Buckert de Jong, M. B. (1996) Increased cadmium excretion in metal-adapted populations of the midge *Chironomus riparius* (Diptera). *Environmental Toxicology and Chemistry* **15**, 332–339.

Prat, N., Toja, J., Solà, C., Burgos, M. D., Plans, M. and Rieradevall, M. (1999) Effect of dumping and cleaning activities on the aquatic ecosystems of the Guadiamar River following a toxic flood. *The Science of the Total Environment* **242**, 231–248.

Rainbow, P. S. (2002) Trace metal concentrations in aquatic invertebrates: why and so what? *Environmental Pollution* **120**, 497–507.

Rasmussen, K. and Lindegaard, C. (1988) Effects of Fe compounds on macroinvertebrate communities in a Danish lowland river stream. *Water Research* **22**, 1101–1109.

Reash, R. J. (2004) Dissolved and total copper in a coal ash effluent and receiving stream: assessment of in situ biological effects. *Environmental Monitoring and Assessment* **96**, 203–220.

Rhea, D. T., Harper, D. D., Farag, A. M. and Brumbaugh, W. G. (2006) Biomonitoring in the Boulder river watershed, Montana, USA: metal concentrations in biofilm and macroinvertebrates, and relations with macroinvertebrate assemblage.

Environmental Monitoring and Assessment **115**, 381–393.

Riba, I., DelValls, T. A., Reynoldson, T. B. and Milani, D. (2006) Sediment quality in Rio Guadiamar (SW, Spain) after a tailing dam collapse: contamination, toxicity and bioavailability. *Environment International* **32**, 891–900.

Riipinen, M. P., Davy-Bowker, J. and Dobson, M. (2008) Comparison of structural and functional stream assessment methods to detect changes in riparian vegetation and water pH. *Freshwater Biology* DOI 10.1111/j.1365-2427.2008.01954.x.

Roberts, D. A., Poore, A. G. and Johnston, E. L. (2006) Ecological consequences of copper contamination in macroalgae: effects on epifauna and associated herbivores. *Environmental Toxicology and Chemistry* **25**, 2470–2479.

Robinson, C. T., Minshall, G. W. and Rushforth, S. R. (1990) Seasonal colonization dynamics of macroinvertebrates in an Idaho stream. *Journal of the North American Benthological Society* **9**, 240–248.

Rothwell, N. (2007) *Parys Mountain and the Lentin Letters.* Awlwch Industrial Heritage Trust, Anglesey, UK.

Rousch, J. M., Simmons, T. W., Kerans, B. L. and Smith, B. P. (1997) Relative acute effects of low pH and high iron on hatching success and survival of the water mite (*Arrenurus manubriator*) and the aquatic insect (*Chironomus riparius*). *Environmental Toxicology and Chemistry* **16**, 2144–2150.

Runnells, D. D., Shepherd, T. A. and Angino, E. E. (1992) Metals in water: determining natural background concentrations in mineralized areas. *Environmental Science and Technology* **26**, 2316–2323.

Rutherford, J. E. and Mellow, R. J. (1994) The effects of an abandoned roast yard on the fish and macroinvertebrate communities surrounding beaver ponds. *Hydrobiologia* **294**, 219–228.

Ryder, G. I. (1989) Experimental studies of the effects of fine sediments on lotic invertebrates. PhD Thesis, University of Otago, Dunedin, New Zealand cited in: Broekhuizen, N., Parkyn, S., Miller, D. 2001. *Hydrobiologia* **457**, 125–132.

Sabater, S., Guasch, H., Ricart, M. *et al.* (2007) Monitoring the effect of chemicals on biological communities: the biofilm as an interface. *Analytical and Bioanalytical Chemistry* **387**, 1425–1434.

Sansalone, J. J. and Buchberger, S. G. (2003) Partitioning and first flush of metals in urban roadway storm water. *Journal of Environmental Engineering* **123**, 134–143.

Sasaki, A., Ito, A., Aizawa, J. and Umita, T. (2005) Influence of water and sediment quality on benthic biota in an acidified river. *Water Research* **39**, 2517–2526.

Schindler, D. W. (1987) Detecting ecosystem responses to anthropogenic stress. *Canadian Journal of Fisheries and Aquatic Sciences* **44**, 6–25.

Schmidt, T. S., Soucek, D. J. and Cherry, D. S. (2002) Integrative assessment of benthic macroinvertebrate community impairment from metal contaminated waters in tributaries of the Upper Powell River, Virginia, USA. *Environmental Toxicology and Chemistry* **21**, 2233–2241.

Schultheis, A. S., Sanchez, M. and Hendricks, A. C. (1997) Structural and functional responses of stream insects to copper pollution. *Hydrobiologia* **346**, 85–93.

Simon, M. L., Cherry, D. S., Currie, R. J. and Zipper, C. E. (2006) The ecotoxicological recovery of Ely Creek and tributaries (Lee County, VA) after remediation of acid mine drainage. *Environmental Monitoring and Assessment* **123**, 109–124.

Smith, J. G. (2003) Recovery of the benthic macroinvertebrate community in a small stream after long-term discharges of fly ash. *Environmental Management* **32**, 77–92.

Smock, L. A. (1983) The influence of feeding habits on whole-body metal concentrations in aquatic insects. *Freshwater Biology* **13**, 301–311.

Solà, C., Burgos, M., Plazuelo, A., Toja, J., Plans, M. and Prat, N. (2004) Heavy metal bioaccumulation and macroinvertebrate community changes in a Mediterranean stream affected by acid mine drainage and an accidental spill (Guadiamar River, SW Spain). *The Science of the Total Environment* **333**, 109–126.

Soucek, D. J., Cherry, D. S., Currie, R. J., Latimer, H. A. and Trent, G. C. (2000) Laboratory to field validation in an integrative assessment of an acid mine drainage – impacted watershed. *Environmental Toxicology and Chemistry* **19**, 1036–1043.

Soucek, D. J., Densen, B. C., Schmidt, T. S., Cherry, D. S. and Zipper, C. E. (2002) Impaired *Acroneuria* sp. (Plecoptera, Perlidae) populations associated with aluminium contamination in neutral pH surface waters. *Archives of Environmental Contamination and Toxicology* **42**, 416–422.

Sridhar, K. R., Krauss, G., Barlocher, F. *et al.* (2001) Decomposition of alder leaves in two heavy metal-polluted streams in central Germany. *Aquatic Microbial Ecology* **26**, 73–80.

Stumm, W. and Morgan, J. J. (1995) *Aquatic Chemistry*. Wiley Blackwell, UK.

Suren, A. M. and Winterbourn, M. J. (1992) The influence of periphyton, detritus and shelter on invertebrate colonisation of aquatic bryophytes. *Freshwater Biology* **27**, 327–340.

Tiefenthaler, L. L., Stein, E. D. and Schiff, K. C. (2008) Watershed and land use-based sources of trace metals in urban storm water. *Environmental Toxicology and Chemistry* **27**, 277–287.

Timmermans, K. R., van Hattum, B., Kraak, M. S. and Davids, C. (1989) Trace metals in a littoral food web: concentrations in organisms, sediment and water. *The Science of the Total Environment* **87/88**, 477–494.

Tripole, S., Gonzalez, P., Vallania, A., Garbagnati, M. and Mallea, M. (2006) Evaluation of the impact of acid mine drainage on the chemistry and the macrobenthos in the Carolina stream (San Luis-Argentina). *Environmental Monitoring and Assessment* **114**, 377–389.

Ullrich, S. M., Ilyushchenko, M. A., Tanton, T. W. and Uskov, G. A. (2007) Mercury contamination in the vicinity of a derelict chlor-alkali plant Part II. Contamination of the aquatic and terrestrial food chain and potential risks to the local population. *The Science of the Total Environment* **381**, 290–306.

Van Damme, P. A., Hamel, C., Ayala, A. and Bervoets, L. (2008) Macroinvertebrate community response to acid mine drainage in rivers of the High Andes (Bolivia). *Environmental Pollution* **156**, 1061–1068.

Van Hattum, B., Timmermans, K. R. and Govers, H. A. (1991) Abiotic and biotic factors influencing in situ trace metal levels in macroinvertebrates in freshwater ecosystems. *Environmental Toxicology and Chemistry* **10**, 275–292.

Vuori, K.-M. (1995) Direct and indirect effects of iron on river ecosystems. *Annales Zoologici Fennici* **32**, 317–329.

Wade, K. R., Ormerod, S. J. and Gee, A. S. (1989) Classification and ordination of macroinvertebrate assemblages to predict stream acidity in upland Wales. *Hydrobiologia* **171**, 59–78.

Wallace, J. B., Huryn, A. D. and Lugthart, G. J. (1991) Colonization of a headwater stream during three years of seasonal insecticidal applications. *Hydrobiologia* **211**, 65–76.

Wang, W. (1987) Factors affecting metal toxicity to (and accumulation by) aquatic organisms – overview. *Environment International* **13**, 437–457.

Warwick, W. F. (1990) Morphological deformities in chironomidae (Diptera) larvae from the Lac St-Louis and Laprairie Basins of the St Lawrence River. *Journal of Great Lakes Research* **16**, 185–208.

Watanabe, K., Monaghan, M. T., Takemon, Y. and Omura, T. (2008) Biodilution of heavy metals in a stream macroinvertebrate food web: evidence from stable isotope analysis. *The Science of the Total Environment* **394**, 57–67.

Watanabe, N. C., Harada, S. and Komai, Y. (2000) Long-term recovery from mine drainage disturbance of a macroinvertebrate community in the Ichi-kawa River, Japan. *Hydrobiologia* **429**, 171–180.

Wiederholm, T. (1984). Response of aquatic insects to environmental pollution. In: *The Ecology of Aquatic Insects* (eds. V. H. Resh and D. M. Rosenberg), pp. 508–557. Greenwood Publishing Group, New York.

Williams, D. D. and Hynes, H. B. N. (1977) Benthic community development in a new stream. *Canadian Journal of Zoology* **55**, 1071–1076.

Winterbourn, M. J., McDiffett, W. F. and Eppley, S. J. (2000) Aluminium and iron burdens of aquatic biota in New Zealand streams contaminated by acid mine drainage: effects of trophic level. *The Science of the Total Environment* **254**, 45–54.

Woodcock, T. S. and Huryn, A. D. (2005) Leaf litter processing and invertebrate assemblages along a pollution gradient in a Maine (USA) headwater stream. *Environmental Pollution* **134**, 363–375.

Yan, N. D., Mackie, G. L. and Gauds, P. (1990) Control of cadmium levels in *Holopedium gibberum* (Crustacea: Cladocera) in Canadian shield lakes. *Environmental Toxicology and Chemistry* **9**, 895–908.

Younger, P. L., Banwart, S. A. and Hedin, R. S. (2002) *Mine Water: Hydrology, Pollution, Remediation*. Kluwer Academic Publishers, Dordrecht, the Netherlands. 442 pp.

CHAPTER FIVE

Impacts of emerging contaminants on the environment

ALISTAIR BOXALL

Introduction

Until very recently, the main focus on chemicals in the environment has been on heavy metals, pesticides and other organic chemicals such as PAHs, PCBs and dioxins. In recent years, there has been increasing concern over the so-called 'emerging contaminants' such as metabolites, transformation products (formed in the environment and treatment processes), human pharmaceuticals, veterinary medicines, nanomaterials, personal care products and flame retardants. These substances have been shown to be released to the environment, or in the case of nanomaterials, will be released to the environment in increasing amounts in the future. In the few monitoring studies that have looked for them, they have been detected in surface waters, groundwaters and drinking waters (e.g., Kolpin *et al.* 1998a, b; 2002; Ferrer *et al.* 2000; Juhler *et al.* 2001; Li *et al.* 2001; Schnoebelen *et al.* 2001; Lagana *et al.* 2002; Zimmerman *et al.* 2002; Battaglin *et al.* 2003).

Alongside the monitoring, studies have been performed to explore the effects of a range of emerging contaminants at the biochemical, cellular, whole organism, population and community levels. While much of the data that have been produced on different classes of emerging contaminants indicate that many pose a small risk to ecosystems and human health, there is some evidence that selected emerging contaminants could affect human and environmental health. For example, the non-steroidal anti-inflammatory drug diclofenac was found to be responsible for the decline in populations of vulture species in Asia (Oaks *et al.* 2004); the antiparasitic drug ivermectin has been shown to affect invertebrates at concentrations lower than those that could occur in the aquatic environment (Garric *et al.* 2007); ethinylestradiol has been associated with endocrine disruption in fish (Lange *et al.* 2001); and there is concern that long-term exposure to antibacterial pharmaceuticals may be contributing to the selection of resistant bacteria (Boxall *et al.* 2003a).

Ecology of Industrial Pollution, eds. Lesley C. Batty and Kevin B. Hallberg. Published by Cambridge University Press. © British Ecological Society 2010.

In this chapter, we provide an overview of the inputs, fate and behaviour and environmental effects of a selection of emerging contaminants: engineered nanomaterials, human and veterinary pharmaceuticals and transformation products. We also discuss possible approaches for identifying those emerging contaminants that are likely to pose the greatest risk to human health and the environment. Finally, recommendations on future research priorities are given.

Engineered nanoparticles

Nanotechnology is a rapidly expanding area, and engineered nanomaterials/ nanoparticles (ENPs) are finding applications in a wide range of areas including use in cosmetics, bioremediation and water treatment (e.g., Kamat & Meisel 2003; Savage & Diallo 2005; Aitken *et al.* 2006). It is therefore inevitable that, during their manufacture and use, ENPs will be released to the environment. NPs may also exist naturally (e.g., Diallo *et al.* 2005) or be formed in water bodies (e.g., Nagy *et al.* 2003) or be released to the environment in mine wastes (e.g., Walker *et al.* 2005). Concerns have therefore been raised over the potential impacts of indirect human exposure to NPs on environmental and human health (e.g., Banfield & Zhang 2001; Biswas & Wu 2005; Boxall *et al.* 2007).

Inputs to the environment

Whilst ENPs may be emitted during the manufacturing process, the route of input to the environment will primarily depend on the end use of the ENP (Boxall *et al.* 2007). For example, pharmaceuticals, cosmetics and sunscreens may be emitted to the sewage system following excretion from the patient or during washing and showering. Once they have passed through the sewer system, they may be released to surface waters. Sunscreens and other cosmetics applied on skin may also enter surface waters directly during swimming or bathing. Waste cosmetics are most likely to be disposed of in household waste that may be landfilled or incinerated. Paints containing ENPs can have both industrial and domestic uses. It is possible that runoff from painted surfaces and domestic use of paints could result in discharges to sewers. In instances where paint is applied to underwater structures or ships, ENPs may be released directly to surface waters. The use of ENPs in fuel and catalysts in vehicles will result in direct aerial emission of particles through vehicle exhaust or emissions to the surface waters and sewers through leakage and spills. Waste lubricants are most likely to be disposed of as special waste that may be landfilled or incinerated. The use of nanoparticles in treatment of polluted water is likely to result in direct emissions to surface and groundwaters or soil. ENPs used to deliver agrochemicals will be released directly to soils and surface waters.

Fate and behaviour of nanomaterials in the environment

Over the past few years, there has been increasing interest in the environmental behaviour of ENPs. A number of studies have investigated the fate and transport of ENPs in environmental systems. There is also a wealth of data available on the behaviour and transport of natural colloids. Available data indicate that, following release to water, nanoparticles (including carbon nanotubes, nanoscale zerovalent iron, titanium dioxide and fullerenes) will aggregate to some degree (e.g., Fortner *et al.* 2005; Guzman *et al.* 2006; Phenrat *et al.* 2007). Aggregates may then settle out (Brant *et al.* 2005). The degree of aggregation and the size range of the aggregates is dependent on the characteristics of the particle (i.e., type, size, surface properties and, for metal particles, the intrinsic magnetic moment) and the characteristics of the environmental system (including pH, ionic strength and dissolved organic carbon content) (Guzman *et al.* 2006; Hyung *et al.* 2007; Phenrat *et al.* 2007). The environmental behaviour of ENPs has also been shown to be modified in the presence of biota (Roberts *et al.* 2007).

Environmental transport studies indicate that NPs will exhibit differing mobilities in the soils and waterbodies and in water treatment processes compared to their corresponding parent form. Selected NPs have been shown to have the potential to contaminate aquifers (Lecoanet *et al.* 2004), and a portion may pass through water treatment processes (e.g., Zhang *et al.* 2008), although the behaviour varies depending on nanomaterial type (Lecoanet *et al.* 2004).

The behaviour of nanoparticles in environmental systems is, therefore, highly complex and appears to be dependent on not only the particle type but also the particle size. A number of modelling approaches have been proposed for predicting behaviour in environmental systems (Mackay *et al.* 2006); however, these have yet to be fully evaluated. Moreover, the studies to date have however generally looked at discrete processes, concentrated on a few nanoparticle types and employed simple test systems.

Effects of ENPs in the environment

Alongside the fate investigations, studies have explored the uptake and effects of nanoparticles on a range of environmental species and endpoints (Oberdorster 2004; Kashiwada 2006; Lovern & Klaper 2006; Oberdorster *et al.* 2006) (Table 5.1). In the laboratory, aquatic organisms appear to rapidly accumulate selected nanoparticles, including carbon black, titanium dioxide and polystyrene (e.g., Lubick 2006, Stone *et al.* 2006).

Laboratory studies with microbes have reported effects of fullerenes on microbial physiology (e.g., Fortner *et al.* 2005; Fang *et al.* 2007), whilst silver nanoparticles have been shown to accumulate in bacterial membranes, ultimately causing cell death (Sondi & Salopek-Sondi 2004). In some cases, there is however a mismatch between laboratory studies and studies to assess impacts in the real environment. For example, under realistic exposure conditions, fullerenes have little impact on the structure and function of the soil microbial communities and microbial processes (Tong *et al.* 2007).

The available data indicate that nanoparticles have low acute toxicity to aquatic organisms (e.g., Lovern & Klaper 2006; Oberdorster *et al.* 2006;

Table 5.1. *Available ecotoxicity data for a range of engineered nanoparticles*

ENP	Test species	Endpoint	Result (mg L^{-1})	Reference
Fullerene C60	*Daphnia magna*	48 h LC50 (mortality)	>35 for water stirred C60; 0.8 for THF applied C60	Zhu *et al.* 2006
	Daphnia magna	Hopping, heart rate appendage movement	Effects observed at 0.26 mg L^{-1}	Lovern *et al.* 2007
	Daphnia magna	21 d mortality, reproduction and moulting	40% mortality observed at 35 mg L^{-1}; effects on moulting and reproduction observed at 2.5 mg L^{-1}	Oberdorster *et al.* 2006
	Hyallela azteca	96 h mortality	No effect at 7 mg L^{-1}	Oberdorster *et al.* 2006
	Copepods	96 h mortality	No effect at 22.5 mg L^{-1}	Oberdorster *et al.* 2006
	Pimephales promelas	96 h mortality and sub-lethal effects	No mortality at 0.5 mg L^{-1}; PMP70 protein Expression suppressed at 0.5 mg L^{-1} but no effect on CYP 1A, 2M1 and 2K1 levels	Oberdorster *et al.* 2006
	Oryzias latipes	96 h mortality and sub-lethal effects	No mortality at 0.5 mg L^{-1}; no effect on CYP 1A, 2M1 and 2K1 PMP 70 protein levels at 0.5 mg L^{-1}	Oberdorster *et al.* 2006
	Escherichia coli	Growth	No growth at 0.4 mg L^{-1}; growth at 0.04 mg L^{-1}	Fortner *et al.* 2005
	Escherichia coli	Respiration	Inhibition at 4 mg L^{-1}; no inhibition at 0.4 mg L^{-1}	Fortner *et al.* 2005
	Bacillus subtilis	Growth	No growth at 0.4 mg L^{-1}; growth at 0.04 mg L^{-1}	Fortner *et al.* 2005

Table 5.1. (cont.)

ENP	Test species	Endpoint	Result (mg L^{-1})	Reference
	Bacillus subtilis	Respiration	Inhibition at 4 mg L^{-1}; no inhibition at 0.4 mg L^{-1}	Fortner *et al.* 2005
	Bacillus subtilis	Phospholipids and membrane phase behaviour	Effects observed at 0.01 mg L^{-1}	Fang *et al.* 2007
	Bacillus subtilis	Minimal inhibitory concentration	nC60 = 0.4–0.6 mg L^{-1}; 'small' nC60 particles = 0.1–0.23 mg L^{-1}; 'large' nC60 particles = 0.75–1.5 mg L^{-1}	Lyon *et al.* 2006
	Pseudomonas putida	Phospholipids and membrane phase behaviour	Effects observed at 0.01 mg L^{-1}	Fang *et al.* 2007
	Soil microbes	Community structure and function	Little effect at 1 mg kg^{-1} (nC60) and 1000 mg kg^{-1} (C60)	Tong *et al.* 2007
TiO$_2$	*Oncorhynchus mykiss*	96 h LC50 (mortality)	>100	Warheit *et al.* 2007
	Daphnia magna	48 h EC50 (immobilisation)	>100	Warheit *et al.* 2007
	Pseudokirchneriella subsapitata	72 h EC50 (growth)	16	Warheit *et al.* 2007
	Daphnia magna	Hopping, heart rate appendage movement	No effects observed at 2.0 mg L^{-1}	Lovern *et al.* 2007
	Bacillus subtilis	Growth inhibition	No inhibition at 500 mg L^{-1}; 75% inhibition at 1000 mg L^{-1}	Adams *et al.* 2006
	Escherichia coli	Growth inhibition	No inhibition at 100 mg L^{-1}; 15% inhibition at 500 mg L^{-1}	Adams *et al.* 2006

Table 5.1. (cont.)

ENP	Test species	Endpoint	Result (mg L^{-1})	Reference
SWNTs	*Amphiascus tenuiremia*	28–35 d EC50 (mortality, development, reproduction)	>10 (effects on mortality, fertilisation and moulting rates observed at 10 mg L^{-1})	Templeton *et al.* 2006
	Oncorhyncus mykiss	10 d sub-lethal effects	Effect on respiration at 0.1 mg L^{-1}; no major disturbance to haematology; changes in brain and gill Zn and Cu; increase in Na^+K^+–ATPase	Smith *et al.* 2007
LPC-SWNTs	*Daphnia magna*	96 h mortality	100% mortality at 20 mg L^{-1}; 20% at 10 mg L^{-1}; no mortality at 5 mg L^{-1}	Roberts *et al.* 2007
C60HxC70Hx	*Daphnia magna*	Hopping, heart rate appendage movement	Effects observed at 0.26 mg L^{-1}	Lovern *et al.* 2007
SiO_2	*Bacillus subtilis*	Growth inhibition	No inhibition at 500 mg L^{-1}; 7% inhibition at 1000 mg L^{-1}	Adams *et al.* 2006
	Escherichia coli	Growth inhibition	No inhibition at 100 mg L^{-1}; 15% inhibition at 500 mg L^{-1}	Adams *et al.* 2006
ZnO	*Bacillus subtilis*	Growth inhibition	90% inhibition at 10 mg L^{-1}	Adams *et al.* 2006
	Escherichia coli	Growth inhibition	14% inhibition at 10 mg L^{-1}	Adams *et al.* 2006

Zhu *et al.* 2006), although they may cause oxidative stress and affect the physiology and reproduction (Lovern & Klaper 2006; Oberdorster *et al.* 2006; Templeton *et al.* 2006). Studies with algae have demonstrated that titanium dioxide nanoparticles inhibit algal photosynthesis (Kim & Lee 2005). Studies with fish have demonstrated oxidative stress in the brains of fish exposed to fullerenes at very low concentrations (Oberdorster 2004), although there is some debate over whether the effects were caused by the fullerenes or the carrier solvent. Studies with plants have shown alumina nanoparticles loaded with phenanthrene to inhibit plant growth (Yang & Watts 2005).

Just like exposure, the factors and processes affecting ecotoxicity seem to be complex. The impacts of ENPs on environmental organisms seem to be determined by a range of characteristics including dissolution potential, aggregation potential, particle surface properties and the characteristics of the exposure environment and the biochemical, physiological and behavioural traits of the organism of interest (e.g., Dhawan *et al.* 2006; Rogers *et al.* 2007). In the future, therefore, we need to bring the exposure and effects studies closer together in order to determine whether ENPs can pose a risk to the environment (SCENIHR 2007; Tiede *et al.* 2009).

Human and veterinary medicines

Medicines play an important role in the treatment and prevention of disease in both humans and animals. Whilst the side effects on human and animal health have been widely documented, only recently have the potential environmental impacts of the manufacture and use of medicines been considered. Much of the data on environmental impacts has been collated in a number of books and review articles (e.g., Halling-Sørensen *et al.* 1998; Daugton & Ternes 1999; Boxall *et al.* 2003a, 2004a; Boxall 2004; Floate *et al.* 2005; Crane *et al.* 2008).

Inputs to the environment

Human and veterinary medicines may be released to the environment by a number of routes. During the manufacturing process, residues may be released from the process and may ultimately enter surface waters. Following administration, human medicines may be absorbed, metabolised and then excreted to the sewer system. They will then enter a treatment works before being released to receiving waters or land during the application of sewage sludge. When used to treat pasture animals, veterinary medicines may be excreted directly to soils or surface waters. Aquaculture treatments will be released directly to surface waters. For intensive livestock treatments, the medicines are likely to enter the environment indirectly through the application of slurry and manure as fertilisers to land. Other minor routes of entry include emissions to air and through the disposal of unused medicines and containers. Once released into the environment, pharmaceuticals will be transported and distributed to air, water, soil or sediment. A range of factors and processes including the physico-chemical properties of the compound and the characteristics of the receiving environment will affect distribution.

Whilst pharmaceuticals will have been released to the environment for decades, it is only recently that attempts have been made to quantify the levels of these compounds in the environment. Using new analytical techniques such as Liquid Chromatography Tandem Mass Spectrometry (LC-MS-MS), low levels of a range of pharmaceuticals, including hormones, steroids, antibiotics and parasiticides are now being detected in soils, surface waters and groundwaters internationally (e.g., Hirsch *et al.* 1999; Kolpin *et al.* 2002). Whilst the reported concentrations are generally low (i.e., sub $\mu g L^{-1}$ in surface waters), the substances have been observed across a wide variety of hydrological, climatic and land-use settings and many of the substances have been detected throughout the year. As a result, questions have been raised over the impacts of veterinary medicines on organisms in the environment and on human health.

Impacts on the environment and human health

In the EU and North America, the environmental risks of human and veterinary medicines now need to be assessed before a product can be marketed and in order to perform the risk assessment, data are often required on the effects on aquatic and terrestrial organisms (Breton & Boxall 2003a). A reasonable body of data is therefore available on the effects of many medicines on aquatic invertebrates, fish and algae, earthworms, plants and soil microbes (e.g., see Boxall *et al.* 2004a). These data have generally been obtained using standard ecotoxicity studies and the studies are often short lived with mortality as the endpoint. Generally, effects occur at much higher concentrations than those measured in the environment.

However, pharmaceutical compounds are either designed to be highly active and interact with receptors in humans and animals or they are toxic towards health, threatening organisms such as bacteria, fungi and parasites. Many lower animals have receptor systems similar to humans and animals, moreover many of the groups of organisms that affect human and animal health and which are targeted by pharmaceuticals play a critical role in the functioning of ecosystems. It is therefore possible that pharmaceuticals may cause subtle effects on aquatic and terrestrial organisms that might not be picked up in standard studies. For human medicines in particular, releases to the environment are likely to be almost continuous so organisms will be exposed for much longer durations than those used in standard tests. Because of this, researchers have begun to investigate some of the more subtle effects caused by long-term low-level exposure to pharmaceuticals. A wide range of subtle impacts have been reported so far (e.g., Table 5.2) including effects on oocytes and testicular maturation; impacts on insect physiology and behaviour; effects on dung

Table 5.2. *Reported subtle effects of pharmaceutical compounds on aquatic and terrestrial organisms*

Substance(s)	Medicine class	Reported effect	Reference
Fenfluramine	Anorexic	Enhances release of serotonin (5-HT) in crayfish which in turn triggers the release of ovary-simulating hormone resulting in larger oocytes with enhanced amounts of vitellin; In fiddler crabs, stimulates the production of gonad-stimulating hormone accelerating testicular maturation	Reported in Daughton & Ternes 1999
17a-ethinyl estradiol	Contraceptive	Endocrine disrupting effects on fish	
Avermectins	Parasiticide	Adults insects – loss of water balance, disruption of feeding and reduced fat accumulation, delayed ovarian development, decreased fecundity and impaired mating; Juvenile insects – delayed development, reduced growth rates, development of physical abnormalities, impairment of pupariation or emergence and a loss of developmental symmetry	Floate *et al.* 2005
Tetracyclines, macrolides and streptomycin	Antibacterials	Antibacterial resistance measured in soil bacteria obtained from sites treated with pig slurry	Sengelov *et al.* 2003
Cypermethrin	Ectoparasiticide	Impact on dung decomposition	Sommer & Bibby 2002
Fenbendazole	Parasiticide	Impact on dung decomposition	Sommer & Bibby 2002
Tylosin	Antibacterial	Impacts on the structure of soil microbial communities	Westergaard *et al.* 2001
Erythromycin	Antibacterial	Inhibition of growth cyanobacteria and aquatic plants	Pomati *et al.* 2004
Tetracylcine	Antibacterial	Inhibition of growth cyanobacteria and aquatic plants	Pomati *et al.* 2004
Ibuprofen	Anti-inflammatory	Stimulation of growth of cyanobacteria and inhibition of growth of aquatic plants	Pomati *et al.* 2004

Table 5.2. (cont.)

Substance(s)	Medicine class	Reported effect	Reference
Fenofibrate	Lipid regulator	Inhibition of basal EROD activity in cultures of rainbow trout hepatocytes	Lavelle et al. 2004
Carbamazepine	Analgesic	Inhibition of basal EROD activity in cultures of rainbow trout hepatocytes	Lavelle et al. 2004
Diclofenac	Analgesic	Inhibition of basal EROD activity in cultures of rainbow trout hepatocytes	Lavelle et al. 2004
Propanolol	Beta blocker	Weak EROD inducer in cultures of rainbow trout hepatocytes	Lavelle et al. 2004
Sulfamethazole	Antibacterial	Inhibition of basal EROD activity in cultures of rainbow trout hepatocytes	Lavelle et al. 2004
Clofibrate	Lipid regulator	Inhibition of basal EROD activity in cultures of rainbow trout hepatocytes	Lavelle et al. 2004
Diazepam	Antianxiety drug	Inhibition in the ability of dissected polyops from the cnidarian Hydra vulgaris to regenerate a hypostome, tentacles and a foot	Pascoe et al. 2003
Digoxin	Cardiac glycoside	Inhibition in the ability of dissected polyops from the cnidarian Hydra vulgaris to regenerate a hypostome, tentacles and a foot	Pascoe et al. 2003
Amlodipine	Calcium channel blocker	Inhibition in the ability of dissected polyops from the cnidarian Hydra vulgaris to regenerate a hypostome, tentacles and a foot	Pascoe et al. 2003

decomposition; inhibition or stimulation of growth in aquatic plant and algae species; and the development of antibacterial resistance in soil microbes. Whilst many of these effects have been seen at environmentally realistic concentrations, the significance in terms of environmental health has yet to be established; this will be one of the challenges in the coming years.

Transformation products

During their use and subsequent release to the environment, synthetic chemicals (pesticides, pharmaceuticals, biocides and industrials) may be degraded by a variety of processes (e.g., Roberts 1998; Roberts & Hutson 1999). Consequently, the environment may be exposed to a mixture of the synthetic chemical itself ('parent' compound) and any resulting degradation products ('degradates').

Recent advances in analytical methodology and greater access to analytical standards have allowed for increased research on the environmental occurrence of degradates. Research on pesticides has shown that degradates can be found in surface water, groundwater, precipitation, air and sediment (Boxall et al. 2004b). Although most studies have been on pesticides, degradates from other chemical classes (e.g., pharmaceuticals and detergents) have also been found in the environment. In some cases, degradates were detected as often or more frequently than the parent compound (Boxall et al. 2004b).

Even though some regulatory schemes require that the impacts of degradates on human and environmental health are considered, for many compounds this information is limited or non-existent. The main exception is for pesticides where not only the risk of the parent compound itself is assessed but also the impacts of major degradates.

Environmental fate of degradates

Once released to or formed in the environment, degradates may be transported and distributed between the major environmental compartments. The concentrations in these compartments depend on a number of factors and processes, including (1) the release scenario of the parent compound, (2) the rate of formation of degradates, (3) the half-lives of the degradates and (4) their distribution behaviour, e.g., their partitioning to sludge, soil and sediment and movement to air and water.

The mechanism, route and rate of degradation of a substance will vary across compartments, for example, if a pesticide is applied to soil, degradation by soil microbes may be important, whereas if the same substance is released directly to a water body, hydrolysis may be a more important mechanism. Climatic conditions (e.g., soil moisture content and sunlight) and the characteristics of the receiving environment (e.g., pH) may also have an impact. For example, photodegradation may be a more important degradation route in tropical conditions than in temperate climates, and for substances released to the sewer, the nature of the treatment process (e.g., aerobic, anaerobic) may affect which products are formed. Many of these routes for pesticides are described in the compilations of Roberts (1998) and Roberts and Hutson (1999).

Comparison of persistence data from laboratory studies for pesticide degradates and parent compounds in soils indicates that selected degradates from a range of chemical classes are often more persistent than their corresponding parent compound (Boxall et al. 2004b). Biotic transformation processes generally result in the formation of degradates that are more polar and water soluble than the parent compound. Hence, the resulting transport behaviour of degradates may also be different from that of the parent compound. A comparison of available data on the sorption potential of degradates and their parents in soils (Boxall et al. 2004b) indicates that around a third of the degradates derived from a range of pesticide types have a K_{oc} value at least an order of magnitude lower than the corresponding parent compound. A small proportion (3%) has a K_{oc} value more than two orders of magnitude lower. These substances might, therefore, be expected to be significantly more mobile in the soils and sewage treatment plants (STPs) than the parent compound, i.e., they are more likely to be released in effluents from treatment works, or to be transported from soils to surface and groundwaters.

The concern is that, if the assessment process focuses only on the environmental compartments of concern to the parent compound, the presence of persistent degradates in other compartments might go unnoticed. The challenge is, therefore, to determine which degradates may be significantly more mobile and which are going to be persistent.

Ecotoxicity of degradates

Data have been generated on the ecotoxicity of pesticide, veterinary medicine and biocide degradates because of the requirement of regulatory schemes (e.g.,the EU pesticide directive 91/414/EEC requires that all major degradates formed at >10% of the applied parent compound are evaluated). Generally, these studies have determined acute effects on organisms used in standard toxicity tests (e.g., Daphnia, rainbow trout, earthworms) – the reason being that, for the vast majority of pesticide metabolites, the acute package demonstrates no risks, so no further testing is triggered. A few studies have, however, assessed sub-lethal and longer term effects (e.g., Osano et al. 2002a, b). The impacts (both acute and longer term) of a few industrial substances have also been investigated, most notably degradates of the non-ionic surfactants (nonylphenol mono- and diethoxylates, nonylphenol carboxylates, nonylphenol ethoxycarboxylates and nonylphenol itself). These substances are believed to have oestrogenic activity due to their ability to mimic the endogenous hormone 17β-estradiol (Jobling et al. 1996).

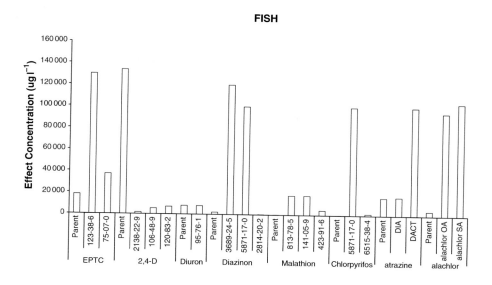

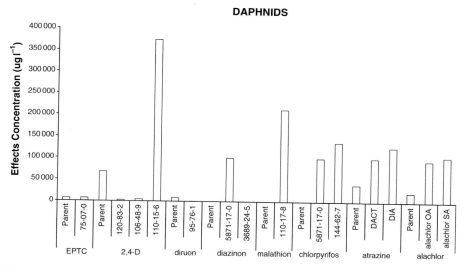

Figure 5.1. Comparison of available ecotoxicity data for some commonly monitored parent pesticides and their transformation products.

The available data demonstrate that, in most cases, degradates have similar toxicity to, or are less toxic than, their parents (Grasso *et al.* 2002; Sinclair & Boxall 2003) (e.g., Fig. 5.1). However, there are instances where degradates can be more toxic (Sinclair & Boxall 2003). For example, of the degradates investigated by Sinclair and Boxall (2003), 41% were less toxic than their parent, 39% had a similar toxicity to their parent (to account for inter-laboratory differences it was assumed that EC50 values within a factor of

3 indicate similar toxicity), 20% of degradates were more than three times more toxic and some degradates (9%) were more than an order of magnitude more toxic. In general, increases in toxicity from parent to degradate were observed for parent compounds that had a low toxicity.

There are a number of possible explanations for these toxicity increases: (1) the active moiety of the parent compound is still present in the degradate; (2) the degradate is the active component of a pro-compound; (3) the bioconcentration factor for the degradate is greater than the parent; and (4) the transformation pathway results in a compound with a different and more potent mode of action than the parent (Sinclair & Boxall 2003).

Identification of future problem contaminants

While we now know a lot more about many emerging contaminants, we have still only detected a small proportion of emerging contaminants that are likely to occur in the environment and we have information on the ecotoxicological effects of an even smaller number of these. It is probably timely, therefore, to begin to tackle the issue of emerging contaminants in the environment in a more systematic way than we are doing at present. In order to do this, we should begin to work towards a better characterisation of (1) what emerging contaminants are released to and formed in the environment; (2) the occurrence, fate and transport of emerging contaminants in the environment; and (3) effects on organisms and the implications in terms of ecosystem functioning. This information could then be used to expand current monitoring programmes to quantify those emerging contaminants that are likely to pose the greatest risk to environmental and human health. The use of ecological monitoring, biomarker assessment and effects directed analysis could also provide valuable information. A potential approach is outlined in Fig. 5.2 and is described in more detail below.

Identifying inputs of emerging contaminants to the environment

A good first step in focusing work on emerging contaminants is to identify what substances are being released to the environment and what substances are being formed in the environment. This information is often very difficult to obtain but can be extremely helpful in prioritising monitoring and testing requirements, steering analytical method development and informing the design of ecotoxicity studies. In the past, we have successfully used a range of sources of information such as usage amounts and patterns, product composition data, usage surveys and life cycle analysis to evaluate the potential for environmental exposure. Other approaches such as remote sensing may also provide valuable information on potential pressures in a catchment. As a result of REACh, our understanding of the use of chemicals in Europe will improve as companies will be required to provide data on amounts used. Metabolites and environmental transformation products are a special case as these are formed

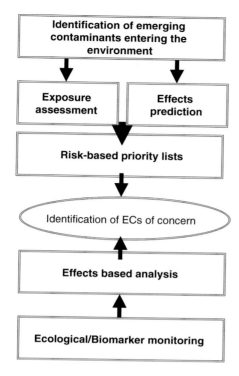

Figure 5.2. A potential scheme, combining, risk prioritisation approaches, ecological monitoring and effects based analyses to identify emerging contaminants posing the largest environmental threat.

either during the use of a product or following release into the environment. The transformation pathways of many substance classes (e.g., pharmaceuticals and plant protection products) have been well-characterised, and information from these types of studies have been used to develop pathway prediction systems for identifying potential transformation pathways of a 'new' molecule. While these approaches are still not yet fully developed and evaluated, in the future they could provide a very valuable tool for identifying transformation products that could occur in a particular system.

Identification of high risk emerging contaminants

The presence of a substance in the environment is not necessarily an indicator of an effect. To determine whether an EC is likely to pose a risk to the environment, we really need to assess the potential for exposure as well as the likely effects. For many emerging contaminants, experimental data on fate and effects in the environment are limited or non-existent although this situation is changing as a result of new regulatory initiatives (e.g., REACh, requirements for environmental risk assessment of pharmaceuticals). We do, however, have a number of simple models available for estimating exposure in different environmental systems and as long as the level of usage of a substance is known, these models often do not perform too badly for certain substance classes. New analytical technologies (e.g., time of flight technologies) also provide a powerful tool to identify and semi-quantify emerging contaminants in real environmental systems. By using the information on potential inputs to a system, it should be possible to focus the analysis on the emerging contaminants of most interest. We also have a number of tools available for estimating the effects of a substance, including read across techniques, QSARs and expert systems. By using exposure models and effects prediction approaches in combination, we can begin to home in on those substances that are likely to pose the greatest risk in a particular situation. The main problem is that for many classes of emerging contaminants, the predictive models are inappropriate. For example, exposure models do not yet exist for assessing the fate of engineered nanoparticles, and good quality QSARs are not always available for predicting ecotoxicity of some chemical classes. However, as REACh begins to have an impact, this situation is likely to improve. Our understanding of risks of ECs would also improve through more sharing of data (e.g., between scientists working in the area and industry), including negative data that may never find its way into a scientific publication.

Previous horizon scanning studies for emerging contaminants

A number of previous studies have been performed to identify priority emerging environmental pollutants (e.g., Boxall et al. 2003b; Sanderson et al. 2003; Thomas & Hilton 2003; Capelton et al. 2006; Sinclair et al. 2006). These have considered a range of classes (veterinary and human medicines and degradates), different exposure pathways and have been aimed at different protection goals (Table 5.3). An outline of the different studies and the outputs are described in more detail below.

A UK Environment Agency funded study identified human pharmaceuticals of most concern to the environment (Thomas & Hilton 2003). Data were obtained from manufacturers on the amounts of all active ingredients used in the UK on a yearly basis. These data were then used to estimate exposure concentrations in the aquatic environment. The relative risk to the environment was then determined by comparing exposure concentrations to either ecotoxicity predictions obtained using QSARs or to the therapeutic dose of the active ingredient. Two priority lists were then developed: one based on the therapeutic dose approach and the other on QSAR predictions. Active ingredients identified as highest priority using both approaches are given in Table 5.4.

Sanderson et al. (2003) ranked approximately 3000 pharmaceuticals in terms of their hazard to algae, daphnids and fish. Modifying additives were the most toxic class, while cardiovascular, gastrointestinal, antiviral, anxiolytic sedatives, hypnotics, antipsychotics, corticosteroid and thyroid pharmaceuticals were identified as the hazardous therapeutic classes.

Table 5.3. *Previous horizon scanning studies for emerging environmental contaminants*

Emerging contaminant class	Protection goal	Exposure pathways covered	Parameters used	Study
Veterinary medicines	Environmental	Soil, surface water	Usage, ecotoxicity	Boxall *et al.* 2003b
Human medicines	Environmental	Surface water	Usage, ecotoxicity, therapeutic dose	Thomas & Hilton 2003
Veterinary medicines	Human	Drinking water, vegetables, meat, fish	Usage, toxicity	Capelton *et al.* 2006
Degradates	Human	Drinking water	Usage, sorption, persistence, toxicity	Sinclair *et al.* 2006
Human medicines	Environmental		Usage, exposure, ecotoxicity	Sanderson *et al.* 2003
Dioxin-like compounds	Human	Sediments, sludge	Chemical analysis, toxicity	Eljarrat & Barcelo 2003

Horizon scanning exercises have been performed to identify veterinary medicines that might pose a risk to both environmental and human health through indirect exposure. Boxall *et al.* (2003b) collated data on the amounts used and patterns of use of veterinary medicines in use in the UK. These data were then used to identify those substances that had the greatest potential to be released to the environment in large amounts. Data on ecotoxicity were then applied to provide a qualitative indication of risk to organisms in the environment. For many substances, insufficient data were available for prioritisation so conservative default values had to be used. Those compounds identified as potential high risk are shown in Table 5.4.

Capelton *et al.* (2006) built upon this prioritisation work and extended it to consider potential indirect risks of veterinary medicines in the environment on human health. The exposure assessment approach was the same as used by Boxall *et al.* (2003a), but mammalian toxicity data were used instead of ecotoxicity data. Data were collated on a wide range of endpoints (including carcinogenicity, mutagenicity, sensitisation, acute toxicity, reproductive toxicity and neurotoxicity), and a simple scoring system was then applied to develop a measure of hazard. The exposure and hazard data were then combined, and

Table 5.4. *Substances that have been identified as a priority for further study (Boxall* et al. *2003b; Thomas & Hilton 2003; Sinclair & Boxall 2003)*

Human medicines	Veterinary medicines			Transformation products
Aminophylline	Amitraz	Flumethrin	Tetracycline	3,5,6-Trifluoro-2-pyridinol
Beclametasone	Amoxicillin	Ivermectin	Tiamulin	Thifensulfuron
Theophylline	Amprolium	Lasalocid Na	Tilmicosin	Krezoxim-methyl acid
Paracetamol	Baquiloprim	Levamisole	Toltrazuril	O-Desmethyl thifensulfuron methyl
Norethisterone	Cephalexin	Lido/lignocaine	Triclabendazole	CGA 321113
Codeine	Chlortetracycline	Lincomycin	Trimethoprim	Carbendazim
Furosemide	Clavulanic acid	Maduramicin	Tylosin	1,2,4, triazole
Atenolol	Clindamycin	Moensin		CL 153815
Bendroflumethiazide	Clopidol	Morantel		Diclofop acid
Chlorphenamine	Cypermethrin	Neomycin		Ethyl-m-hydroxyphenyl carbamate
Lofepramine	Cyromazine	Nicarbazin		Triazine amine A
Dextropropoxyphene	Decoquinate	Nitroxynil		BTS 27919
Procyclidene	Deltamethrin	Oxolinic acid		Desmethyl isoproturon
Tramadol	Diazinon	Oxytetracycline		Deethyl atrazine
Clotrimazole	Diclazuril	Phosmet		De-isopropylatrazine
Thiridazine	Dihydrostreptomycin	Piperonyl butoxide		Thiophene-sulphonamide
Mebeverine	Dimethicone	Poloxalene		Propachlor oxanilic acid
Terbinafine	Emamectin benzoate	Procaine benzylpenicillin		Propachlor ethane sulphonic acid
tamoxifen	Enrofloxacin	Procaine penicillin		4-Hydroxy-2,5,6-trichloro isophthalonitrile
Trimethoprim	Fenbendazole	Robenidine HCL		Triazamate metabolite II
Sulfamethoxazole	Flavomycin	Salinomycin Na		1-Methyl-3-(4-isopropylphenyl) urea
Fenofibrate	Flavophospholipol	Sarafloxicin		TCPSA
Diclofenac	Florfenicol	Sulphadiazine		3-Carbary-2,4,5-trichloro benzoic acid

compounds were classified as high, medium, low or very low risk. Around 30 active ingredients were classified as high, medium or low risk.

Sinclair *et al.* (2006) described a screening approach for identifying pesticide degradates that might pose a risk to drinking water supplies. The approach combines data on pesticide usage, degradate formation, properties and persistence with a simple screening model to determine the potential exposure of drinking water to different degradates. Exposure estimates are then compared to Acceptable Daily Intakes for the associated parent pesticide to determine relative risk. The approach has been applied to the UK and California. Degradates identified as highest priority in UK drinking water sources are given in Table 5.4.

Monitoring of the environment

As ecosystems are so complex, it is probably impossible to predict the impact of a substance on the actual environment based on laboratory studies and modelling alone. The approaches described above should therefore go hand-in-hand with biological based monitoring programmes in the real environment. For example, ecological monitoring will provide an indication of the impacts on structure and functioning of a system. This approach is already employed in surface water assessment in many countries, but perhaps we should begin to expand the monitoring to other systems (e.g., soil systems). Monitoring of biomarker responses might also be appropriate although the link between biomarker response and ecological functioning is often not clear. In instances where impacts on an ecosystem are observed, the use of effects directed analysis may help to identify stressors – both emerged and emerging.

Research needs

It is clear that, over the past few years, there has been increasing interest in the risks of emerging contaminants in the environment. There are, however, still many questions that need to be addressed before we can say whether residues in the environment are a threat to human and environmental health:

1. What are the risks of substances that have yet to be studied? Due to resource limitations only a small proportion of substances in use today have been investigated. There is, therefore, a need to develop an understanding of how other substances will affect the environment and for the further development of approaches for identifying substances of most concern.
2. How can we analyse certain emerging contaminants in environmental media? While there have been significant advances in analytical technology over the past decade which now allows us to detect many classes of emerging contaminants at low levels in complex media, for selected contaminants (e.g., engineered nanoparticles), method development is still in its infancy (e.g., Tiede *et al.* 2008).

3. How can we better assess ecotoxicity? It is possible that current standard ecotoxicity tests are not appropriate for assessing the impacts of many emerging contaminants (e.g., engineered nanoparticles and pharmaceuticals). The use of more subtle endpoints such as impacts on behaviour, physiology and biochemistry appear to show some merit. Further work should be performed to identify these subtle impacts.

4. What do the ecotoxicity data mean? A number of subtle effects have been demonstrated following exposure to selected emerging contaminants at environmentally realistic concentrations. We need to establish what these data mean in terms of effects on ecosystem functioning.

5. What are the mechanisms determining and fate and behaviour of emerging contaminants? For many traditional contaminants, our understanding of those factors and processes affecting fate and behaviour in the environment is well-developed and models are available for predicting a range of important fate parameters (e.g., sorption, bioaccumulation). However, for many emerging chemical contaminant classes, other fate mechanisms appear to be important. In the future, we need to try and further understand these mechanisms in order to develop improved models for use in environmental risk assessment.

6. How can we mitigate against any identified risks? In the event that a risk of an emerging contaminant to the environment is identified, it may be necessary to introduce treatment and mitigation options. By better understanding those factors controlling the fate and behaviour or different classes of emerging contaminant (see recommendation 5), we should be better placed to optimise existing remediation technologies or develop new approaches to reduce risks.

References

Adams, L. K., Lyon, D. Y. and Alvareez, P. J. J. (2006) Comparative ecotoxicity of nanoscale TiO$_2$, SiO$_2$ and ZnO water suspensions. *Water Research* **40**, 3527–3532.

Aitken, R. J., Chaudhry, M. Q., Boxall, A. B. A. and Hull, M. (2006) In-depth review: manufacture and use of nanomaterials – current status in the UK and global trends. *Occupational Medicine-Oxford* **56**, 300–306.

Banfield, J. F. and Zhang, H. Z. (2001) Nanoparticles in the environment. *Reviews in Mineralogy and Geochemistry* **44**, 1–58.

Battaglin, W. A., Thurman, E. M., Kalkhoff, S. J. and Porter, S. D. (2003) Herbicides and transformation products in surface waters of the Midwestern United States, *Journal of the American Water Resources Association* **39**, 743–756.

Biswas, P. and Wu, C. Y. (2005) Critical review: nanoparticles and the environment. *Journal of Air and Waste Management Association* **55**, 708–746.

Boxall, A. B. A. (2004) The environmental side effects of medication. *EMBO Reports.* **5**, 1110–1116.

Boxall, A. B. A., Fogg, L. A., Kay, P., Blackwell, P. A., Pemberton, E. J. and Croxford, A. (2003b) Prioritisation of veterinary medicines in the UK environment. *Toxicology Letters* **142**, 399–409.

Boxall, A. B. A., Fogg, L. A., Kay, P., Blackwell, P. A., Pemberton, E. J. and Croxford, A. (2004a) Veterinary medicines in the environment. *Reviews in Environmental Contamination and Toxicology* **180**, 1–91.

Boxall, A. B. A., Kolpin, D. W., Halling-Sorensen, B. and Tolls, J. (2003a) Are veterinary medicines causing environmental risks? *Environmental Science and Technology* **37**, 286A–294A.

Boxall, A. B. A., Sinclair, C. J., Fenner, K., Kolpin, D. W. and Maund, S. (2004b) When synthetic chemicals degrade in the environment. *Environmental Science and Technology* **38**, 369A–375A.

Boxall, A. B. A., Tiede, K. and Chaudhry, M. Q. (2007) Engineered nanomaterials in soils and water: how do they behave and could they pose a risk to human health? *Nanomedicine* **2**, 919–927.

Brant, J., Lecoanet, H. and Weisner, M. R. (2005) Aggregation and deposition characteristics of fullerene nanoparticles in aqueous systems. *Journal of Nanoparticle Research* **7**, 545–553.

Breton, R. and Boxall, A. (2003) Pharmaceuticals and personal care products in the environment: regulatory drivers and research needs. *QSAR and Cominatorial Science* **22**, 399–409.

Capelton, A., Courage, C., Rumsby, P., Stutt, E., Boxall, A. and Levy, L. (2006) Prioritising veterinary medicines according to their potential indirect human exposure and toxicity profile. *Toxicology Letters* **163**, 213–223.

Crane, M., Boxall, A. B. A. and Barrett, K. (2008) *Veterinary Medicines in the Environment*. CRC Press, Boca Raton. 240 pp.

Daughton, C. C. and Ternes, T. A. (1999) Pharmaceuticals and personal care products in the environment: agents of subtle change? *Environmental Health Perspectives* **107**, 907–938.

Dhawan, A., Taurozzi, J. S., Pandey, A. K. *et al.* (2006) Stable colloidal dispersion of C60 fullerenes in water: evidence for

genotoxicity. *Environmental Science and Technology* **40**, 7394–7401.

Diallo, M. S, Glinka, C. J., Goddard, W. A. and Johnson, J. H. (2005) Characterisation of nanoparticles and colloids in aquatic systems 1. Small angle neutron scattering investigations of Suwanee River fulvic acid aggregates in aqueous solutions. *Journal of Nanoparticle Research* **7**, 435–448.

Eljarrat, E. and Barcelo, D. (2003) Priority lists for persistent organic pollutants and emerging contaminants based on their relative toxic potency in environmental samples. *Trends in Analytical Chemistry* **22**, 655–665.

Fang, J., Lyon, D. Y., Wiesner, M. R., Dong, J. and Alvarez, P. J. J. (2007) Effect of a fullerene water suspension on bacterial phospholipids and membrane phase behaviour. *Environmental Science and Technology* **41**, 2636–2642.

Ferrer, I., Thurman, E. M. and Barcelo, D. (2000) First LC/MS determination of cyanazine amide, cyanazine acid and cyananzine in groundwater. *Environmental Science and Technology* **34**, 714–718.

Floate, K., Wardaugh, K., Boxall, A. B. A. and Sherratt, T. (2005) Faecal residues of veterinary parasiticides: non-target effects in the pasture environment. *Annual Reviews in Entomology* **50**, 153–179.

Fortner, J. D., Lyon, D. Y., Sayes, C. M. *et al.* (2005) C-60 in water: nanocrystal formation and microbial response. *Environmental Science and Technology* **39**, 4307–4316.

Garric, J., Vollat, B., Duis, K. *et al.* (2007) Effects of the parasiticide ivermectin on the cladoceran *Daphnia magna* and the green alga *Pseudokirchneriella subcapitata*. *Chemosphere* **69**, 903–910.

Grasso, P., Alberio, P., Redolfi, E., Azimonti, G. and Giarei, C. (2002) *Statistical Evaluation of Available Ecotoxicology Data on Plant Protection Products and their Metabolites*. Report to the European Commission, ICPHRP, Milan, Italy.

Guzman, K. A., Finnegan, M. P. and Banfield, J. F. (2006) Influence of surface potential on aggregation and transport of titania nanoparticles. *Environmental Science and Technology* **40**, 7688–7693.

Halling-Sørensen, B., Nors Nielsen, S., Lanzky, P. F., Ingerslev, F., Holten Lützhøft, H. C. and Jørgensen, S. E. (1998) Occurrence, fate and effects of pharmaceutical substances in the environment – a review, *Chemosphere* **36**, 357–393.

Hirsch, R., Ternes, T., Haberer, K. and Kratz, K. L. (1999) Occurrence of antibiotics in the aquatic environment. *The Science of the Total Environment* **225**, 109–118.

Hyung, H., Fortner, J. D., Hughes, J. B. and Kim, J. H. (2007) Natural organic matter stabilizes carbon nanotubes in the aqueous phase. *Environmental Science and Technology* **41**, 179–184.

Jobling, S., Sheahan, D., Osborne, J. A., Matthiessen, P. and Sumpter, J. (1996) Inhibition of testicular growth in rainbow trout (*Oncorhynchus mykiss*) exposed to estrogenic alkylphenolic compounds. *Environmental Toxicology and Chemistry* **15**, 194–202.

Juhler, R. K., Sorensen, S. R. and Larsen, L. (2001) Analysing transformation products of herbicide residues in environmental samples, *Water Research* **35**, 1371–1378.

Kamat, P. V. and Meisel, D. (2003) Nanoscience opportunities in environmental remediation. *Comptes Rendus Chimie* **6**, 999–1007.

Kashiwada, S. (2006) Distribution of nanoparticles in see-through medaka (*Oryzias latipes*). *Environmental Health Perspectives* **114**, 1697–1702.

Kim, S. C. and Lee, D. K. (2005) Preparation of TiO_2-coated hollow glass beads and their application to the control of algal growth in eutrophic water. *Microchemical Journal* **80**, 227–232.

Kolpin, D. W., Barbash, J. E. and Gilliom, R. J. (1998a) Occurrence of pesticides in shallow groundwater of the United States: initial results from the National Water-Quality Assessment Program. *Environmental Science and Technology* **32**, 558–566.

Kolpin, D. W., Furlong, E. T., Meyer, M. T. *et al.* (2002) Pharmaceuticals, hormones, and other organic wastewater contaminants in US streams, 1999–2000: a national reconnaissance. *Environmental Science and Technology* **36**, 1202–1211.

Kolpin, D. W., Thurman, E. M. and Linhart, S. M. (1998b) The environmental occurrence of herbicides: the importance of degradates in ground water. *Archives of Environmental Contamination Toxicology* **35**, 385–390.

Lagana, A., Bacaloni, A., De Leva, I., Faberi, A., Giovanna, F. and Marino, A. (2002) Occurrence and determination of herbicides and their major transformation products in environmental waters. *Analytica Chimica Acta* **462**, 187–198.

Lange, R., Hutchinson, T. H., Croudace, C. P. *et al.* (2001) Effects of the synthetic estrogen 17 alpha-ethinylestradiol on the life cycle of the fathead minnow. *Environmental Toxicology and Chemistry* **20**, 1216–1227.

Lavelle, N., Ait-Aissa, S., Gomez, E., Casellas, C. and Porcher, J. M. (2004) Effects of human pharmaceuticals on cytotoxicity, EROD activity and ROS production in fish hepatocytes. *Toxicology* **196**, 41–55.

Lecoanet, H. F., Bottero, J. Y. and Wiesner, M. R. (2004) Laboratory assessment of the mobility of nanomaterials in porous media. *Environmental Science and Technology* **38**, 5164–5169.

Li, P. C., Swanson, E. J. and Gobas, F. A. (2002) Diazinon and its degradation products in agricultural water courses in British Columbia, Canada. *Bulletin of Environmental Contamination and Toxicology* **69**, 59–65.

Lovern, S. B. and Klaper, R. (2006) *Daphnia magna* mortality when exposed to titanium dioxide and fullerene (C60) nanoparticles. *Environmental Toxicology and Chemistry* **25**, 1132–1137.

Lovern, S. B., Strickler, J. R. and Klaper, R. (2007) Behavioural and physiological changes in

Daphnia magna when exposed to nanoparticle suspensions (titanium dioxide, nano C-60, and C(60)HxC(70)Hx). *Environmental Science and Technology* **41**, 4465–4470.

Lubick, N. (2006) Still life with nanoparticles. *Environmental Science and Technology* **40**, 4328.

Lyon, D. Y., Adams, L. K., Falkner, J. C. and Alvareez, P. J. (2006) Antibacterial activity of fullerene water suspensions: effects of preparation method and particle size. *Environmental Science and Technology* **40**, 4360–4366.

Mackay, C. E., Johns, M., Salatas, J. H., Bessinger, B. and Perri, M. (2006) Stochastic probability modelling to predict the environmental stability of nanoparticles in aqueous suspension. *Integrated Environmental Assessment and Management* **2**, 293–298.

Nagy, N. M., Konya, J., Beszeda, M. *et al.* (2003) Physical and chemical formations of lead contaminants in clay and sediment. *Journal of Colloid and Interface Science* **263**, 13–22.

Oaks, J. L., Gilbert, M., Virani, M. Z. *et al.* (2004) Diclofenac residues as the cause of vulture population decline in Pakistan. *Nature* **427**, 630–633.

Oberdorster, E. (2004) Manufactured nanomaterials (fullerenes, C60) induce oxidative stress in the brain of juvenile largemouth bass. *Environmental Health Perspectives* **112**, 1058–1062.

Oberdorster, E., Zhu, S., Blickley, T. M., McClellan-Green, P. and Haasch, M. L. (2006) Ecotoxicology of carbon-based engineered nanoparticles: effect of fullerene (C60) on aquatic organisms. *Carbon* **44**, 1112–1120.

Osano, O., Admiraal, W. and Otieno, D. (2002b) Developmental disorders in embryos of the frog *Xenopus laevis* induced by chloroacetanilide herbicides and their degradation products. *Environmental Toxicology and Chemistry* **21**, 375–379.

Osano, O., Admiraal, W., Klamer, H. J., Pastor, D. and Bleeker, E. A. (2002a) Comparative toxic and genotoxic effects of chloroacetanilides, formamidines and their degradation products on *Vibrio fischeri* and *Chironomus riparius*. *Environmental Pollution* **119**, 195–202.

Pascoe, D., Karntanut, W. and Mueller, C. T. (2003) Do pharmaceuticals affect freshwater invertebrates? A study with the cnidarian *Hydra vulgaris*. *Chemosphere* **51**, 521–528.

Phenrat, T., Saleh, N., Sirk, K., Tilton, R. D. and Lowry, G. V. (2007) Aggregation and sedimentation of aqueous nanoscale zerovalent iron dispersions. *Environmental Science and Technology* **41**, 284–290.

Pomati, F., Netting, A. G., Calamari, D. and Neilan, B. A. (2004) Effects of erythromycin and ibuprofen on the growth of *Synechocystis* sp. and *Lemna minor*. *Aquatic Toxicology* **67**, 387–396.

Roberts, A. P., Mount, A. S., Seda, B. *et al.* (2007) In-vivo biomodification of lipid-coated nanotubes by *Daphnia magna*. *Environmental Science and Technology* **41**, 3025–3029.

Roberts, T. (1998) *Metabolic Pathways of Agrochemicals, Part One: Herbicides and Plant Growth Regulators*. The Royal Society of Chemistry, Cambridge, UK.

Roberts, T. and Hutson, D. (1999) *Metabolic Pathways of Agrochemicals, Part Two: Insecticides and Fungicides*. The Royal Society of Chemistry, Cambridge, UK.

Rogers, N. J., Franklin, N. M., Apte, S. C. and Batley, G. E. (2007) The importance of physical and chemical characterization in nanoparticle toxicity studies. *Integrated Environmental Assessment and Management* **3**, 303–304.

Sanderson, H., Johnson, D. I., Reitsma, T., Brain, R. A., Wilson, C. J. and Solomon, K. R. (2003) Ranking and prioritisation of environmental risks of pharmaceuticals in surface waters. *Regulatory Toxicology and Pharmacology* **39**, 158–183.

Savage, N. and Diallo, M. S. (2005) Nanomaterials and water purification: opportunities and challenges. *Journal of Nanoparticle Research* **7**, 331–342.

SCENIHR. (2007) *Opinion on the Appropriateness of the Risk Assessment Methodology in Accordance with the Technical Guidance Documents for New and Existing Substances for Assessing the Risks of Nanomaterials.* EC Health and Consumer Protection DG.

Schnoebelen, D. J., Kalkhoff, S. J. and Becher, K. D. (2001) *Occurrence and Distribution of Pesticides in Streams of the Eastern Iowa Basins 1996–1998.* USGS Report, Iowa, USA.

Sengelov, G., Agerso, Y., Halling-Sorensen, B., Baloda, S. B., Andersen, J. S. and Jensen, L. B. (2003) Bacterial antibiotic resistance levels in Danish farmland as a result of treatment with pig manure slurry. *Environment International* **28**, 587–595.

Sinclair, C. J. and Boxall, A. B. A. (2003) Assessing the ecotoxicity of pesticide transformation products. *Environmental Science and Technology* **37**, 4617–4625.

Sinclair, C. J., Boxall, A. B. A., Parsons, S. A. and Thomas, M. (2006) Prioritization of pesticide environmental transformation products in drinking water supplies. *Environmental Science and Technology* **40**, 7283–7289.

Smith, C. J., Shaw, B. J. and Handy, R. D. (2007) Toxicity of single walled carbon nanotubes to rainbow trout (*Oncorhyncus mykiss*): respiratory toxicity, organ pathologies and other physiological effects. *Aquatic Toxicology* **82**, 94–109.

Sommer, C. and Bibby, B. M. (2002) The influence of veterinary medicines on the decomposition of dung organic matter in soil. *European Journal of Soil Biology* **38**, 155–159.

Sondi, I. and Salopek-Sondi, B. (2004) Silver nanoparticles as antimicrobial agent: a case study on *E. coli* as a model for gram-negative bacteria. *Journal of Colloid Interface Science* **275**, 177–182.

Stone, V., Fernandes, T. F., Ford, A. and Cristofi, N. (2006) *Suggested Strategies for the Ecotoxicological Testing of New Nanomaterials.* Materials Research Society, Symposium Proceedings v.895.

Templeton, R. C., Ferguson, P. L., Washburn, K. M., Scrivens, W. A. and Chandler, G. T. (2006) Life-cycle effects of single-walled carbon nanotubes (SWNTs) on an estuarine meiobenthic copepod. *Environmental Science and Technology* **40**, 7387–7393.

Thomas, K. V. and Hilton, M. J. (2003) *Targeted Monitoring Programme for Pharmaceuticals in the Aquatic Environment.* Environement Agency R&D Technical Report P6–012/6. Environment Agency, UK.

Tiede, K., Boxall, A. B. A., Tear, S. P., Lewis, J., David, H. and Hassellov, M. (2008) Detection and characterisation of engineered nanoparticles in food and the environment. *Food Additives and Contaminants* **25**, 795–821.

Tiede, K., Hassellov, M., Breitbarth, E., Chaudhry, Q. and Boxall, A. B. A. (2009) Considerations for environmental fate and ecotoxicity testing to support environmental risk assessment for engineered nanoparticles. *Journal of Chromatography A* **1216**, 503–509.

Tong, Z., Bischoff, M., Nies, L., Applegate, B. and Turco, R. F. (2007) Impact of fullerene (C60) on a soil microbial community. *Environmental Science and Technology* **41**, 2985–2991.

Walker, S. R., Jamieson, H. E., Lanzirotti, A., Andrade, C. F. and Hall, G. E. M. (2005) The speciation of arsenic in iron oxides in mine wastes from the Giant gold mine, NWT: application of synchrotron micro-XRD and micro-XANES at the grain scale. *Canadian Mineralogist* **43**, 1205–1224.

Warheit, D. B., Hoke, R., Finlay, C., Doner, E. M., Reed, K. L. and Sayes, C. M. (2007) Development of a base set of toxicity tests using ultrafine TiO$_2$ particles as a component of nanoparticle risk management. *Toxicology Letters* **171**, 99–110.

Westergaard, K., Muller, A. K., Christensen, S., Bloem, J. and Sorensen, S. J. (2001) Effects of tylosin on the soil microbial community. *Journal of Soil Biochemistry* **33**, 2061–2071.

Yang, L. and Watts, D. J. (2005) Particle surface characteristics may play an important role in phytotoxicity of alumina nanoparticles. *Toxicology Letters* **158**, 122–132.

Zhang, Y., Chen, Y. S., Westerhoff, P., Hristovski, K. and Crittenden, J. C. (2008) Stability of commercial metal oxide nanoparticles in water. *Water Research* **42**, 2204–2212.

Zhu, S., Oberdoerster, E. and Haasch, M. L. (2006) Toxicity of an engineered nanoparticle (fullerene C60) in two aquatic species, *Daphnia* and fathead minnow. *Marine Environmental Research* **62**, S5–S9.

Zimmerman, L. R., Schneider, R. J. and Thurman, E. M. (2002) Analysis and detection of the herbicides dimethenamid and flufenacet and their sulfonic and oxanilic acid degradates in natural water. *Journal of Agricultural and Food Chemistry* **50**, 1045–1052.

CHAPTER SIX

Ecological monitoring and assessment of pollution in rivers

J. IWAN JONES, JOHN DAVY-BOWKER,
JOHN F. MURPHY AND JAMES L. PRETTY

Introduction

Many organisms respond to pollution in a predictable way, and it has long been realised that the biota can be used to determine the extent of pollution at a site, a technique termed biomonitoring. Much of the science of biomonitoring developed in aquatic systems, driven by concerns about the impact of industrial and domestic pollution on potable water resources. Over the past century, aquatic biomonitoring has travelled a long way from the early methodologies, and much about the pitfalls and benefits of using biota to assess pollution or other stressors has been discovered. Here we describe the history of biomonitoring and how our understanding has developed, with particular focus on RIVPACS (River InVertebrate Prediction And Classification System). This system marked a major advance in biomonitoring techniques, introducing the reference condition approach, where the physical and geographical characteristics of the river were taken into account when determining what taxa would be expected to be present if the site were not polluted. Assessment of a site was then based on a comparison of the observed community and derived scores, to that expected if the site were not polluted. RIVPACS was also the first biomonitoring tool to incorporate a measure of uncertainty; any assessment is based on spatially and temporally variable samples and it is necessary to calculate the confidence associated with the quality class derived using these samples.

We are now in an era where the Water Framework Directive places a legal obligation on European nations to use the biota to assess the ecological quality of their rivers, lakes, coastal and transitional (brackish) water bodies. This legislation marks a move away from assessing the influence of a single pressure (organic pollution) on the water body, and now elements of the biota, other than just macroinvertebrates, are used to assess a wide range of pressures on ecological quality (e.g., acidification, low flows, hydromorphology, heavy

Ecology of Industrial Pollution, eds. Lesley C. Batty and Kevin B. Hallberg. Published by Cambridge University Press. © British Ecological Society 2010.

metals). As the Water Framework Directive requires member states to achieve 'Good' ecological status by 2015, techniques are now urgently required that can predict the consequences and cost-effectiveness of potentially difficult and expensive catchment management measures. There is now an unprecedented drive by water managers and freshwater biologists to develop the wide range of biomonitoring tools needed to predict and support the return of ecological integrity to European waters after decades of industrial, agricultural and domestic impacts; and this against an uncertain background of global change.

Early biomonitoring techniques

Historically the link between water and the development of industry has always been strong. Water has been used as a power source, for transport, in processing and as a conduit for waste. As a consequence, industry and associated urban areas developed along waterways, with profound consequences for rivers: historically pollution of rivers has been chronic in many industrial areas. However, the strong link between industrial development and the development of urban areas to house the workforce, coupled with a common usage of waterways for waste disposal, has always made it difficult to determine the source of pollution, industrial or urban, impacting upon waterways. Initial concerns were for public health and the security of potable water supplies; increasingly regular outbreaks of cholera and other waterborne diseases occurred in the expanding cities, more likely a consequence of domestic sewage rather than industrial pollution. These public health concerns led to the implementation of combined sewerage systems in several developing industrial cities, with the aim of removing waste, both industrial and urban, to a safe distance from water supplies. It wasn't until late in the nineteenth century that the impact of pollution on the biological community of freshwaters was first noted, but again organic pollution and public health were paramount in these observations.

Concern over the purity of potable water supplies and safety of fish for human consumption led to the development of simple environmental monitoring systems to determine the extent of organic pollution, based on the protozoan fauna (Cohn 1853; Mez 1898). Later workers noted the impact of gross (organic) pollution on a wider range of organisms. In particular, the 'saprobic system' was proposed by Kolkwitz and Marsson (1902, 1908, 1909) to estimate the extent of organic pollution, particularly from urban sewage, on streams and rivers using benthic invertebrates, microbes and plants. In this system saprobity is defined as the sum total of all those metabolic processes within the bioactivity of a body of water which are the antithesis of primary production. Four conditions of saprobity were defined, using either chemical or biological methods, namely,

Polysaprobic (p)   Reduction processes predominate
Alpha-mesosaprobic (á)   Reduction processes diminish and

Beta-mesosaprobic (β)   oxidation processes predominate
Oligosaprobic (o)   Complete oxidation

The saprobic system was refined by various workers through the early twentieth century and was formalised in the Prague Convention of 1966 (Sládeček 1967). With further refinements (see Rolauffs *et al.* 2003), the index still continues to be used in several European countries, in two ways that differ in calculation method (i.e., the formula of Pantle and Buck (1955) or the formula of Zelinka and Marvan (1961)) and in applied species indicative values (i.e., the list of Sládeček (1973) or the revised list given in the latest German standard (DIN 38410)). Whilst sensitive to all pollution, the saprobic system is primarily an assessment of the extent of organic pollution, and as with all early bioassessment techniques, a method to determine the quality of water resources for potable supply.

However, many workers wanted a numerical score, rather than one of four conditions, to enable the extent of pollution (or recovery) to be measured more precisely. Thus, numerical indices were adopted, such as the Trent Biotic Index (TBI; Woodiwiss 1964), which gave a site a score of 1 to 10 based on the presence and diversity of a variety of indicator taxa. As the name suggests the index was developed for assessment of the running waters of the English Midlands, where rivers were used to both discharge industrial and urban waste, and as a water source for domestic and industrial use. Although the impacts of domestic and industrial pollution on these rivers were not separable, due to combined outfall, the index was essentially based on the sensitivity of invertebrates to the low oxygen stress associated with organic pollution. The TBI in turn spawned a number of indices across the globe, each adapted to their own fauna (e.g., Indice Biotique in France (Tuffery & Verneaux 1968), Chutter's Biotic Index in South Africa (Chutter 1972), Hilsenhoff's Biotic Index in the USA (Hilsenhoff 1988)), and into other regions of the UK where the rivers are of a different character to those of the Midlands (e.g., Chandler's Score in Scotland (Chandler 1970)). In 1976 the Water Data Unit of the UK Department of the Environment convened a working party to provide a biological classification scheme to report the biological quality of rivers across the UK. This Biological Monitoring Working Party (BMWP) produced a scoring system that relied heavily on the perceived tolerance of organisms to pollution (typically sewage pollution but with no ability to discriminate between industrial and other sources) as assessed by expert judgement, where invertebrate families were given a score from 1 to 10 (Biological Monitoring Working Party 1978). Initially, the total score for all taxa present at a site was used as an unofficial raw measure of the extent of pollution at a site, but it soon became apparent that taxon richness and sampling effort had an influence on this value (Armitage *et al.* 1983). To rectify this problem, the total score was replaced by the average score assigned to all taxa recorded (ASPT) which, together with the number of scoring families (Ntaxa), was adopted as the standard measures of quality.

Whilst most biomonitoring techniques were developed based on the response of invertebrates, other biological groups were incorporated into some systems. In the USA, Patrick et al. (1954) developed a methodology based on the structure of diatom assemblages, and European workers developed ideas from the saprobic system using algae (Kelly 1998). The Quality Rating System (also known as the 'Q-Value' system), which has been in use in the Republic of Ireland since 1970 (Flanagan & Toner 1972), uses a combination of benthic algae, macrophytes and invertebrates. As well as having different sensitivities to toxins (particularly herbicides, but also acidity, heavy metals and other pollutants), phototrophs respond readily to nutrient pollution (Kelly 1998; Holmes et al. 1999) and, thus, are better capable of detecting nutrient impacts than invertebrates where any effect is indirect via the effects on phototrophs. At the other end of the food web, fish are particularly sensitive to organic and inorganic materials that bioaccumulate (e.g., mercury, dioxins, radionucleo-tides), and growth anomalies (deformities, eroded fins, lesions and tumours) can be used to detect specific toxins, as well as indices based on community integrity (e.g., Index of Biotic Integrity (Fausch et al. 1990; Scardi et al. 2006)).

RIVPACS and the reference condition approach

Despite continued improvements in quantifying the response of invertebrates and other biological quality elements to pollution, none of the above systems confronted the fundamental problem that physical and chemical characteristics of rivers vary naturally. Different rivers contain different fauna and flora and the likelihood of capture of a taxon at an unpolluted site is largely dependent upon the natural environmental character of that site. As many of the characteristics of the taxa tolerant of pollution are associated with low oxygen stress, sluggish rivers have a fauna that produces a low score even where sites are unpolluted. As a consequence, the practical application of early approaches often relied heavily on expert judgement to provide an assessment of the deviation from the expected community and, thus, the extent of pollution. By adopting the 'reference condition' approach, the River InVertebrate Prediction And Classifica-tion System (RIVPACS) represented a major step forward in the bioassessment of rivers (Wright et al. 1984). In this system, it was assumed that in order to judge the impact of pollution on the fauna of a site, that site should be compared with the fauna of similar sites that are not subject to any apparent environmental stress. These sites are called reference sites. In this way, site quality is measured as a ratio, the observed/expected score, where the observed score is that of the test site and the expected score is that predicted by RIVPACS based on the fauna at similar reference sites. In RIVPACS the test site is matched to the reference sites by the pollution insensitive physical, chemical and geographical characteristics of the site.

To define the reference condition, an extensive dataset was compiled of macroinvertebrate assemblages, collected using a standardised methodology (Furse *et al.* 1981; Murray-Bligh *et al.* 1997) and identified to species level, from representative sites across the UK which were not subject to pollution or other environmental stress. These 268 reference sites from England, Scotland and Wales (later further reference sites were added from Northern Ireland, as well as additional ones from England, Scotland and Wales) contained 642 species, including species groups for taxa where it was not possible to iden-tify individuals to species accurately. Using these data, a biological classifica-tion was derived with TWINSPAN, which successively divided the reference sites into groups in a hierarchical manner based on the similarity of their fauna (Moss 1997). Initially, 16 such end groups were identified varying in their physical characteristics from naturally acidic, high gradient, upland streams to large, sluggish rivers draining lowland basins on sedimentary geologies (Wright *et al.* 1984; Wright 1997). Multiple discriminant analysis was then used to identify the best set of physical, chemical and geographical predictor variables to separate the different biological end groups (Moss 1997). Multiple discriminant analysis does not assume that any individual characteristic is capable of separating any two or more classes, but that combinations of characteristics in multidimensional space will produce a plane of separation that can be used to discriminate between classes (Fig. 6.1a). There was an initial stepwise selection procedure to eliminate those variables that did not make a statistically significant contribution to the explanatory power of the model; the original list of 28 physico-chemical variables was reduced to an optimal subset of 11 variables in the final model (Furse *et al.* 1984; Moss *et al.* 1987). It is these pollution insensitive predictor variables that are then used to match the test site to the most appropriate reference biological end groups.

In undertaking this process, it was recognised that pigeonholing rivers into a prescriptive classification was an artificial process as, in reality, rivers repre-sent a continuous gradient of change from one type to another and do not fall into distinct biological types (Furse *et al.* 1984; Wright *et al.* 1984). Thus, in defining the reference condition, RIVPACS does not assume a perfect match between the test site and the reference biological end groups, but uses a probabilistic assignment of a test site to determine end group membership (based on similarity to test site and frequency of occurrence of that class) to produce a predicted community based on several end groups, weighted in proportion to the probability of end group membership (Fig. 6.1b). This makes the predictions robust with respect to the method of classifying reference sites (Clarke *et al.* 2003). Furthermore, this approach provides a means of automati-cally identifying test sites that are inadequately represented by the reference sites, thus preventing unreliable predictions.

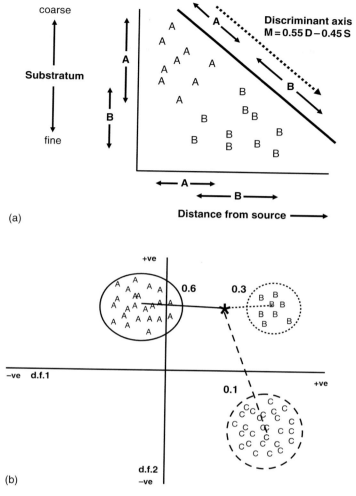

Figure 6.1. (a) Illustration of discriminant analysis. Groups A and B are not distinct in terms of either distance from source or substratum, but it is possible to separate the two groups along the discriminant axis M using a combination of both parameters where $M = 0.55\,D - 0.45\,S$. It is possible to separate multiple groups and parameters using this approach. (b) Illustration of how discriminant functions (d.f.1 and 2) are used to determine probabilities of end group inclusion. The probability that the site * belongs to each group is proportional to its distance from the group centroid and the number of sites in the group.

   The probabilities of the test site belonging to each biological end group are used to calculate the expected fauna. For each taxon, the proportion of sites within each end group that contain that taxon is combined with the probability of the test site belonging to that end group (Table 6.1). The contributions to the probability of the taxon occurring are summed across all probable end groups,

Table 6.1. *Calculating the expected probability of capture of a taxon at a test site. See Fig. 6.1 for determination of probable end group membership*

End group	Probability site belongs to end group	% of reference sites in end group with taxon *i* present	Contribution of the end group to the likelihood of capture of taxon *i*
A	0.6	30	0.18
B	0.3	40	0.12
C	0.1	60	0.06

Note: Expected likelihood of capture (P_E) of taxon $i = 0.36$.

to provide an overall predicted probability of occurrence for the test site. In the example illustrated in Table 6.1 and Fig. 6.1b, the test site has a probability of 0.6 of being in biological end group A and taxon *i* occurs in 30% of group A reference sites, giving a contribution of 0.18 (0.6×0.3) to the expected probability. Using the same process, groups B and C contribute 0.12 (0.3×0.4) and 0.06 (0.1×0.6) to the expected probability, respectively, giving an overall predicted probability of taxon *i* occurring at the test site (P_E) of 0.36 ($0.18 + 0.12 + 0.06$; see Table 6.1). This calculation is done for each of the 642 taxa in turn. The expected abundance is also calculated in a similar manner, using the average observed abundance of the taxon for the reference sites in each end group. Using the same approach and site group probabilities ensure consistency in the predictions of the expected probability of occurrence and abundance of taxa (Moss *et al.* 1987; Clarke *et al.* 2003). Thus, as part of the basic output from the RIVPACS software, the macroinvertebrate taxa are listed in decreasing order of expected probability of occurrence. The expected probabilities of occurrence and expected abundance are then used to derive estimates of expected values for a wide range of indices, which are compared to the values derived from sampling the test site to give an observed/expected measure of site quality.

An essential feature of the RIVPACS approach is the classification of reference sites using the invertebrate fauna. Thus, sites with similar invertebrate assemblages are brought together and no assumptions are made about the environmental features that influence species occurrence. Furthermore, as rivers represent a continuum of variation, RIVPACS does not assume exact matching to a discrete classification of reference site types. This differs from other reference condition approaches, such as the Rapid Biological Assessment method used by the United States Environmental Protection Agency (US EPA; Barbour *et al.* 1999), in which environmental attributes are used to a priori define the typology of reference sites and it is assumed that the biological assemblages of these sites are similar within types. Norris and Hawkins (2000) provides a thorough critique of the relative merits of the RIVPACS and a-priori typological approaches. They concluded that a RIVPACS approach was superior to the approach adopted by the US EPA. Indeed, RIVPACS-type predictive models have now been developed for many regions within the USA (Hawkins *et al.* 2000; Ostermiller & Hawkins 2004; Yuan 2006; Carlisle *et al.* 2008) and were used throughout the USA in the national Wadeable Stream Assessment (Hawkins *et al.* 2008), as well as in Australia (Turak *et al.* 1999, Simpson & Norris 2000), Sweden (Johnson 2003), Czech Republic (Kokeš *et al.* 2006) and Canada (Reynoldson *et al.* 1995).

Statistical approaches other than TWINSPAN have been used to achieve the biological classification of the reference sites, such as Bray-Curtis unweighted pair-group mean arithmetic averaging (UPGMA) (Davies 1994; Reynoldson *et al.* 1995; Hawkins *et al.* 2000) and recently the application of artificial intelligence techniques, e.g., Kohonen self-organising maps, has been demonstrated (Giraudel & Lek 2001; Walley & Conner 2001; Céréghino *et al.* 2003, Gevrey *et al.* 2004). Irrespective of the methods used, it must be stressed that biological variation is continuous and that sites do not fall into completely distinct biological types. It is also important to consider the evenness, and absolute size of the end groups: end groups should contain a sufficient number of reference sites to generate reliable predictions of taxon occurrence (Moss *et al.* 1999).

Since its first introduction, RIVPACS has become the main tool used by the regulatory authorities in the UK for the biological monitoring of rivers, with later versions having an improved coverage of reference sites, both geographically and biologically. For practical use, an expected score is derived for average BMWP score per taxon (ASPT) and number of BMWP scoring families for the test site (Ntaxa), and compared to the observed scores derived from a family level identification of samples taken from the test site. Although all types of pollution are likely to affect observed/expected scores, as previously, the purpose of this monitoring was essentially the assessment of organic pollution, particularly in terms of sewage effluent. It was not until the advent of the Water Framework Directive (2000/60/E) that the focus of biological monitoring changed.

The Water Framework Directive, a sea change in thinking

The Water Framework Directive (WFD) is the most substantial piece of legislation relating to water quality passed by the European Commission to date, and marked a shift from largely chemical to ecological standards. The legislation stipulated that several different biological quality elements had to be assessed, not just invertebrates; now fish, phytoplankton, and macrophytes and phytobenthos had to be considered also, dependent upon the water body type. Furthermore, this legislation represented a sea change in thinking: the main driver of the legislation was not concern about public health, but of sustainable use of water resources. Naturally, public health still featured highly in the thinking behind the directive (Article 24, 2000/60/E), but the impact of pressures other than sewage pollution on the quality and quantity of water resources were given far more weight than previously. The aim of the directive is to maintain and improve the quality of all substantial inland and coastal water bodies, namely rivers, lakes, groundwater, and transitional (brackish) and coastal waters. Under this legislation, a variety of biological, hydromorphological and physico-chemical quality elements must be assessed and compared to a reference condition where there are no, or only very minor, anthropogenic alterations to the values of the elements for the surface water body type from those normally associated with that type under undisturbed conditions.

Clearly, the idea of reference condition that forms the basis of RIVPACS heavily influenced the drafting of WFD legislation but, in turn, the WFD set new requirements for RIVPACS and other existing biomonitoring tools in use by member states. A description of what constitutes 'reference condition' was now clearly defined by legislation, and a process of screening had to be undertaken to determine if the RIVPACS reference sites complied with these definitions. Hydromorphological, physico-chemical and land-use characteristics of the sites and their catchments were compiled to ensure that there were no, or very minor, anthropogenic pressures acting upon the sites at the time they were sampled. For point source pressures, the process was relatively straightforward (e.g., are significant obstructions to natural flows present?), but for parameters influenced by diffuse pressures the process was somewhat more difficult. For most river types we do not have a measure of what the 'natural' phosphorus concentration would be in the absence of, or with very minor, anthropogenic pressure. Here a statistical approach was taken where deviation from end group mean values for that variable were compared to deviations from end group mean community response (ASPT and Ntaxa), thus correlating cause and impact (Davy-Bowker *et al.* 2007a). Those sites that indicated significant deviation from end group mean in both the variable of interest and the community response were deemed to be impacted. Sites that failed to pass this screening process were removed from the reference site database, and a new classification was derived with 835 sites covering the whole of the UK divided into 54 end groups. As it is likely that large budgets will be riding on the assessment of ecological quality, in the interests of transparency this database of reference sites is now freely available (Davy-Bowker *et al.* 2007b; http://www.sniffer.org.uk; http://www.ceh.ac.uk/sections/re/rivpacs_database_1.html).

Provision to classify biological communities and match to test sites using a variety of pollution insensitive parameters is given in the WFD (System B), although a simpler fixed typology based only on altitude, catchment size and geology (System A) which produces a type classification for categorising test sites is also permissible (Table 6.2). However, the predictions of reference

Table 6.2. *Physical, chemical and geographical characteristics that can be used in the Water Framework Directive Systems A and B to define typologies for matching test sites to reference sites*

SYSTEM A

Altitude typology	Size typology based on catchment area	Geology typology
High > 800 m Mid altitude 200–800 m Lowland < 200 m	Small 10–100 km² Medium 100–1000 km² Large 1000–10 000 km² Very large > 10 000 km²	Calcareous Siliceous Organic

SYSTEM B

Obligatory factors	Optional factors
Altitude Latitude Longitude Geology Size	Distance from river source Energy of flow (function of flow & slope) Mean water width Mean water depth Mean water slope Form and shape of the main riverbed River discharge (flow) category Valley shape Transport of solids Acid neutralising capacity Mean substratum composition Chloride Air temperature range Mean air temperature Precipitation

condition produced by pre-existing RIVPACS-type System B models (RIVPACS, UK; SWEPAC(SRI), Sweden; PERLA, Czech Republic) were more effective (i.e., produced lower standard deviations of O/E ratios where O is an individual reference site and E is the expected end group mean) than models based solely on the WFD System A variables or null models based solely on a single expectation of the average across all reference sites (Davy-Bowker *et al.* 2006). This is primarily because System A uses very broad categories of map-derived variables that have limited ecological relevance, whereas RIVPACS-type System B models are typically based on continuous variables selected for their ecological significance.

It should be stressed that the reference condition as defined by the WFD does not represent a desire to revert to pre-industrial conditions by removing human influence from the catchment, but a desire to achieve high quality and sustainable water resources for use by both humans and other organisms that persist within the catchment. Importantly, by inference, the WFD defines reference conditions as having to be statistically comparable to the test sites. Thus, the measures of the biological community used to produce the required observed and expected scores must be derived without any differences in the measure at either reference or test sites arising from unaccounted differences in sampling methodology. There must be an equal probability of taxa being recorded in the data used to derive the reference conditions as in a test site that does not suffer any stress, or if there is bias between the sampling methodologies, any effect must be quantified and included in the calculated O/E. Historic data rarely achieve these requirements, which precludes long distance hind-casting of reference conditions to some pre-industrial utopia.

For some biological quality elements, it is possible to produce a description of historic communities using palaeolimnological techniques (identifying sub-fossil remains from sediments) or other historic data. However, if different techniques are used to define reference condition from those used to assess sites, any methodological bias (due to sampling, preservation or technique differences) between the representation of taxa in samples used to define reference condition and those derived from sampling test sites must be taken into account. Any difference in uncertainty will also have to be accounted for (see below).

This strict definition of reference condition also differs markedly from a target condition as would be applied to the restoration of an impacted site. Whilst the target of restoration may be to achieve reference condition, a restoration target is more likely to include other characteristics of the community, either as individual species or other aspects of community composition, which would be required at a site to indicate that a restoration had been successful (e.g., presence of a Biodiversity Action Plan species). Furthermore, the methods used to derive these requirements do not have to be comparable to those used to assess the level of success. Hence, an anecdotal historic record of a rare species may be sufficient to include its recovery as a target. Having said this, the WFD does aim to restore all water bodies to 'good' ecological status, where there is little impact upon the structure and function of the biological community. This is an admirable aim and one that will bring considerable conservation benefit. If a functionally equivalent community is restored to a site, it is likely that the appropriate niches for species of conservation value will be present, and there is a high probability of them being able to persist, even if not initially present. Conversely, if the community is not functioning in the appropriate way, species of conservation value are unlikely to persist, even if present, as their niches are lacking.

Knowledge transfer, new indices and uncertainty

Due to its comprehensive nature, covering a wide range of water body types and biological quality elements, the WFD has precipitated an unprecedented development of new bioassessment tools across the EU. This process of development has benefited greatly from the knowledge gained through existing tools such as RIVPACS. Correspondingly, existing tools have been enhanced further to deal with the issues raised by the WFD. Following the move towards the sustainable use of water resources, many new indices are being developed for pressures other than sewage pollution, including, for the first time, indices to assess pollutants from primarily industrial sources. Examples include pH, heavy metals, thermal pollution and pesticides, the consequences of industrial and intensive agricultural processes. Rather than relying on expert judgement, as had been done previously (e.g., BMWP), new objective statistical techniques, such as artificial intelligence (e.g., Walley & O'Conner 2001) and partial ordination (e.g., Davy-Bowker et al. 2005) are being used to establish the response of taxa to specific stressors. By removing the influence of variables that characterise river type, it is possible to indicate the response of the taxa to the gradient of stress that is of interest; the rank order of sensitivity and the tolerance of taxa can be used to identify the taxa most likely to be impacted first, and the point of significant impact, along the gradient of increasing stress (Fig. 6.2). An example of this objective statistical approach is the Acid Water Indicator Community (AWIC, see Table 6.3), which relates the invertebrate community to stream pH (Davy-Bowker et al. 2005), thus enabling the effects of low pH to be assessed. This index is a considerable improvement on previous attempts (Rutt et al. 1990). Tests indicate that AWIC has general applicability and compares favourably with direct pH measurement, where annual samples at fortnightly – monthly intervals are typically required to estimate mean pH with comparable precision to a single AWIC sample (Ormerod et al. 2006). As with all measures of condition, the precision of the measure used has great implications in terms of confidence in the assessment, and the cost involved in achieving that level of confidence.

Irrespective of the measure used, in comparing a test site to reference condition an environmental quality ratio (EQR) is derived from observed/expected scores, which ranges from 0 to 1. For reporting, this range is divided into five quality classes, namely high, good, moderate, poor and bad, and member states must report the proportion of water bodies in each of these classes, together with the confidence of the assessments. Any assessment is intended to represent the status of the water body as a whole over a period of time, but is obtained by sampling/surveying one or more biological quality elements at one or more spatial locations within the water body at one or more points in time during the period for which the assessment is intended to apply. The results of all bioassessment techniques are influenced by variation in the observed fauna, which affects the certainty of the estimate of quality.

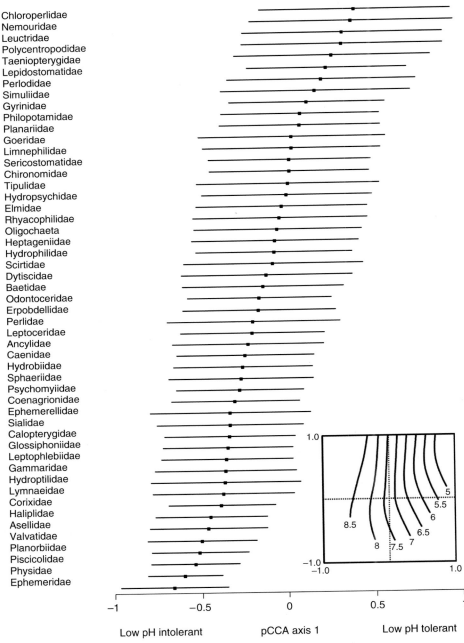

Figure 6.2. The position of AWIC taxa along axis 1 of a partial CCA with mean pH as the sole explanatory variable and altitude (m), slope (m km^{-1}) and distance from source (km) as co-variables. Horizontal bars indicate the width of the distribution (standard deviation) around each taxon score. Aggregated families are shown by their primary name (see Table 6.3). Inset indicates the pH gradient in the pCCA ordination space (reproduced from Davy-Bowker *et al.* 2005).

Table 6.3. *Acid Water Indicator Community (AWIC) Index scores of invertebrate families (reproduced from Davy-Bowker* et al. *2005)*

Taxon	Score	Taxon	Score	Taxon	Score
Ephemeridae	6	Psychomyiidae (inc. Ecnomidae)	6	Hydropsychidae	4
Physidae	6	Sphaeriidae	6	Tipulidae	4
Piscicolidae	6	Hydrobiidae (inc. Bithyniidae)	6	Chironomidae	4
Planorbiidae	6	Caenidae	6	Sericostomatidae	4
Valvatidae	6	Ancylidae (inc. Acroloxidae)	6	Limnephilidae	4
Asellidae	6	Leptoceridae	6	Goeridae	4
Haliplidae	6	Perlidae	6	Planariidae (inc. Dugesiidae)	4
Corixidae	6	Erpobdellidae	6	Philopotamidae	3
Lymnaeidae	6	Odontoceridae	6	Gyrinidae	3
Hydroptilidae	6	Baetidae	6	Simuliidae	3
Gammaridae (inc. Crongonyctidae & Niphargidae)	6	Dytiscidae (inc. Noteridae)	6	Perlodidae	2
Leptophlebiidae	6	Scirtidae	6	Lepidosto-matidae	2
Glossiphoniidae	6	Hydrophilidae (inc. Hydraenidae)	6	Taenioptery-gidae	2
Calopterygidae	6	Heptageniidae	6	Polycentro-podidae	1
Sialidae	6	Oligochaeta	6	Leuctridae	1
Ephemerellidae	6	Rhyacophilidae (inc. Glossosomatidae)	6	Nemouridae	1
Coenagrionidae	6	Elmidae	6	Chloroperlidae	1

By dividing assessment quality into classes, there is a possibility that sites will be misclassified, and the probability of misclassification occurring is influenced by how much variation there is in the observed biota relative to the width of the classes (Fig. 6.3). This variation is attributable to:

1. sampling variation due to random or systematic differences between samples collected
2. sample processing errors caused by sub-sampling, sorting and identification
3. 'natural' temporal variation
4. the effects of pollution, environmental stress or remediation.

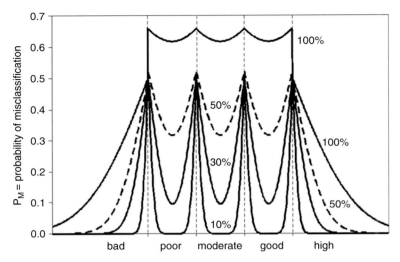

Figure 6.3. Plot of the probability of misclassifying a site into a different status class versus its true environmental quality ratio (EQR) for a range of error/uncertainty standard deviations in the observed EQR value. The EQR range has been divided into the five WFD classes, with the middle three of equal width, W. Plots are shown for standard deviation equal to 10, 30 and 50% of W. Uncertainty is derived from various sources associated with sampling sites at fixed locations in space and time (see text for fuller explanation). The less uncertainty there is associated with an assessment, the more likely a site will be correctly classified. Note that uncertainty has the largest influence near class boundaries: those sites that have a score that corresponds with a class boundary have a probability of misclassification of at least 0.5 as they could be either side of the boundary.

Hence, it is necessary to identify how much of the variation can actually be attributed to directional change as a result of environmental stress or remediation (item 4 above). Put simply, how much of a difference is required before the alarm bells should be rung, or the glasses raised and a success toasted. The importance of this uncertainty is now paramount, as the WFD requires that all inland and coastal waters within defined river basin districts must reach at least 'good' status by 2015. Any water bodies failing to reach this level must be improved by the implementation of what are often likely to be expensive and politically difficult catchment management plans. There is a lot of money riding on how sure we are that water bodies have failed to reach good status.

Considerable effort was put into quantifying the uncertainty associated with RIVPACS assessments both prior and subsequent to the implementation of the WFD (Clarke 1997), and we use the understanding developed for this advanced tool to help illustrate the requirements for other tools.

In the UK, CEH conduct regular quality assurance audits of the environment agencies, quantifying sample processing errors, where samples are re-sorted and identified by a highly skilled team (Dines & Murray-Bligh 1997). This process provides a measure of this component (2 above) of uncertainty and enables quality scores to be corrected for the bias caused by sorting and identification errors (missed taxa tend to be more sensitive increasing both metrics used, Ntaxa and ASPT) as well as identifying areas where further training is required (Dines & Murray-Bligh 1997; Haase et al. 2006). Sampling uncertainty (1 above) was quantified by a programme of nested, spatially replicated sampling across a range of river types and qualities (Clarke et al. 2002). Long-term 'natural' variation (3 above) has recently been quantified using temporally (and spatially) replicated datasets, taking care to separate sampling and spatial variation from within-year variation and from between-year variation (Davy-Bowker et al. 2007c). The importance of natural temporal variation varies, dependent on the time period that the assessment is intended to represent and the time interval from reference to test sample: temporal variation is particularly important when using fossil remains to define reference condition. The extent of temporal variation is likely to vary between river types, with some types having more long-term temporal variation than others.

The variation from all these sources has been combined within RIVPACS to provide simulations of the probable quality classes for any measured value, and an output is given of the probability of the test site being in those quality classes (Clarke 1997; Davy-Bowker et al. 2007c). It should be remembered that, if a site has a score that corresponds exactly with a class boundary, it could be either class and, thus, there is at least a 50% probability of misclassifying it.

The WFD includes several different biological quality elements for which there may be several metrics, each of which has associated uncertainty. A methodology has been developed to assess the effect of the various sources of variation and errors in the observed and the reference condition values of one or more metrics on the overall uncertainty in assignment of water bodies to ecological status class. The software package STARBUGS (STAR Bioassessment Uncertainty Guidance Software (Clarke 2004); see Software at http://www.eu-star.at or www.ceh.ac.uk/products/software/water.html) uses the various estimates of the components of uncertainty to generate many random simulations of the potential metric values for a site, from which predefined metric-based classification rules and class boundaries are used repeatedly on each simulation to build up estimates of the probabilities that a particular water body belongs to each of the WFD ecological status classes.

Whilst uncertainty may seem to be an embarrassment to biomonitoring techniques that we should be able to do without, it should be remembered that it affects all assessments of quality, whether they be physical, chemical or biological, as it is not possible to sample the whole water body all the time. A full understanding of the factors influencing uncertainty, however, can help us design tools and sampling strategies that reduce uncertainty in the most cost-effective manner. Once the influence of temporal and spatial variance has been quantified relative to sampling variance, sampling strategies can be designed to reduce that component that is having the largest effect. Uncertainty can be combined with an estimate of monetary cost per sample to derive a cost-effective precision based on the number of samples required to achieve a set level of precision, and thus enable comparisons of different techniques (or sampling strategies). It can be the case that superficially expensive techniques (per sample) are, in fact, more cost-effective than superficially cheaper ones as overall fewer samples need to be collected (Neale *et al.* 2006). Such an approach is vital to compare quick-and-dirty approaches (e.g., the use of aerial photographs) to more traditional labour intensive techniques (e.g., field survey).

A further aspect of uncertainty that needs to be addressed is the response of the water body to programmes of measures. According to the directive, appropriate management must be put into place to ensure that all water bodies have achieved 'good' status by 2015. Such management techniques are likely to be costly and difficult, and managers will have to be confident that the measures put into place will not only reduce the level of stress that the water body is suffering (e.g., resulting in reduced nutrient loading and hence lower concentrations in the water), but that the biota will respond to the improved conditions to a sufficient extent that the measured O/E will achieve 'good' status. Issues of cost-effectiveness will complicate these decisions further. As well as knowledge of the effectiveness of the management technique in controlling the pressure, a full understanding of the response of the biota to a reduction in the pressure and the sensitivity of the assessment model to changes in the biota, together with any uncertainty associated with each level of response will be required to achieve informed and defensible decisions. Although the information used to develop tools such as RIVPACS are available to inform many steps in this decision process (e.g., biological response to pressures, uncertainty), for most biological quality elements and pressures this knowledge is still lacking and, until better estimates of the required values are available, other approaches such as Bayesian belief networks may have to be used to achieve the required goals within the available time frame.

Conclusions

Although in recent years the WFD has precipitated an unprecedented drive by water managers and freshwater biologists to develop a wide range of biomonitoring tools (Furse *et al.* 2006), continued effort will be required if society is to achieve the goal of sustainable use of water resources. The world faces a future where water resources will come under increasing pressure from climate change and population growth. Bioassessment techniques will be needed to assess the impact of newly identified pressures (e.g., pharmaceutical residues), and separate these impacts from those of climate change. Over time, reference condition may have to change, as communities will change; eventually unimpacted test sites will no longer hold communities that are comparable to 'historic' reference sites. The pressure to build new tools is particularly intense for developing nations where effective, cheap techniques are urgently required to provide their populations with clean, sustainable water supplies, such that industrial and urban development can continue without being compromised by poor quality natural resources. Even well-developed tools such as RIVACS continue to be improved and adapted, and a new updated version (RIVPACS 4 – River Invertebrate Classification Tool) has recently been built (Davy-Bowker *et al.* 2007c). Biomonitoring has travelled a long way in the last century, but the next century will present many new challenges that will require new techniques and tools if we are not to let the legacy of our past compromise the future.

References

Armitage, P.D., Moss, D., Wright, J.F. and Furse, M.T. (1983) The performance of a new biological water quality score system based on macroinvertebrates over a wide range of unpolluted running water sites. *Water Research* **17**, 333–347.

Barbour, M.T., Gerritsen, J., Snyder, B.D. and Stribling, J.B. (1999) *Revision to Rapid Bioassessment Protocols for Use in Streams and Rivers: Periphyton, Benthic Macroinvertebrates and Fish.* 2nd edn. EPA 841-B-99-002, US Environmental Protection Agency, Office of Water, Washington, DC.

Biological Monitoring Working Party (1978) *Final Report: Assessment: A Presentation of the Quality of Rivers in Great Britain.* Unpublished report, Department of the Environment, Water Data Unit.

Carlisle, D.M., Hawkins, C.P., Meador, M.R., Potapova, M. and Falcone, J. (2008) Biological assessments of Appalachian streams based on predictive models for fish, macroinvertebrate, and diatom assemblages. *Journal of the North American Benthological Society* **27**, 16–37.

Céréghino, R., Park, Y.-S., Compin, A. and Lek, S. (2003) Predicting the species richness of aquatic insects in streams using a limited number of environmental variables. *Journal*

of the North American Benthological Society **22**, 442–456.

Chandler, J.R. (1970) A biological approach to water quality management. *Water Pollution Control* **69**, 415–422.

Chutter, F.M. (1972) An empirical biotic index of the quality of water in South African streams and rivers. *Water Research* **6**, 19–30.

Clarke, R.T. (1997) Uncertainty in estimates of biological quality based on RIVPACS. In: *Assessing the Biological Quality of Fresh Waters* (eds. J.F. Wright, D.W. Sutcliffe and M.T. Furse), pp. 39–54. Freshwater Biological Association, Ambleside, UK.

Clarke, R.T. (2004) *STAR Deliverable. Error/ Uncertainty Module Software STARBUGS (STAR Bioassessment Uncertainty Guidance Software. User Manual.* http://www.eu-star.at.

Clarke, R.T. and Davy-Bowker, J. (2006) *Development of the Scientific Rationale and Formulae for Altering RIVPACS Predicted Indices for WFD Reference Condition.* Scotland & Northern Ireland Forum for Environmental Research. Project WFD72B report.

Clarke, R.T., Furse, M.T., Gunn, R.J.M., Winder, J.M. and Wright, J.F. (2002) Sampling variation in macroinvertebrate data and implications for river quality indices. *Freshwater Biology* **47**, 1735–1751.

Clarke, R.T., Wright, J.F. and Furse, M.T. (2003) RIVPACS models for predicting the expected macroinvertebrate fauna and assessing the ecological quality of rivers. *Ecological Modelling* **160**, 219–233.

Cohn, F. (1853) Über lebende Organismen im Trinkwasser. *Günzberg Z Klin Med,* **4**, 229–237.

Davies, P.E. (1994) *River Bioassessment Manual: Monitoring River Health Initiative.* Department of the Environment, Sport, Territories, Land and Water Resources R and D Corporation, Commonwealth Environment Protection Agency, Canberra, Australia.

Davy-Bowker, J., Clarke, R.T., Corbin, T. *et al.* (2007c) *River Invertebrate Classification Tool.* Scotland & Northern Ireland Forum for Environmental Research. Project WFD72C report.

Davy-Bowker, J., Clarke, R.T., Furse, M.T. *et al.* (2007a) *RIVPACS Pressure Data Analysis.* Scotland & Northern Ireland Forum for Environmental Research).

Davy-Bowker, J., Clarke, R.T., Furse, M.T. *et al.* (2007b) *RIVPACS Database Documentation.* Scotland & Northern Ireland Forum for Environmental Research. Project WFD46 report.

Davy-Bowker, J., Clarke, R.T., Johnson, R.K., Kokes, J., Murphy, J.F. and Zahradkova, S. (2006) A comparison of the European Water Framework Directive physical typology and RIVPACS-type models as alternative methods of establishing reference conditions for benthic macroinvertebrates. *Hydrobiologia* **566**, 91–105.

Davy-Bowker, J., Murphy, J.F., Rutt, G.R., Steel, J.E.C. and Furse, M.T. (2005) The development and testing of a macroinvertebrate biotic index for detecting the impact of acidity on streams. *Archiv für Hydrobiologie* **163**, 383–403.

DIN 38410 *(Deutsche Einheitsverfahren zur Wasser- und Abwasser- und Schlamm-untersuchung) T.2,* 1990. Bestimmung des Saprobien index, Berlin.

Dines, R.A. and Murray-Bligh, J.A.D. (1997) Quality assurance and RIVPACS. In: *Assessing the Biological Quality of Fresh Waters* (eds. J.F. Wright, D.W. Sutcliffe and M.T. Furse), pp. 70–78. Freshwater Biological Association, Ambleside, UK.

Fausch, K.D., Lyons, J., Karr, J.R. and Angermeier, P.L. (1990) Fish communities as indicators of environmental degradation. *American Fisheries Society Symposium* **8**, 123–144.

Flanagan, P.J. and Toner, P.F. (1972) *The National Survey of Irish Rivers. A Report on Water Quality.* An Foras forbartha, WR/R1, Dublin.

Furse, M.T., Herring, D., Brabec, K., Buffagni, A., Sandin, L. and Verdonschot, P.F.M. (2006) *The Ecological Status of European Rivers: Evaluation and Intercalibration of Assessment Methods.* Developments in Hydrobiology 188. Springer, Dordrecht, the Netherlands.

Furse, M.T., Moss, D., Wright, J.F. and Armitage, P.D. (1984) The influence of seasonal and taxonomic factors on the ordination and classification of running-water sites in Great Britain and on the prediction of their macroinvertebrate communities. *Freshwater Biology* **14**, 257–280.

Furse, M.T., Wright, J.F., Armitage, P.D. and Moss, D. (1981) An appraisal of pond-net samples for biological monitoring of lotic macro-invertebrates. *Water Research* **15**, 679–689.

Gevrey, M., Rimet, F., Park, Y.-S., Giraudel, J.-L., Ector, L. and Lek, S. (2004) Water quality assessment using diatom assemblages and advanced modelling techniques. *Freshwater Biology* **49**, 208–220.

Giraudel, J.L. and Lek, S. (2001) A comparison of self-organizing map algorithm and some conventional statistical methods for ecological community ordination. *Ecological Modelling* **146**, 329–339.

Haase, P., Murray-Bligh, J., Lohse, S. *et al.* (2006) Assessing the impact of errors in sorting and identifying macroinvertebrate samples. *Hydrobiologia* **566**, 505–521.

Hawkins, C.P., Norris, R.H., Hogue, J.N. and Feminella, J.W. (2000) Development and evaluation of predictive models for measuring the biological integrity of streams. *Ecological Applications* **10**, 1456–1477.

Hawkins, C. P., Paulsen, S. G., Van Sickle, J. and Yuan, L. L. (2008) Regional assessments of stream ecological condition: scientific challenges associated with the USA's national Wadeable Stream Assessment. *Journal of the North American Benthological Society* **27**, 805–807.

Hilsenhoff, W. L. (1988) Rapid field assessment of organic pollution with a family-level biotic index. *Journal of the North American Benthological Society* **7**, 65–68.

Holmes, N. T. H., Newman, J. R., Chadd, J. R., Rouen, K. J., Saint, L. and Dawson, F. H. (1999) *Mean Trophic Rank: A User's Manual*. Research and Development, Technical Report E38. Environment Agency, Bristol, UK.

Johnson, R. K. (2003) Development of a prediction system for lake stony-bottom littoral macroinvertebrate communities. *Archiv für Hydrobiologie* **158**, 517–540.

Kelly, M. G. (1998) Use of community-based indices to monitor eutrophication in European rivers. *Environmental Conservation* **25**, 22–29.

Kokeš, J., Zahradkova, N. D., Hodovsky, J., Jarkovsky, J. and Soldan, T. (2006) The PERLA system in the Czech Republic: a multivariate approach for assessing the ecological status of running waters. *Hydrobiologia* **566**, 343–354.

Kolkwitz, R. and Marsson, M. (1902) Grundsätze für die biologische Beurteilung des Wassers nach seiner Flora und Fauna. *Mitt. aus d. Kgl. Prüfungsanstalt für Wasser versorgung u. Abwässerbeseitigung* **1**, 33–72.

Kolkwitz, R. and Marsson, M. (1908) Ökologie der pflanzlichen Saprobien. *Ber Deutsch Bot Ges* **26a**, 505–519.

Kolkwitz, R. and Marsson, M. (1909) Ökologie der tierischen Saprobien. *International Reviews in Hydrobiology* **2**, 126–152.

Mez, C. (1898) Mikroskopische Wasseranalyse. *Anleitung zur Untersuchung des Wassers mit besonderer Berücksichtigung von Trink- und Abwasser J.* Springer, Berlin.

Moss, D. (1997) Evolution of statistical methods in RIVPACS. In: *Assessing the Biological Quality of Fresh Waters* (eds. J. F. Wright, D. W. Sutcliffe and M. T. Furse), pp. 25–37. Freshwater Biological Association, Ambleside, UK.

Moss, D., Furse, M. T., Wright, J. F. and Armitage, P. D. (1987) The prediction of the macro-invertebrate fauna of unpolluted running-water sites in Great Britain using environmental data. *Freshwater Biology* **17**, 41–52.

Moss, D., Wright, J. F., Furse, M. T. and Clarke, R. T. (1999) A comparison of alternative techniques for the protection of the fauna of running water sites in Great Britain. *Freshwater Biology* **41**, 167–181.

Murphy, J. F. and Davy-Bowker, J. (2005) Spatial structure in lotic macroinvertebrate communities in England and Wales: relationship with physicochemical and anthropogenic stress variables. *Hydrobiologia* **534**, 151–164.

Murphy, J. F. and Davy-Bowker, J. (2006) The predictive modelling approach to biomonitoring: taking river quality assessment forward. In: *Biological Monitoring of Rivers: Applications and Perspectives*. (eds. M. Ziglio, M. Siligardi and G. Flaim), pp. 383–399. Wiley & Sons, Chichester, UK.

Murray-Bligh, J. A. D., Furse, M. T., Jones, F. H., Gunn, R. J. M., Dines, R. A. and Wright, J. F. (1997) *Procedure for Collecting and Analysing Macroinvertebrate Samples for RIVPACS*. Environment Agency, Bristol and IFE, Wareham, UK.

Neale, M. W., Kneebone, N. T., Bass, J. A. B. *et al.* (2006) *Assessment of the Effectiveness and Suitability of Available Techniques for Sampling Invertebrates in Deep Rivers*. EU INTERREG IIIA Ireland/Northern Ireland, North South Share River Basin Management Project Report.

Norris, R. H and Hawkins, C. P (2000) Monitoring river health. *Hydrobiologia* **435**, 5–17.

Ormerod, S. J., Lewis, B. R., Kowalik, R. A., Murphy, J. F. and Davy-Bowker, J. (2006)

Field testing the AWIC index for detecting acidification in British streams. *Archiv für Hydrobiologie* **166**, 99–115.

Ostermiller, J. D. and Hawkins, C. P. (2004) Effects of sampling error on bioassessments of stream ecosystems: application to RIVPACS-type models. *Journal of the North American Benthological Society* **23**, 363–382.

Pantle, E. and Buck, H. (1955) Die biologische Überwachung der Gewässer und die Darstellung der Ergebnisse, *Gas und Wasserfach* **96**, 604.

Patrick, R., Hohn, M. H. and Wallace, J. H. (1954) A new method for determining the pattern of the diatom flora. *Notulae Naturae* **259**, 2–12.

Reynoldson, T. B, Bailey, R. C., Day, K. E. and Norris, R. H. (1995) Biological guidelines for freshwater sediment based on BEnthic Assessment of SedimenT (the BEAST) using a multivariate approach for predicting biological state. *Australian Journal of Ecology* **20**, 198–219.

Rolauffs, P., Hering, D., Sommerhäuser, M., Rödiger, S. and Jähnig, S. (2003) *Entwicklung eines leitbildorientierten Saprobienindexes für die biologische Fließgewässerbewertung.* Umweltbundesamt Texte 11/03, pp. 1–137.

Rutt, G. P., Weatherley, N. S. and Ormerod, S. J. (1990) Relationships between the physicochemistry and macroinvertebrates of British upland streams – the development of modelling and indicator systems for predicting fauna and detecting acidity. *Freshwater Biology* **24**, 463–480.

Scardi, M., Tancioni, L. and Cataudella, S. (2006) Monitoring methods based on fish. In: *Biological Monitoring of Rivers: Applications and Perspectives* (eds. M. Ziglio, M. Siligardi and G. Flaim), pp. 135–153. Wiley & Sons, Chichester, UK.

Simpson, J. C. and Norris, R. H. (1997) Biological assessment of river quality: development of AusRivAS models and outputs. In: *Assessing the Biological Quality of Fresh Waters* (eds. J. F. Wright, D. W. Sutcliffe and

M. T. Furse), pp. 125–142. Freshwater Biological Association, Ambleside, UK.

Sládeček, V. (1967) The ecological and physiological trends in the saprobiology. *Hydrobiologia* **30**, 513–526.

Sládeček, V. (1973) System of water quality from the biological point of view, *Ergebnisse der Limnologi* **7**, 1–128.

Tuffery, G. and Verneaux, J. (1968) *Méthode de détermination de la qualité biologique des eaux courantes. Exploitation codifiée des inventaires de fauna du fond.* Ministère de l'Agriculture (France). 23 pp.

Turak, E., Flack, L. K., Norris, R. H., Simpson, J. and Waddell, N. (1999) Assessment of river condition at a large spatial scale using predictive models. *Freshwater Biology* **41**, 283–298.

Walley, W. J. and O'Connor, M. A. (2001) Unsupervised pattern recognition for the interpretation of ecological data. *Ecological Modelling* **146**, 219–230.

Woodiwiss, F. S. (1964) The biological system of stream classification used by the Trent River Board. *Chemistry Industry* **11**, 443–447.

Wright, J. F.(1997) An introduction to RIVPACS. In: *Assessing the Biological Quality of Fresh Waters* (eds. J. F. Wright, D. W. Sutcliffe and M. T. Furse), pp. 1–24. Freshwater Biological Association, Ambleside, UK.

Wright, J. F., Moss, D., Armitage, P. D. and Furse, M. T. (1984) A preliminary classification of running-water sites in Great Britain based on macroinvertebrate species and the prediction of community type using environmental data. *Freshwater Biology* **14**, 221–256.

Yuan, L. (2006) Theoretical predictions of observed to expected ratios in RIVPACS-type predictive model assessments of stream biological condition. *Journal of the North American Benthological Society* **25**, 841–850.

Zelinka, M. and Marvan, P. (1961) Zur Präzisierung der biologischen Klassifikation der Reinheit fließender Gewässer. *Archiv für Hydrobiologie* **57**, 389–407.

CHAPTER SEVEN

Detecting ecological effects of pollutants in the aquatic environment

ALASTAIR GRANT

Introduction

In the marine environment a widely used definition of pollution is the 'intro-duction by man, directly or indirectly, of substances or energy into the marine environment (including estuaries) *resulting in deleterious effects* such as harm to living resources, hazards to human health, hindrance of marine activities, including fishing, impairing quality for use of sea-water and reduction of amenities' (GESAMP 1982 – my emphasis). The increasing sophistication of analytical chemical methods means that we can detect contamination by a wide range of chemicals in almost any aquatic environment (see Chapter 5). But the rational regulation of direct contaminant discharges to the environment and the setting of priorities for dealing with contaminants arising from diffuse sources requires us to be able to identify the subset of cases of contamination where deleterious effects are, or may be, occurring. There has been consider-able recent improvement in methods for *prospective* risk assessment – methods that allow an assessment of whether particular concentrations of a substance might cause ecological effects in the field. For example, the development of species sensitivity distributions has given greatly improved information on whether the sensitivity of standard laboratory test organisms reflects the sensitivity of the much wider range of organisms that occur in the field (see, e.g., Maltby *et al.* 2005), and there is discussion of risk assessment methods elsewhere in this volume (Chapters 5 and 9). However, determining whether an individual substance *actually is* having deleterious effects on the ecology at any particular location remains a major challenge for ecotoxicology.

The aim of this chapter is to review recent advances in methods for detecting ecological effects of pollutants at field sites, and discuss the reasons why it is often difficult to do better than detecting areas where there is severe ecological damage. I will also review some examples from the literature, including my own work, where it has proved possible to detect subtle ecological changes due

Ecology of Industrial Pollution, eds. Lesley C. Batty and Kevin B. Hallberg. Published by Cambridge University Press. © British Ecological Society 2010.

to pollution or subtle effects of contaminants on individual organisms that, if continued in the medium and long term, will feed through into ecological effects at a population level. So there are circumstances where we *can* detect subtle effects of pollution on field populations. An understanding of the conditions necessary to achieve this, combined with an appreciation of the reasons why this is sometimes not possible, leads us to the conclusion that it is likely that there are many other aquatic locations where pollution is having adverse ecological effects but where we are not able to detect them using currently available methods.

What effects do we expect pollution to have on ecosystems?

It is helpful to begin by identifying the sorts of ecological change that we expect to observe when an ecosystem is exposed to a damaging pollutant. Fig. 7.1 shows a model for the effects of pollution on a community. Except in extreme environments, where few species can survive because of physical stress, we would expect a reasonably diverse community of organisms to be present at an unpolluted site. If this area becomes polluted, some or all of the species present may migrate out of the area, be killed or fail to reproduce, leading to reduced diversity. If the pollution event is transitory, the community will recover in time. But if the pollution source is continuous, or the pollutant persists in the environment (in sediments, for example), then these pollution sensitive species will be unable to recolonise. In the case of severe pollution,

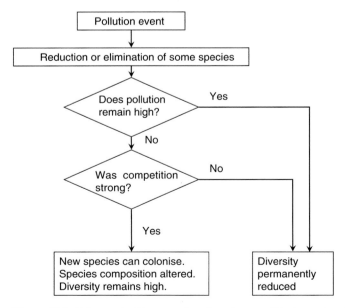

Figure 7.1. Schematic diagram showing the possible impacts of pollution on a community. If interspecific competition was strong, the result can be altered community composition but unchanged diversity.

diversity will remain low. However, it is possible that, with more moderate pollution, tolerant species may replace those that have been lost. This is possible if these more tolerant species are poor competitors that are normally excluded from the site by competitively superior but less stress-tolerant species (this would also be consistent with the observation that pollution can increase the success of invasive species, c.f. Piola & Johnston 2008). If there are sufficient tolerant species available in the local species pool to colonise, this may produce a community with similar diversity but a different composition to that present before the pollution occurred. A similar change in community composition may occur even if the pollutant is not acutely toxic. Chronic toxic effects in which a pollutant reduces survival, growth or reproduction of sensitive species, may reduce the population size of these sensitive species, relaxing the intensity of competition on less sensitive species and allowing them to become established or increase.

Does this pattern occur in reality?

If this conceptual model is correct, we can classify sites into three categories:

1. Pristine sites which may be contaminated, but where there are negligible effects of the contaminants on the species that are present;
2. Severely polluted sites at which diversity shows a significant reduction relative to that at pristine sites, as a result of adverse effects of contaminants
3. Moderately polluted sites where diversity is comparable to that at pristine sites, but where adverse effects of contaminants on more sensitive species have altered community composition to favour pollution-tolerant species.

What evidence is there that this corresponds to what actually does occur in the field? At the pristine end of the continuum, we hope that there are still sites where effects of contamination are negligible, although proving this would be difficult and there is evidence that virtually none of the world's coral reefs are unaffected by human activity (Pandolfi *et al.* 2003). At the other end of the spectrum, there are many examples where pollution has reduced the diversity of communities. Detecting pollution that is sufficiently severe to cause considerable reduction in diversity is straightforward, and I do not intend to review examples of this here. Detecting subtle changes in community composition that are not accompanied by a reduction in diversity presents a much greater challenge. Differences in community composition between sites can be caused by changes in environmental characteristics other than pollution. So detecting changes due to pollution requires the availability of good control sites, which are identical to the impacted sites in every respect apart from the level of contamination (Grant & Millward 1997). In studies where this condition is satisfied, subtle ecological changes associated with impacts of pollution may be present over quite extensive areas. In a classic study of the ecological effects of the discharge of oil-based drilling muds from the Ekofisk oil platform, John Gray (Gray *et al.* 1990) showed that, although the area of reduced diversity extends only for about 250 m out from the platform, more subtle ecological changes were detectable across an area of sea bed with a diameter of around 5 km. Subsequently, Olsgard and Gray (1995) carried out the same multivariate statistical analyses on data from a number of other oil and gas fields in the Norwegian sector of the North Sea. They showed that, when environmental conditions were heterogeneous, the only adverse effect that could be detected around oil and gas platforms in the North Sea was a small area where diversity was reduced. However, when the environment was homogeneous, a much larger area of altered faunal composition could be detected. At more heterogeneous sites, presumably the oil-based drilling muds were causing similar changes to the fauna outside of the area of reduced diversity, but these effects were not indistinguishable from differences produced by the environmental heterogeneity. Other studies of oil and gas platforms have failed to demonstrate subtle effects on community composition as a result of environmental heterogeneity (e.g., Terlizzi *et al.* 2008). Changes in community composition not associated with reduced diversity have been replicated in an experimental mesocosm study, in which sediment cores were amended with a 3-mm layer of drill cuttings (Schaanning *et al.* 2008). After 3 months, there were no significant differences in diversity between treatments, but community composition did differ between clean and cuttings treated cores. However, the authors suggested that the differences in community composition were due to altered sediment characteristics, rather than toxicity, as the cuttings contained low toxicity water- and olefin-based drilling muds.

There is also evidence for this hypothesised pattern in biofilms in rivers subject to herbicide runoff from vineyards (Dorigo *et al.* 2007). These authors found significant differences in the community structure of both eukaryotic and prokaryotic micro-organisms between pristine and contaminated sites, but no associated differences in diversity. Evidence that the differences were due to herbicides rather than other environmental differences between sites came from the fact that the sensitivity of photosynthesis to diuron exposure in laboratory conditions was lower in contaminated sites than it was in pristine sites. And in a study of herbivorous insects along an air pollution gradient in California, Jones and Paine (2006) found that particular groups of insects were associated with the pollution gradient, although there were no marked relationships between pollutant concentrations and insect abundance, species richness or diversity.

Is a change in community composition a deleterious effect?

So a number of studies have been able to identify sites where community composition is altered by exposure to pollutants without diversity being reduced. Whether this alteration in community composition represents a deleterious effect is still subject to debate. An extreme position would be that any alteration of species composition from a 'natural' state is undesirable. But other assessors will wish to take into account some measure of the relative desirability of the species being lost and gained. On land and in freshwater, we have sufficient information about the distribution and abundance of individual species to be able to set conservation priorities, and could use these to judge what constituted a deleterious change. In the marine environment, we cannot yet do this, although there would probably be agreement that the presence of known opportunistic species is a deleterious effect (implicit in the biological index used by Ugland *et al.* 2008) and some progress is being made in classifying substantial numbers of marine species according to their sensitivity to disturbance (Borja *et al.* 2004; Borja & Muxika 2005). In a study of temporal changes in the benthos of the Tees estuary, Warwick *et al.* (2002) interpret a reduction in taxonomic diversity as a 'detrimental' change, despite the fact that it was accompanied by an *increase* in Shannon diversity. In terms of the underlying species changes, the increase in Shannon diversity was caused by a reduction in the abundance of some common species, particularly the polychaete *Spiophanes bombyx*, whereas the reduction in taxonomic distinctiveness was visible only in the presence/absence data so reflects a reduction in phylogenetic diversity in the rarer species. By some criteria, the changes caused by pollution might be viewed as desirable changes. Some studies of atmospheric deposition have found higher diversity at 'polluted' sites. Mechanisms that may lead to this include direct effects of particulate deposition adding nutrients and/or raising the pH of acidic or nutrient poor soils (Brandle *et al.* 2001) and indirect effects whereby negative effects on vegetation lead to a change in leaf litter quality (Fenn & Dunn 1989). Even where there is a reduction in diversity, severe metal contamination may produce communities dominated by species or ecotypes that do not occur elsewhere. Microbial communities in freshwater affected by acid mine drainage such as the Rio Tinto in Spain and Iron Mountain in California are intriguing, supporting populations of metal- and acid-tolerant extremophiles, although in the case of the Rio Tinto may pre-date mining activity (Bond *et al.* 2000; Edwards *et al.* 2000; Gonzalez-Toril *et al.* 2003; Baker *et al.* 2004). Anthropogenic contamination that allows the persistence of metal-tolerant higher plants such as the calamine violet *Viola calaminaria* (Bizoux & Mahy 2007) could be viewed as a beneficial conservation measure (see Chapter 2).

However, species with wider environmental tolerances are likely to be more widespread than those with narrower tolerances. So changes in community composition due to pollution may well represent reductions in rare species, and their replacement by more common species. So if we are concerned to protect aquatic biodiversity, we should seek to avoid levels of pollution that produce even these relatively subtle changes.

A case study of polluted estuaries in South West England

A number of independent studies have been carried out on severely contaminated estuaries in South West England, and these give some interesting insights into the difficulties that can be faced by ecological monitoring. Restronguet Creek, one of the tributaries of the Fal, and the Hayle estuary (Fig. 7.2), receive

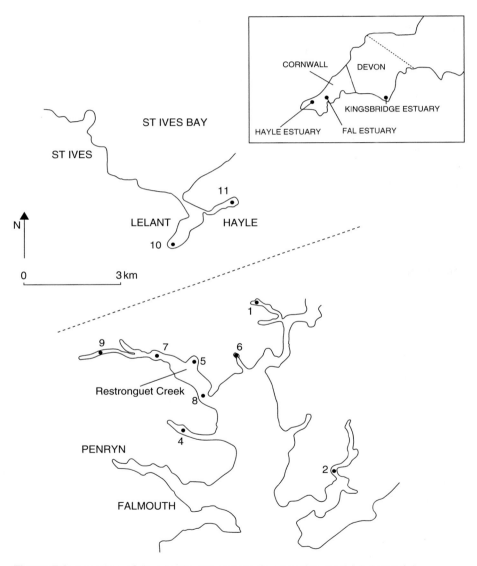

Figure 7.2. Location of sites in the Fal and Hayle estuaries at which bacterial communities have been characterised using T-RFLP analysis, and location of these two estuaries within South West England. Further details are given by Ogilvie and Grant (2008).

drainage water from disused tin and copper mines and are grossly contaminated as a result. The highest sediment metal concentrations occur in the upper reaches of Restronguet Creek, with total Cu often in excess of $3000 \, \mu g \, g^{-1}$ and total Zn as high as $4500 \, \mu g \, g^{-1}$ (personal observation). Metal concentrations are lower in the sandier sediments of the Hayle estuary ($450–850 \, \mu g \, g^{-1}$ Cu, $450–2500 \, \mu g \, g^{-1}$ Zn), but metals are more readily bioavailable there. Pore water Cu concentrations in Hayle sediments are regularly in excess of $500 \, \mu g \, L^{-1}$, and we have measured concentrations of up to $3400 \, \mu g \, L^{-1}$ Cu (in comparison with a maximum of $65 \, \mu g \, L^{-1}$ Cu in Restronguet Creek sediments and a UK EQS for marine waters of only $5 \, \mu g \, L^{-1}$). The most marked ecological effects can be attributed to copper contamination. Populations of several invertebrates from these estuaries can only survive there because they have evolved enhanced copper tolerance (Grant et al. 1989). Individuals of the common estuarine polychaete *Nereis diversicolor* from the Hayle and upper Restronguet Creek are able to tolerate Cu concentrations that are five times higher than those that can be tolerated by animals from clean estuaries. Nematode communities from both estuaries show a similar increase in Cu tolerance while microbial communities from the most contaminated parts of the Hayle are 2000 times less sensitive to Cu than those from uncontaminated sites (Millward & Grant 1995, 2000; Ogilvie & Grant 2008).

Work on the nematode communities in the Fal is consistent with the conceptual picture described above. Nematode diversity is significantly reduced only in the most contaminated tributary, Restronguet Creek, but community composition is altered in other less contaminated tributaries in a pattern that correlates with severity of contamination (Somerfield et al. 1994). These data suggest that there is an alteration of the nematode communities in all tributaries accept for Percuil River. The nematode communities in tributaries other than Percuil River also show increased tolerance to copper, indicating that the changes in community composition reflect a replacement of sensitive species by more tolerant species (Millward & Grant 2000), and providing strong evidence that there is an effect of pollution on these nematode communities. Meiofaunal copepods appear to be less sensitive than the nematodes, and changes in copepod communities can only be found in Restronguet Creek (Somerfield et al. 1994).

In contrast, the microbial communities in these estuaries provide an extreme example of how community response to environmental heterogeneity can mask the effects of a pollutant. There is no relationship between bacterial community composition, as quantified using T-restriction fragment length polymorphism (T-RFLP) analysis on bacterial 16S rDNA, and the extent of metal contamination, despite the fact that sediment metal concentrations vary by a factor of 20, and pore water metal concentrations by three orders of magnitude (Ogilvie & Grant 2008; Fig. 7.3). The data presented in Fig. 7.3 are for two

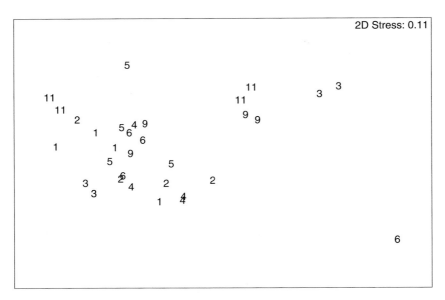

Figure 7.3. Multidimensional scaling of data on bacterial community structure in the Hayle and Fal estuaries (from Ogilvie & Grant 2008). Site numbers as in Fig. 7.2, with higher numbers indicating higher sediment pore water Cu concentrations.

replicate T-RFLP analyses on each of two sediment samples collected from each site. Replicate analyses on the same sample are generally close together in Fig. 7.3, but replicate samples from the same site may have bacterial communities that are almost as different from each other as the differences between the least and most contaminated sites. Despite this, the communities show strong differentiation in metal tolerance that is closely correlated with sediment metal concentrations. Thymidine and leucine incorporation rates at the least contaminated site are more than four orders of magnitude more sensitive to Cu than those from the most contaminated site in the Hayle estuary. So there is a strong ecological impact of Cu on the microbial communities present, but conventional ecological monitoring methods completely fail to detect it.

It is difficult to reach any clear conclusions as to the extent of ecological impacts on the macrofauna in these systems. Only very occasional individuals of the bivalve *Scrobicularia plana* are found in Restronguet Creek, although this does occur in the other tributaries (Bryan & Gibbs 1983). Warwick (2001) shows that the macrofauna in the Fal is different from that in other muddy estuaries in South West Britain, with over-representation of small polychaetes relative to other muddy estuaries. He notes that the isopod *Cyathura carinata* is absent from the Fal. He also suggests that the amphipod *Corophium volutator* is absent from the Fal, but in fact it does occur sporadically in the most contaminated parts of Restronguet Creek (Bryan & Gibbs 1983) and the Hayle estuary (personal observation). Warwick included the least contaminated tributary of the Fal, Percuil River, in this comparison. Sediments here still contain around 200 μg g^{-1} Cu, which is about an order of magnitude higher than background concentrations. So it is possible that macrofauna in *all* the tributaries of the Fal estuary, including Percuil River, are adversely affected by metal pollution. If so, the difference in macrofauna between the Fal and other muddy estuaries in South West Britain could be a consequence of metal contamination. However, the low diversity of macrofauna and the limited number of uncontaminated control sites limits the extent to which these conclusions can be drawn, and it is entirely possible that the differences reflect other environmental character-istics, as the sediments in the Fal system are rather finer grained than other estuaries in South West Britain. In the case of nematode communities, we do have evidence that there is no alteration in community composition in Percuil River, as community composition in the Helford River is very similar (Millward 1995). So the conclusion from all these studies is that there are probably effects of metal contamination on the fauna of the Hayle and most of the Fal system. But these effects are only apparent on some components of the biota. Multi-variate statistical analyses detect changes in nematode composition in moder-ately contaminated sites, but this is only possible because environmental conditions are sufficiently homogeneous. For bacteria, ecological monitoring methods fail to detect impacts that are clearly demonstrated by the enhanced metal tolerance of the same microbial communities.

Have we reached the methodological limits of detecting effects of pollution in the field?

There were considerable successes in the development and validation of multi-variate statistical methods to detect ecological change in the 1980s and early 1990s (see Warwick & Clarke 1991; Clarke & Warwick 2001b). However, since then, there seems to have been only limited progress in the development of ecological monitoring methods in the marine environment. This limited pro-gress appears to result from the difficulty of controlling for nuisance variables other than the severity of pollution rather than any intrinsic limitation of the statistical methods. Unless some novel solutions can be identified to deal with this limit, it may be that we cannot progress further. Methods such as the extent of taxonomic distinctiveness (Clarke & Warwick 2001a) show some signs of an association with pollution effects, but the association has not proved to be strong enough for these methods to have entered into widespread use as monitoring tools (Warwick *et al.* 2002). Freshwater faunas are rather species poor and in rivers water velocity is the strongest control on which species occur. It has therefore proved possible to identify pollution-sensitive and pollu-tion-tolerant taxa and construct indices such as the BMWP (or NWC) score (National Water Council 1981). It has also been possible to predict the probability of individual species being present at a site with particular physical and chemical characteristics. The RIVPACS system makes these predictions for the UK, allowing the fauna at a particular site to be compared with that predicted for a site with those same physical characteristics (Clarke *et al.* 2003). The methodology is transferable to other biogeographic regions if a suitable training dataset from uncontaminated rivers is available. A site is designated as polluted if there are appreciable numbers of absences. This methodology is unlikely to detect changes in community composition not accompanied by a reduction in diversity. But this phenomenon may not be common in freshwater, as the depauperate nature of freshwater faunas means that there will often not be a large species pool of more tolerant species that can replace sensitive species eliminated by pollution. Recent attempts have been made to construct similar indices of environmental quality in the marine environment, driven in part by the need to be able to quantify 'good' ecological status as part of the implementation of the EU Water Framework Directive. The AZTI or AMBI index (Borja *et al.* 2000, 2007) is based on a classification of species into five categories ranging from disturbance sensitive species to primary opportunists. We do not yet have a clear picture of how this index relates to other measures of pollution impact, but some studies have found that the value of the AZTI index is not always closely related to other measures of disturbance (e.g., Labrune *et al.* 2006).

These methods may make some further progress, but dealing with the effects of environmental differences other than pollution presents a fundamental barrier to progress, and one that is difficult to solve using multivariate statistical analyses of community composition. A potential solution may come from the examination of relationships between pollutant concentrations and the abundance and/or occurrence of individual species. Some studies have applied the concept of species sensitivity distributions to understand the relationship between contaminant concentrations and reduction in the abundance of individual species (Leung *et al.* 2005; Kwok *et al.* 2008). The result of these studies is thresholds below which ecological effects are unlikely to be occurring, rather than assessment of whether or not effects are occurring at any particular location. Inter-correlation of pollutant concentrations with other environmental variables is a difficulty in these studies, but recent developments in dealing with co-linearity in multiple regression, particularly hierarchical partitioning (Chevan & Sutherland 1991; Mac Nally 2002), may provide a solution to this (see Brown *et al.* in preparation, for a worked example in a different marine context).

Direct toxicity testing of environmental samples
Given these apparent limits to ecological monitoring, are there alternative approaches which might be more sensitive or might be less affected by

environmental heterogeneity? We have already touched upon Pollution Induced Community Tolerance and shown that this is as sensitive as conventional monitoring for nematodes and much more sensitive for microbes (see above). Another alternative approach is to expose organisms to environmental samples and examine their responses, dealing with environmental heterogeneity by exposing the same species to all samples, although this does not directly give information on what is happening in the field. If environmental samples are toxic to a widespread organism, it gives cause for concern. With this end in mind, considerable effort has gone into developing methods of sediment toxicity testing, using both whole sediments and pore waters (e.g., Matthiessen *et al.* 1998; Chapman *et al.* 2002; Thomas *et al.* 2003; Borgmann *et al.* 2005; McCready *et al.* 2006; Shipp & Grant 2006). In particular, the development of sensitive sub-lethal sediment toxicity tests have allowed detection of effects of contaminant concentrations lower than those that are acutely toxic (Thain *et al.* 1997; Shipp & Grant 2006; Allen *et al.* 2007). What is apparent from published studies using sediment toxicity testing is that toxic sediments may be remarkably widespread. Matthiessen *et al.* (1998) found acute and chronic toxicity of sediments from a number of sites in the Tyne estuary. Tests on more than 100 sites in and around Sydney Arbour, Australia, found that nearly a quarter of sediments showed a marked effect on amphipod reburial and over half showed a more than 20% reduction of sea urchin egg fertilisation (McCready *et al.* 2004). This study also reported that more than 80% of sediments were highly toxic to the Microtox test, but this overestimates toxicity as the authors did not remove elemental sulphur from solvent extracts of contaminants before testing (and see also Pardos *et al.* 1999; McCready *et al.* 2006). The performance of sediment toxicity testing in the Fal and Hayle systems shows that it is of similar sensitivity to the best available ecological monitoring. Shipp and Grant (2006) found that growth rate of the gastropod *Hydrobia ulvae* in a 28-day test was reduced when exposed to sediment from all the tributaries of the Fal estuary except Percuil River.

A conceptually related approach is the use of Scope for Growth (SFG) in *Mytilus edulis* as an indicator of environmental quality. Other evidence suggests that marine organisms and communities can be extremely sensitive to contaminants. Widdows *et al.* (1995, 2002) collected individuals of the mussel *Mytilus edulis* from an uncontaminated site, deployed them at field sites around the UK, then returned them to the laboratory and measured Scope for Growth by quantifying feeding and respiration rates. SFG was reduced at many sites in the southern North Sea and at a number of locations in western Scotland, the Irish Sea and South West England. At some locations, surplus energy was less than 25% of that in unstressed animals. Tissue body burdens of PAHs at the end of the exposure were sufficient to explain between 50 and 80% of the reduction in SFG. The ecological significance of these results is not known, but if continued for a prolonged period would result in reduced growth rate, and presumably translate into ecological effects.

Conclusions

It is clear from the examples outlined that pollution can produce subtle changes in community composition which can be detected in some circumstances, but in other cases are undetectable because of environmental heterogeneity. Widely used methods of ecological monitoring are particularly susceptible to environmental heterogeneity, so are almost certainly failing to detect extensive effects of pollution in the marine environment, although this is less likely to be occurring in freshwater. This implies that there may be substantial parts of the marine environment where undetected effects of pollution are occurring, and evidence of global effects of humans on coral reefs, and evidence of stress on mussels throughout large parts of the UK's coastal waters give some cause for concern. New statistical methodologies may improve the ability of monitoring to detect effects, but it may be that the problem is addressed more effectively by alternative approaches. Sediment toxicity testing, pollution induced community tolerance and biomarkers such as Scope for Growth that are linked closely to ecological effects are proving to be as sensitive as ecological monitoring but are less affected by environmental heterogeneity.

References

Allen, Y. T., Thain, J. E., Haworth, S. and Barry, J. (2007) Development and application of long-term sublethal whole sediment tests with *Arenicola marina* and *Corophium volutator* using Ivermectin as the test compound. *Environmental Pollution* **146**, 92–99.

Baker, B. J., Lutz, M. A., Dawson, S. C., Bond, P. L. and Banfield, J. F. (2004) Metabolically active eukaryotic communities in extremely acidic mine drainage. *Applied and Environmental Microbiology* **70**, 6264–6271.

Bizoux, J. P. and Mahy, G. (2007) Within-population genetic structure and clonal diversity of a threatened endemic metallophyte, *Viola calaminaria* (Violaceae). *American Journal of Botany* **94**, 887–895.

Bond, P. L., Druschel, G. K. and Banfield, J. F. (2000) Comparison of acid mine drainage microbial communities in physically and geochemically distinct ecosystems. *Applied*

and *Environmental Microbiology* **66**, 4962–4971.

Borgmann, U., Grapentine, L., Norwood, W. P., Bird, G., Dixon, D. G. and Lindeman, D. (2005) Sediment toxicity testing with the freshwater amphipod *Hyalella azteca*: relevance and application. *Chemosphere* **61**, 1740–1743.

Borja, A. and Muxika, H. (2005) Guidelines for the use of AMBI (AZTI's Marine Biotic Index) in the assessment of the benthic ecological quality. *Marine Pollution Bulletin* **50**, 787–789.

Borja, A., Franco, J. and Muxika, I. (2004) The biotic indices and the Water Framework Directive: the required consensus in the new benthic monitoring tools. *Marine Pollution Bulletin* **48**, 405–408.

Borja, A., Franco, J. and Perez, V. (2000) A marine Biotic Index to establish the ecological quality of soft-bottom benthos within European estuarine and coastal

environments. *Marine Pollution Bulletin*
40, 1100–1114.

Borja, A., Josefson, A. B., Miles, A., *et al.*
(2007) An approach to the intercalibration
of benthic ecological status assessment
in the North Atlantic ecoregion, according
to the European Water Framework
Directive. *Marine Pollution Bulletin*
55, 42–52.

Brandle, M., Amarell, U., Auge, H., Klotz, S. and
Brandl, R. (2001) Plant and insect diversity
along a pollution gradient: understanding
species richness across trophic levels.
Biodiversity and Conservation **10**, 1497–1511.

Brown, M. J. H., Mossman, H. L., Davy, A. J.
and Grant, A. (In prep.) Elevation,
drainage and sediment oxygenation
constrain halophyte colonisation
on a salt marsh created by managed
realignment.

Bryan, G. W. and Gibbs, P. E. (1983) Heavy metals
in the Fal Estuary (Cornwall): a study of
long term contamination by mining waste
and its effects on estuarine organisms.
*Marine Biological Association of the United
Kingdom, Occasional Publication* **2**, 1–112.

Chapman, P. M., Ho, K. T., Munns, W. R.,
Solomon, K. and Weinstein, M. P. (2002)
Issues in sediment toxicity and ecological
risk assessment. *Marine Pollution Bulletin* **44**,
271–278.

Chevan, A. and Sutherland, M. (1991)
Hierarchical partitioning. *American
Statistician* **45**, 90–96.

Clarke, K. R. and Warwick, R. M. (2001a)
A further biodiversity index applicable to
species lists: variation in taxonomic
distinctness. *Marine Ecology-Progress Series*
216, 265–278.

Clarke, K. R. and Warwick, R. M. (2001b) *Change in
Marine Communities*. 2nd edn. PRIMER-E Ltd,
Plymouth, UK.

Clarke, R. T., Wright, J. F. and Furse, M. T. (2003)
RIVPACS models for predicting the
expected macroinvertebrate fauna and
assessing the ecological quality of rivers.
Ecological Modelling **160**, 219–233.

Dorigo, U., Leboulanger, C., Berard, A., Bouchez,
A., Humbert, J. F. and Montuelle, B. (2007)
Lotic biofilm community structure and
pesticide tolerance along a contamination
gradient in a vineyard area. *Aquatic Microbial
Ecology* **50**, 91–102.

Edwards, K. J., Bond, P. L., Gihring, T. M. and
Banfield, J. F. (2000) An archaeal iron-
oxidizing extreme acidophile important in
acid mine drainage. *Science* **287**, 1796–1799.

Fenn, M. E. and Dunn, P. H. (1989) Litter
decomposition across an air-pollution
gradient in the San Bernardino Mountains.
Soil Science Society of America Journal **53**,
1560–1567.

GESAMP. (1982) *The Health of the Oceans*, UNEP
Regional Seas Reports and Studies. United
Nations Environment Programme, Regional
Seas Programme Activity Centre, Geneva,
Switzerland, 111 pp.

Gonzalez-Toril, E., Llobet-Brossa, E., Casamayor,
E. O., Amann, R. and Amils, R. (2003)
Microbial ecology of an extreme acidic
environment, the Tinto River. *Applied and
Environmental Microbiology* **69**, 4853–4865.

Grant, A. and Millward, R. N. (1997) Detecting
community responses to pollution.
In: *Responses of Marine Organisms to their
Environment* (eds. L. E. Hawkins and
S. Hutchinson), pp. 201–209. Proceedings
of the 30th European Marine Biology
Symposium, University of Southampton,
Southampton, UK.

Grant, A., Hateley, J. G. and Jones, N. V. (1989)
Mapping the ecological impact of
heavy-metals on the estuarine polychaete
Nereis-Diversicolor using inherited metal
tolerance. *Marine Pollution Bulletin*
20, 235–238.

Gray, J. S., Clarke, K. R., Warwick, R. M. and
Hobbs, G. (1990) Detection of initial effects
of pollution on marine benthos – an
example from the Ekofisk and Eldfisk
oilfields, North Sea. *Marine Ecology-Progress
Series* **66**, 285–299.

Jones, M. E. and Paine, T. D. (2006) Detecting
changes in insect herbivore communities

along a pollution gradient. *Environmental Pollution* **143**, 377–387.

Kwok, K. W., Bjorgesaeter, A., Leung, K. M. et al. (2008) Deriving site-specific sediment quality guidelines for Hong Kong marine environments using field-based species sensitivity distributions. *Environmental Toxicology and Chemistry* **27**, 226–234.

Labrune, C., Amouroux, J. M., Sarda, R. et al. (2006) Characterization of the ecological quality of the coastal Gulf of Lions (NW Mediterranean). A comparative approach based on three biotic indices. *Marine Pollution Bulletin* **52**, 34–47.

Leung, K. M., Bjorgesaeter, A., Gray, J. S. et al. (2005) Deriving sediment quality guidelines from field-based species sensitivity distributions. *Environmental Science and Technology* **39**, 5148–5156.

Mac Nally, R. (2002) Multiple regression and inference in ecology and conservation biology: further comments on identifying important predictor variables. *Biodiversity and Conservation* **11**, 1397–1401.

Maltby, L., Blake, N., Brock, T. C. M. and Van Den Brink, P. J. (2005) Insecticide species sensitivity distributions: importance of test species selection and relevance to aquatic ecosystems. *Environmental Toxicology and Chemistry* **24**, 379–388.

Matthiessen, P., Bifield, S., Jarrett, F. et al. (1998) An assessment of sediment toxicity in the River Tyne Estuary, UK by means of bioassays. *Marine Environmental Research* **45**, 1–15.

McCready, S., Birch, G. F., Long, E. R., Spyrakis, G. and Greely, C. R. (2006) Relationships between toxicity and concentrations of chemical contaminants in sediments from Sydney Harbour, Australia, and vicinity. *Environmental Monitoring and Assessment* **120**, 187–220.

McCready, S., Spyrakis, G., Greely, C. R., Birch, G. F. and Long, E. R. (2004) Toxicity of surficial sediments from Sydney Harbour and vicinity, Australia. *Environmental Monitoring and Assessment* **96**, 53–83.

Millward, R. (1995) The effects of chronic and acute metal-enrichment on the nematode community structure, composition and function in Restronguet Creek, *SW England*. University of East Anglia.

Millward, R. N. and Grant, A. (1995) Assessing the impact of copper on nematode communities from a chronically metal enriched estuary using pollution-induced community tolerance. *Marine Pollution Bulletin* **30**, 701–706.

Millward, R. N. and Grant, A. (2000) Pollution-induced tolerance to copper of nematode communities in the severely contaminated Restronguet Creek and adjacent estuaries, Cornwall, United Kingdom. *Environmental Toxicology and Chemistry* **19**, 454–461.

National Water Council. (1981) *River Quality – The 1980 Survey and Future Outlook*. National Water Council, London.

Ogilvie, L. A. and Grant, A. (2008) Linking pollution induced community tolerance (PICT) and microbial community structure in chronically metal polluted estuarine sediments. *Marine Environmental Research* **65**, 187–198.

Olsgard, F. and Gray, J. S. (1995) A comprehensive analysis of the effects of offshore oil and gas exploration and production on the benthic communities of the Norwegian continental shelf. *Marine Ecology-Progress Series* **122**, 277–306.

Pandolfi, J. M., Bradbury, R. H., Sala, E. et al. (2003) Global trajectories of the long-term decline of coral reef ecosystems. *Science* **301**, 955–958.

Pardos, M., Benninghoff, C., Thomas, R. L. and Khim-Heang, S. (1999) Confirmation of elemental sulfur toxicity in the Microtox assay during organic extracts assessment of freshwater sediments. *Environmental Toxicology and Chemistry* **18**, 188–193.

Piola, R. F. and Johnston, E. L. (2008) Pollution reduces native diversity and increases invader dominance in marine hard-substrate communities. *Diversity and Distributions* **14**, 329–342.

Schaanning, M. T., Trannum, H. C., Oxnevad, S., Carroll, J. and Bakke, T. (2008) Effects of drill cuttings on biogeochemical fluxes and macrobenthos of marine sediments. *Journal of Experimental Marine Biology and Ecology* **361**, 49–57.

Shipp, E. and Grant, A. (2006) *Hydrobia ulvae* feeding rates: a novel way to assess sediment toxicity. *Environmental Toxicology and Chemistry* **25**, 3246–3252.

Somerfield, P. J., Gee, J. M. and Warwick, R. M. (1994) Soft-sediment meiofaunal community structure in relation to a long-term heavy-metal gradient in the Fal Estuary System. *Marine Ecology-Progress Series* **105**, 79–88.

Terlizzi, A., Bevilacqua, S., Scuderi, D. *et al.* (2008) Effects of offshore platforms on soft-bottom macro-benthic assemblages: a case study in a Mediterranean gas field. *Marine Pollution Bulletin* **56**, 1303–1309.

Thain, J. E., Davies, I. M., Rae, G. H. and Allen, Y. T. (1997) Acute toxicity of ivermectin to the lugworm *Arenicola marina*. *Aquaculture* **159**, 47–52.

Thomas, K. V., Barnard, N., Collins, K. and Eggleton, J. (2003) Toxicity characterisation of sediment porewaters collected from UK estuaries using a *Tisbe battagliai* bioassay. *Chemosphere* **53**, 1105–1111.

Ugland, K. I., Bjorgesaeter, A., Bakke, T., Fredheim, B. and Gray, J. S. (2008) Assessment of environmental stress with a biological index based on opportunistic species. *Journal of Experimental Marine Biology and Ecology* **366**, 169–174.

Warwick, R. M. (2001) Evidence for the effects of metal contamination on the intertidal macrobenthic assemblages of the Fal Estuary. *Marine Pollution Bulletin* **42**, 145–148.

Warwick, R. M. and Clarke, K. R. (1991) A comparison of some methods for analyzing changes in benthic community structure. *Journal of the Marine Biological Association of the United Kingdom* **71**, 225–244.

Warwick, R. M., Ashman, C. M., Brown, A. R. *et al.* (2002) Inter-annual changes in the biodiversity and community structure of the macrobenthos in Tees Bay and the Tees estuary, UK, associated with local and regional environmental events. *Marine Ecology-Progress Series* **234**, 1–13.

Widdows, J., Donkin, P., Brinsley, M. D. *et al.* (1995) Scope for growth and contaminant levels in North-Sea mussels *Mytilus edulis*. *Marine Ecology-Progress Series* **127**, 131–148.

Widdows, J., Donkin, P., Staff, F. J. *et al.* (2002) Measurement of stress effects (scope for growth) and contaminant levels in mussels (*Mytilus edulis*) collected from the Irish Sea. *Marine Environmental Research* **53**, 327–356.

CHAPTER EIGHT

With the benefit of hindsight: the utility of palaeoecology in wetland condition assessment and identification of restoration targets

PETER GELL

Introduction

Pollution sources to aquatic ecosystems can be categorised as point (or direct), those derived from identifiable sources such as sewage treatment plant out-falls, or diffuse, where the source of pollutant is more difficult to identify, such as surface erosion. In the former case, the effluent loads can be high; however, by virtue of a more clear relationship between source and impact, cause is more readily identifiable and solutions more readily encouraged or directed (Smol 2008). Diffuse pollution sources often create chronic symptoms of elevated pollution loads that are more difficult to establish experimentally and more difficult to identify spatially. In many instances, the drivers of these heightened releases of pollutants to receiving waters have a long history and originated from settlements and developments that extend beyond the memory of modern societies. The widespread and deep-in-time nature of diffuse sources of pollution, coupled with their nature as being, effectively, multiple point sources, renders the identification of the causes of diffuse pollution uncertain and so poses a greater challenge in terms of mitigation.

Diffuse pollutants are most often represented by sediments and solutes. Widespread vegetation clearance, catchment settlement, intensive tilling and cropping and excessive stocking rates of grazing animals all contribute to exposing surface soils to erosive forces that increase sediment loads to aquatic systems. This acts to increase sedimentation rates in streams and lakes and to increase the turbidity of the water itself. The chronic increase in supply of clays and colloids leads to a reduction in light penetration removing benthic substrates from the photic zone and advantaging phytoplankton over more productive attached macrophytes (Reid *et al.* 2007). In some systems, floodplain

Ecology of Industrial Pollution, eds. Lesley C. Batty and Kevin B. Hallberg. Published by Cambridge University Press. © British Ecological Society 2010.

sediments are bound with native phosphorus (Olley & Wallbrink 2004), which is remobilised when riparian surfaces are exposed or when changes to the hydrological response of the catchment elevates the erosive potential of storm runoff. This change acts to further advantage productive phytoplankton reinforcing the state shift from a previous clear-water condition. The same catchment changes driving erosion also alter the surface-groundwater balance and can lead to the increased accession of precipitation to recharge zones. The resultant increased hydraulic pressure can expand upslope the region of groundwater discharge (Macumber 1991) and, where this is overlain by strata rich in connate salts, can lead to surface secondary salinisation. This leads to increased flux of solutes to aquatic systems elevating surface water salinity. The associated increased sodicity leads to soil efflorescence and makes the soil surface more dispersive and erodable (Neave & Rayburg 2006). Furthermore, in provinces rich in sulphate salts, the burial of sulphates in sites of sediment accumulation can lead to the rapid production of sulphides. These, if exposed under conditions of drought, for example, can lead to the release of sulphurous acids in the manner of acid sulphate soils in coastal contexts where sediments and salts have accumulated over millennia. In all, the sum of many diffuse sources of pollution, delivered at very low loads over long periods, can result in a cocktail of acute salinity, acidity, turbidity or eutrophication. Ultimately, due to the interrelatedness of the drivers of pollution, several pollution symptoms can emerge synchronously in wetland systems (Gell *et al.* 2007a). Here, the multifaceted suite of drivers in action makes cause and effect particularly difficult to unravel.

Commitments to restore wetlands

The restoration of wetland systems is now a high profile ideal of many developed and developing nations. Restoration or rehabilitation of wetlands affected by direct sources of pollution is more easily affected as the degraded condition is often attributable to few pollution sources. Depending on the nature of the system, rehabilitation can occur rapidly, e.g., in fast flowing streams, or can take decades, e.g., endorheic (closed) crater lakes (Gell, unpublished data) and estuaries (Chapter 14). The restoration of wetlands subject mostly to diffuse pollution sources, particularly if situated within large catchments such as the Murray Darling Basin (MDB) of Australia or the Mississippi River basin in the USA, is a vastly greater challenge as the degradation may have occurred over decades, is driven by a myriad of causes and sources of pollution and continues to be subjected to a pervasive shift in hydrological, ecological and socio-political regime.

South Australia's wetland strategy

There are now emerging strategies and policies to preserve, rehabilitate and restore wetlands, including rivers and lakes. Across Australia, the Federal

Government has supported the generation of state-based wetland strategies. In South Australia, at the base of the MDB, the 2001 Select Committee of the South Australian Parliament stated that 'wetlands are essential to the maintenance of hydrological, physical and ecological health of the riverine environment and provide economic, social and cultural benefits to the broader community' (DEH & DWLBC 2003, p. 9). One of the cornerstones of the 2003 *Wetlands Strategy for South Australia*, was a 'recognition that action is needed to see water-dependent ecosystems understood, protected and restored or rehabilitated as part of addressing South Australia's degradation of land and water resources and loss of biological diversity' (DEH & DWLBC 2003, p. 11). Furthermore, of the seven key research priorities identified under this strategy were:

- defining, for differing wetland types and proposed uses, their limits or thresholds of acceptable change;
- identifying natural wetting and drying cycles of wetlands and the biological and other responses of these;
- environmental water needs of different wetland types and their related biota;
- better understanding of the macro- and microinvertebrates of wetlands;
- implications of long-term alterations/changes in salinity and climate change on wetlands;
- inventory tools including minimum datasets, classification systems, wetland condition indices . . .; and
- early warning indicators of impacts on wetlands and monitoring protocols (DEH & DWLBC 2003, p. 25).

These priorities recognised the need to understand wetland change, particularly the drivers of change and the wetland responses, to build the state's capacity to restore or rehabilitate its water-dependent ecosystems.

The Water Framework Directive

Perhaps an even more ambitious approach to wetland restoration is the Water Framework Directive (WFD) or 'Directive 2000/60/EC of the European Parliament and of the council of 23 October 2000; establishing a framework for community action in the field of water policy'. It identifies the need to manage European water resources in an integrated fashion by understanding all components of the ecosystem at the catchment scale (Bennion & Battarbee 2007). The principal aim of the WFD is to 'achieve good ecological quality in all relevant waters by 2015'. High ecological quality is deemed to be a wetland that is entirely or largely undisturbed relative to a reference or baseline condition identified mostly on the basis of the ecological structure and function of its aquatic ecosystem. The condition of good condition is satisfied if the site has only slightly deviated from the reference. This equates to the measure of biological integrity – a measure of human impact – used in the US Clean Water Act (Barbour *et al.* 2000). The first iteration of the River Basin Planning process within the WFD will be to identify sites that are at risk of being classified as moderate, poor or bad status and to establish a status review and monitoring programme to identify the catchment drivers of the water condition. Here the WFD has similar goals to the Wetland Strategy of South Australia with the 'natural' baseline the reference against which condition is assessed.

Identifying targets for restoration

Critical to the assessment of the condition or biological integrity of a wetland is the identification of this reference condition or baseline. Most of this work has focused on lotic (flowing) waters. Relative to the timing of settlement and the onset of industrialisation, the monitoring of waterways has only been a very recent feature of management. In some locations, there are long records of water quality which, despite being patchy in time, can be used to assess past condition and recent recovery (Chapter 13). The comparison of these data with modern monitoring, however, is problematic due to changes in monitoring approaches and technologies over time. Furthermore, the variability of these systems questions whether the data from one sampling event are representative of the average condition of the waterway. While the vouchering of a specimen provides incontravertible evidence for the presence of an organism, uncertainty as to the sampling intensity of the operator in the past questions whether absence of evidence is evidence for absence. To overcome this, reference conditions have been modelled based on non-chemical variables in a space-for-time assessment of change or impact. Here, the deviation in invertebrate species assemblages from a modelled reference condition underpins the river condition assessment process within RIVPACS (Chapter 6) in the UK and AusRivas (Reynoldson *et al.* 1997) across Australia. This too has proven to represent a particular challenge, and it is recognised that, if the body of 'reference' sites used to build the model are in fact impacted, then the inferred condition score of the test sites will be artificially raised. In this instance, some impacted sites may qualify as less impacted and they will not be highly rated for restoration need. This has been particularly the case in lowland streams in Australia where few sites are considered to be in reference condition (Thoms *et al.* 1999).

Given this focus on running waters, it stands that assessment of the condition of lentic (still water) waters is based on even less data. In 2004, in recognition of this, and the limited knowledge of the nature of these systems, the South Australian Government initiated *The River Murray Wetland Baseline Study*. This is intended to provide a twenty-first century baseline to ensure that these wetlands deteriorate no further. While this may provide sufficient evidence against which to compare future change, the absence of past monitoring data exposes modern wetlands to mismanagement as managers cannot be called to task if evidence for change is weak or absent. It was under this paradigm that a catchment manager in northwestern Victoria was able to claim that Psyche Bend Lagoon, a shallow floodplain lake lying alongside the River Murray, was in its natural condition after irrigation waters were diverted away from the lagoon to arrest salt fluxes to the main river. Long-term monitoring data (Gell et al. 2002), however, revealed that the wetland salinity was steady at around $2.0 \, ms \, cm^{-1}$ for most of 1995–1997, but rapidly salinised to $60 \, ms \, cm^{-1}$ within 18 months of the diversion. Clearly, facing liability for the impact on the wetland, the manager chose to misrepresent the wetland's natural condition. The lack of a clearly identified baseline condition of the river system more broadly also leads to a weakening of the government's commitment to restore it with environmental flows. A scientific reference group identified that $1500 \, GL \, a^{-1}$ of flow was required for the River Murray to have any likelihood of achieving good ecological condition (Jones et al. 2002a). However, Benson et al. (2003) were able to successfully claim that the degraded state of the river had been exaggerated by the scientists and that it would not benefit greatly from the widespread allocation of increased flow. Clearly, where issues of liability or livelihoods are at the core of an issue, the evidence for a baseline condition needs to be objective and defensible.

Wetlands and sediment records

Wetlands act as sediment sinks in catchments and so can be used as a means to establish mean sedimentation rates and the flux and accumulation of pollutants at the end of the system. They represent, therefore, a classic receptor, at a catchment scale, in the sense of the source-pathway-receptor principal of the ecological risk assessment framework of the UK Environment Agency (see Chapter 9). Care should be taken, however, as they have the capacity to also act as a source (e.g., Farmer 1991). The sediment input to lowland wetlands is potentially greater than their upland counterparts owing to the amplification of fluxes into higher order streams. The contribution of sediments carried by the main river to floodplain wetlands is influenced by the degree of connectivity between the river and the wetland and the frequency of floods of heights sufficient to connect the two. These allochthonous (external) sources combine with endogenic (internal) production of sediments and organic matter that can further accelerate gross sediment accretion. The trajectory towards complete infilling may be balanced by declining sediment trapping efficiency as a wetland fills and sediment scouring occurs with high flow events. Increased flux of allochthonous sources down large rivers may come from surface erosion from the catchment, river bank collapse and soil surface sodicity and this may, at least in part, be balanced, in regulated systems, by sediment trapping behind impoundments (Olley & Wallbrink 2004). Endogenic drivers that accelerate accretion rates include wave-driven erosion of exposed littoral zones and increased sediment trapping by aquatic macrophytes, themselves promoted by accelerated nutrient fluxes. The separation of these allochthonous, river-derived indicators from those produced within the wetland enables assessment of both the condition of the river and the individual wetland to be made.

The chronic nature and cause of diffuse pollution often allows it to be underrated as a driver of change. Humans experiencing pollution are, naturally, alert to changes that occur over short time frames such as black water or anoxia events that lead to fish kills. In many instances, this ecological catastrophe is predisposed due to the antecedent conditions brought on by diffuse sources over decades. This degraded state is often not recognised as the chemical shifts are difficult to perceive and the short-term focus of the observer denies them a credible baseline against which to assess condition. Rarely were ecological monitoring programmes established early enough, and equally as rarely were early monitoring programmes funded for long enough, for a strong temporal perspective to be acquired to put present conditions in context. This record of change can be accessed retrospectively however. By virtue of the retention, in archives of wetland sediment, of biological indicators of wetland condition over time, an extended record, even beyond intensive human settlement, can be generated. These palaeoecological studies can provide lessons of the past lost to memory and provide the benefits of hindsight otherwise unavailable to restoration scientists and practitioners. The strongest knowledge base, as Davis (1989) identified, and Bennion and Battarbee (2007) reiterated recently, is the splicing of long-term studies with palaeoecological studies, so that the past and the present can cooperate to enable us to identify baseline condition(s) and the natural variation around it, the timing and nature of impact, the responsiveness of the system and the present trajectory of change.

Study areas

Two study regions are selected here to reveal the insights gained from palaeo-limnological approaches to wetland condition assessment. The first, the MDB, is subject to the impacts of long-term land development for primary industry. Sites in the lower catchment are illustrated here to reveal the complex interaction of increases in the flux of salts, sediments and nutrients in a highly abstracted water resource under a drying climate. Two coastal wetlands outside the basin are discussed to reveal the impact of water diversions on wetland condition. The second region, the upper Mississippi River Basin, focuses on two in-channel lakes that have accumulated sediments since the last deglaciation. Held up by natural impoundments, these wetlands represent a unique opportunity to evaluate the changing condition of the river per se without having to distinguish allochthonous from autochthonous evidence.

The MDB covers $1.073 \times 10^6 \, km^2$ and ranges from montane landscapes in the southeast that, in winter-spring, provide for most of the catchment runoff, to semi-arid and arid conditions in the southwest. The northernmost parts of the catchment extend into sub-tropical climates that can provide sporadic high flows in summer-autumn. The catchment was settled from the 1840s and now provides 40% of Australia's agricultural gross domestic product; however, this has come at considerable environmental cost. In excess of 80% of the 14×10^9 GL mean annual flow has been diverted for irrigation agriculture and domestic use (Commonwealth of Australia 2001). In many lowland parts of the catchment, saline regional groundwater tables are perennially within the capillary zone and large areas of the catchment have been impacted by secondary salinisation. Nutrients are released from urban sources, irrigation return drains and clays mobilised from exposed river banks. Stock grazing rates were high within a few years of settlement and the controlling influence of ENSO cycles has exacerbated erosion through extreme droughts and flood events. The wetlands under study range from levee and oxbow lakes along the river floodplain to terminal lakes that owe their origin to the reduction in gradient and formation of barrier spits after the stabilisation of sea levels in the mid Holocene. Most floodplain wetlands are shallow (<1 m), and they tend to dry regularly while some have been maintained by connection to a series of weirs or locks along the rivers and so are deeper (1–3 m). The terminal lakes are variously influenced by tides. Of the wetlands discussed outside the MDB, Lake Curlip is a shallow lake on the Brodribb River near the mouth of the Snowy River catchment in eastern Victoria. Lastly, Mullins Swamp is a coastal lagoon in southeast South Australia, hemmed in between the modern coastal foredune and the last interglacial dune system, the Woakwine Range.

The Upper Mississippi wetlands were formed by natural barrages of alluvial material deposited at the junctions of smaller rivers with the larger Mississippi River channel. These represent a unique geomorphic context in that large, linear lakes have formed in the main river channels. Accumulating material derived from the St Croix River catchment is Lake St Croix that was effectively dammed by deposits accumulated at its junction with the Mississippi River (Blumentritt *et al.* 2009). This catchment, of 19 900 km^2, remains mostly forested with some agricultural development. Average inflow in 2004 was 165 m^3 s^{-1} (Triplett *et al.* 2008). Lake St Croix is 35 km^2 in area and has a residence time of 20–50 days. Lake Pepin, in contrast, is on the main Mississippi River stem and was formed by sediments deposited at its junction with the Chippewa River. Here, there is considerable catchment development for intensive crops such as corn and soybean and the twin cities of Minneapolis St Paul lie 80 km upstream in the Minnesota catchment. The Mississippi catchment above the lake is 12 000 km^2 and the mean flow is 600 m^3 s^{-1}. Lake Pepin receives sediments and other material from the Mississippi and Minnesota River catchments (Engstrom *et al.* 2009). Lake Pepin is 103 km^2 and has an average residence time of 19 days.

Methods

Over the past decade, over 40 floodplain wetlands within the MDB have been cored for palaeolimnological research (Barnett 1994; Thoms *et al.* 1999; Ogden 2000; Reid *et al.* 2002; Gell *et al.* 2005a, b, 2006, 2007b, unpublished data; Fluin *et al.* 2007). Here, the records from a range of wetlands across the basin are discussed from its middle to upper reaches nearer the Australian alpine region to the Coorong at the mouth of the River Murray. The sites outside the MDB are in coastal contexts that reveal the impact of water diversions. All wetlands are selected to reveal wetland sedimentation rates and to demonstrate the use of microfossil indicators to reveal changes to river and wetland condition. In all sites, a Russian (d-section) corer (after Jowsey 1966) was used to retain the depth/age profile but often a field piston corer (after Chambers & Cameron 2001) was used to increase sediment volume for analysis. In several wetlands, cores were taken from 2 or 3, and in the Coorong 30, locations. At most sites, samples were taken for ^{14}C AMS dating. Elsewhere samples were taken for luminescence dating. In most instances, recent samples were taken for alpha and gamma spectrometry to generate a ^{210}Pb decay profile and to gain evidence for ^{137}Cs activity. The upper sediments of most cores were subsampled to identify the arrival of exotic *Pinus* pollen (after Ogden 1996). The changing nature of the sediments in many cores was analysed utilising magnetic susceptibility analysis by passing unopened PVC cores through a Bartington BS2 loop. Subsamples were collected from all cores to extract fossil diatom algae to reconstruct changes in wetland condition according to the known preferences of species for salinity, pH and nutrients. Samples were prepared and enumerated using standard methods (after Battarbee *et al.* 2001). Species identification was supported by reference to Krammer and Lange-Bertalot (1986, 1988, 1991a, b) and Witkowski *et al.* (2001). The ecological inferences from each species are based on Van Dam *et al.* (1994), Gell (1997), Sonneman *et al.* (2000), Gell *et al.* (2002), Tibby (2004) and Tibby and Reid (2004). The timing of increases in river plankton in wetlands closely linked to the main channel is used to aid the identification of the onset of river regulation in the mid 1920s.

A network of 25 cores along five, equally spaced shore-to-shore transects were collected by Engstrom and co-workers from Lakes Pepin and St Croix in 1995–1996. Cores were taken with portable piston and Livingstone corers to depths of 3.5 to 4.0 m. The magnetic susceptibility of all cores was determined across 2-cm intervals with a Bartington core logging sensor. Dry density data were gained from loss-on-ignition techniques at 100° C, 550° C and 1000° C. Small pieces of sub-fossil terrestrial woody material were sieved from selected cores for ^{14}C dating using AMS. There were 20 sediment cores from Lake Pepin that were analysed for ^{137}Cs and 10 for ^{210}Pb. In addition, 20 sediment cores were analysed for total phosphorous (Engstrom *et al.* 2009). Samples from five of these cores were prepared for pollen analysis to assist with chronology, and ten samples from core V.1 were analysed for sub-fossil diatoms to infer changing water quality conditions using the transfer function of Bennion (1994). Similar analyses were undertaken on 24 cores taken across five transects in Lake St Croix. Eight cores from Lake St Croix were used for ^{137}Cs, ^{210}Pb and ^{14}C AMS dating. All sediments were analysed for mercury, silica, sedimentary phosphorus and diatoms for inferring phosphorus.

Results and discussion

As the sediments of catchment wetlands integrate the fluxes of materials moving down a watershed, their sediments accumulate a wide variety of chemical, physical and biological indicators of a wide range of pollutants. In most instances, each site records evidence for multiple drivers of degradation (e.g., salinisation and eutrophication). The evidence from the Murray and Mississippi study regions for sediment and metal pollution, salinisation, eutrophication and drivers of pH change are discussed in order below.

Sediment pollution
Sedimentation rates

Direct evidence for an increased flux in the sediments from a catchment can be derived from estimates of sedimentation rate. Sediment can be carried down a main river channel and deposited in the calmer waters of an impounded section such as a reservoir, or natural impoundments such as Lakes Pepin and St Croix. These impoundments can trap much of the sediment travelling down the channel and lead to a reduction in the flux of particles down the system. Depending on the residence time, finer, colloidal material can pass impoundments and be transferred down the system. In times of flood, there can be a lateral exchange of suspended and dissolved material between the channel and the floodplain and its wetlands. Additionally, sediments can also be washed or blown directly into any wetland from neighbouring slopes and floodplains.

Sedimentation rates are best described as mass accumulated per unit area of sediment to allow for empirical comparisons of change down core so avoiding differential moisture content or compaction. Where they are expressed as accumulation rates (mm in sediment depth per unit time) then a qualitative comparison can be made mindful of these factors.

Lowland, limestone-rich landscapes such as the lower River Murray are challenging sites to derive robust sediment chronologies (Gell *et al.* 2005a). This is due to the episodic transport of sediment, elevated wetland salinity concentrations and natural and artificial wetting and drying regimes. Despite this, plausible chronologies can be established by utilising a range of dating techniques. These chronologies can be supported by the assembly of a set of palaeorecords that identify regional patterns in microfossil assemblage changes. Once established, the arrival of exotic pollen biomarkers and any switch to river plankton-dominated assemblages can be used as corroborative evidence to support independent radiometric-based time lines.

The sediment accumulation rates in many wetlands down the Murray River system are summarised in Gell et al. (2006, 2009). The pre-industrial rates of accumulation were typically in the order of $0.1-1.0 \, \mathrm{mm \, yr^{-1}}$ with most sites in the lower end of this range. Pre-industrial rates in excess of $1.0 \, \mathrm{mm \, yr^{-1}}$ occurred in large meanders that were scoured deeply in the mid Holocene. In all sites with the dating strength to allow comparison, the post-European sedimentation rates increased, in some instances by almost two orders of magnitude. Early post-European sedimentation rates were as high as $20 \, \mathrm{mm \, yr^{-1}}$ in sites near to heavily stocked sheep runs (Gell et al. 2005b). Here they declined post regulation owing to the trapping of coarser sediments behind weirs as suggested by Olley and Wallbrink (2004). Recent rates as high as $10-30 \, \mathrm{mm \, yr^{-1}}$ are occurring, possibly driven by the increased sensitivity of floodplain surfaces to erosion through drought and increased sodicity. Exceptionally high values were returned from Pikes River wetland ($30 \, \mathrm{mm \, yr^{-1}}$) and Coorong South Lagoon ($18 \, \mathrm{mm \, yr^{-1}}$). The rates returned for the estuarine Coorong sites are considered to be net sedimentation rates as the interaction of tides can remove settled sediments in these systems (Lawler 2005). Along the Mississippi the pre-industrial (c. 1830) sediment accumulation rate for Lake Pepin was reported as ranging from as low as $0.5 \, \mathrm{kg \, m^{-2} \, yr^{-1}}$ in the lower part of the basin to $2.1 \, \mathrm{kg \, m^{-2} \, yr^{-1}}$ in the uppermost transect (Engstrom et al. 2009). The post settlement rates throughout most of the basin are typically $7-15 \, \mathrm{kg \, m^{-2} \, yr^{-1}}$ but as high as $20-30 \, \mathrm{kg \, m^{-2} \, yr^{-1}}$ in the upper transect. In all these values represent a 15-fold increase relative to the pre-1830 baseline.

High sedimentation rates reflect the flux of sediment that can impact on wetland biota through changing light regime (discussed below) and directly by changing benthic habitat and impacting upon the function of filter feeding organisms. Geomorphically, high sedimentation rates can directly truncate the life of the wetland itself. About 17% of the 1830 volume of Lake Pepin is now filled with sediment. In both the Murray and Mississippi situations, flow regulation has reduced the prospect of flood-induced scouring and wetland renovation. This recent net increase in sedimentation rate, thereby, represents a clear risk to the longevity of the wetland. Given the shallow nature of many River Murray wetlands, and the elimination of medium level floods (Jones et al. 2002b), the terrestrialisation of wetlands is likely to proceed at a pace greater than wetland renewal. At pre-industrial sediment accumulation rates Pepin Lake would have expected to have another 4000 years of life as a lake, however, current rates allow for a projection of only 340 years until complete filling by sediment. Many of the floodplain wetlands in the Lower Murray River system are even shallower ($<1-2 \, \mathrm{m}$), so sedimentation rates in the order of $10 \, \mathrm{mm \, yr^{-1}}$

or more forecast complete wetland filling within decades. The outcome of this infilling is the shallowing of one wetland near Swanport adjacent to the Lower River Murray. Here, this wetland is celebrated as a complex and diverse mosaic of aquatic macrophyte communities. It is evident that the pre-industrial condition was of open water and the upper 2.0 m or so of sediment is post-industrial. So the present condition is derived, and an artefact of accelerated sediment flux. The sediment accumulation trajectory of this wetland, widely considered to be in good ecological condition, is for complete infilling in the near future. In contrast, Murrundi Wetland to its south attained a swamp-like condition about 2000 years ago (Gell *et al.* 2005a). To reinstate its perceived open water condition, as recollected during the extreme floods of the 1950s, this wetland has been dredged and, in part, is now an open water environment.

Metal pollutants

Metals are often transported with sediments and can accumulate in sediment sequences. Fluorescence spectroscopy techniques were applied to sediment digestates from a range of levels from ten cores from Lake Pepin to reconstruct changes in mercury flux over time (Balogh *et al.* 1999). They identified that the natural, pre-industrial whole basin mercury accumulation rate was $3 \, \mathrm{kg \, yr}^{-1}$, believed to be typical of lakes sampled across the region. From *c.* 1830 this increased rapidly to a maximum of $357 \, \mathrm{kg \, yr}^{-1}$ in the 1960s. Since the 1800s, 18.1 tons of Hg were deposited into the lake, with half contributed between 1940 and 1970 when the rate of catchment development increased well ahead of the implementation of mitigation measures. This increase from baseline is attributed, initially, to direct industrial uses of the metal and to landscape disturbance enhancing the transmission of sediments and Hg. In addition, coal combustion increased atmospheric deposition of Hg to waterways later in settlement. The commissioning of a wastewater treatment plant in 1938 merely kept pace with accelerating development, until a secondary treatment capability was added and industrial use of Hg declined in the 1960s. These changes, and the advent of the Clean Water Act in 1972, have seen a 70% decline in Hg accumulation in the uppermost sediments relative to the 1960s peak. The successful control of releases from the treatment plant is mostly responsible and, now, diffuse sources are again the principal source, attesting to the relative difficulty in controlling these sources relative to the more readily identifiable point sources. This is further reinforced by the observations of ongoing increases in the global, atmospheric contribution of Hg and recent, regional declines (Engstrom & Swain 1997). The timing of this recovery is not unlike that evident from long-term observations from another stream subject to heavy industrialisation, the River Tame in the UK (Chapter 13). The modern Hg accumulation rate of $110 \, \mathrm{kg \, yr}^{-1}$ remains well above baseline, but sediment-based approaches have identified that the treatment plant has been responsible, allowing managers to demonstrate that they are working to improve the ecological condition of the system with respect to mercury accumulation.

Turbidity and its biological impacts

In large, slow moving river systems, a large proportion of the suspended sediment load is fine to colloidal. These particles often remain in suspension and increase that water's turbidity even in quite still waters. They tend to absorb or reflect light and so increase the attenuation of light. This shallowing of the photic zone limits primary production within the benthic or littoral zones and advantages planktonic primary producers as they can remain near light at the water surface. Palaeoecological evidence for this has now been generated for the upper (Ogden 2000; Reid *et al.* 2007), middle (Gell *et al.* 2006) and lower (Fluin *et al.* 2007) reaches of the River Murray through shifts from benthic to planktonic or tychoplanktonic taxa. The record from Sinclair Flat on the lowland River Murray in South Australia (Fig. 8.1) provides definitive evidence for a clear water baseline condition reflected in the abundance of benthic diatom taxa (*Epithemia adnata, Eunotia serpentina*). While undated, this record replicates shifts to planktonic forms associated with river regulation from 1922 at 44 cm. From 33 cm, facultative planktonic taxa (*Staurosira elliptica, Staurosirella pinnata*), virtually absent in the state of the wetland at its origin, become codominant. Diatom evidence across the basin is supported by that of sub-fossil cladoceran remains where assemblages shifted, coincident with the algal changes, from benthic species usually associated with aquatic macrophytes to planktonic forms (Ogden 2000; Reid *et al.* 2007).

Eutrophication

The derived diatom communities of many wetlands across southeastern Australia supported many taxa known to be associated with nutrient enrichment. In the UK, small species of *Stephanodiscus* and *Cyclostephanos* are clear indicators of eutrophication in shallow water systems (Sayer 2001). In Australia, perhaps due to contrasting ionic composition or nutrient ratios, stream-wetland interaction or salinity levels, these small planktonic taxa are often subdominant. As such, the nutrient enrichment message is often blurred by indicators of other changes to the wetland. One with very clear evidence for cause and effect is that of Willsmere Lagoon (Gell *et al.* 2005a). The pre-industrial baseline condition of this lagoon is reflected in the dominance of the oligotrophic plankton *Cyclotella stelligera*. Increased sediment accumulation, and a switch to sediment borne and sediment dwelling taxa, mark the industrial period along the Yarra River near the city of Melbourne, settled in 1835. Intensification of settlement around the lagoon and diversion of drainage water from urban surfaces and a constructed freeway in the 1970s mark a shift to eutrophic taxa and a diatom inferred TP

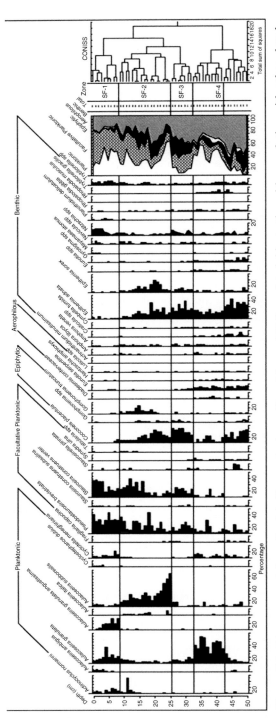

Figure 8.1. Diatom record from Sinclair Flat on the lower River Murray. While undated, biostratigraphic correlation with other, dated records suggest the lowermost sediments to be pre-settlement and those above 25 cm (*Aulacoseira subborealis* [= *A. subarctica*] rise) to be post regulation. The pre-industrialisation diatom flora is dominated by the epiphyte *Cocconeis placentula* and shallow, clear water taxa *Eunotia serpentina* and *Epithemia adnata*. River plankters replace these and the nutrient and turbidity indicators *Pseudostaurosira brevistriata*, *Staurosira contruens* forma *venter* and *Nitzschia* spp. come to dominate the upper sedimentary levels. The plankters *Actinocyclus normanii* and *Cyclotella meneghiniana* increase from 12 cm marking increased salinity and/or nutrients concentration. The record shows remarkable parallels with that from Lake Pepin (Engstrom *et al.* 2009) from the upper Mississippi River showing similar responses to industrialisation and regulation in the respective catchments. CONISS is the programme within TILIA (Grimm 1993) that generates the dendrogram to guide placement of zone boundaries.

concentrations of $500\,\mu g\,L^{-1}$ contrasting markedly with early industrial levels of $\sim 100\,\mu g\,L^{-1}$.

An understanding of nutrient flux in floodplain wetlands is confounded by the unknown frequency of flooding and so level of hydrological and sedimentary exchange between the wetland and the channel. The in-channel, natural impoundments of lakes St Croix and Pepin overcome this limitation and so provide a unique opportunity to assess nutrient flux using mass balance approaches. Engstrom et al. (2009) undertook such mass balance analyses for phosphorus on Lake Pepin. Phosphorus accumulation in the lake's sediments increased 15-fold since 1830 from a baseline of 60 tonnes p.a. to 900 tonnes presently, attributed evenly between increased loadings and increased phosphorus retention. Total phosphorus accumulation rate increased most rapidly after 1940 and decreased after the 1970s, correlated with changing sediment fluxes. Organic phosphorus increased to the present with maximum values in the 1990s. This sedimentary phosphorus record was compared with diatom-inferred total phosphorus that reflects principally the nutrient's bioavailable concentration in the water column. Again, as in Europe, small eutraphentic, planktonic forms including Stephanodiscus spp., Cyclotella meneghiniana and Cyclostephanos invisitatus dominate the upper sediments that represent the post-1960s period, some entering the record earlier than others. The inferred TP reinforces the sediment measurements showing increased concentrations from early in settlement, increasing markedly from the 1940s, and attaining maxima in the uppermost layers. This record parallels that of some in the lowland River Murray with declines in benthic taxa soon after settlement and a peak in Aulacoseira subarctica in the early to mid 1900s. These lowland systems from different continents, stimulated coincidentally by sediment and other releases associated with industrial development from a range of sources, show a remarkably similar record of change reinforcing the diatom palaeorecord as a sensitive measure of eutrophication from industrialisation.

Salinity

Salts represent a less well-appreciated form of industrial pollution but are recognised as a part of the cocktail released from the coal and textile industries. Diffuse pollution sources of salts are best known from saline seepage associated with hydrological change and dryland salinisation through the removal of vegetation cover or wetland or irrigation salinity. These have a long history with evidence for salinisation impacting on productivity in the Tigris-Euphrates, and Indus River catchments from early in settlement 4500–4000 years ago (Goudie 1993). In the MDB too, salinity has a long history relative to the duration of settlement. From as early as 1880, wetland salinity was evident, as indicated by rises in the brackish water diatom taxa Amphora veneta, Gyrosigma acuminatum and Tryblionella hungarica (Gell et al. 2005b). While their

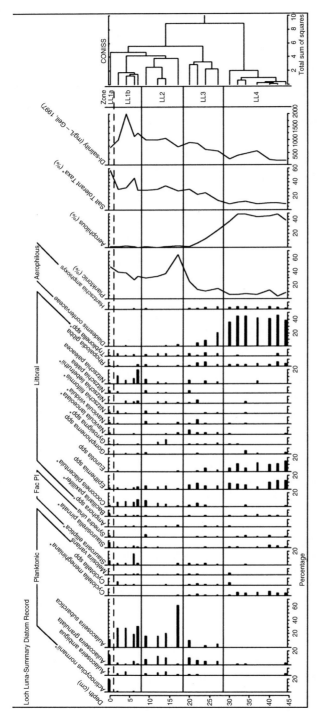

Figure 8.2. The diatom record from Loch Luna in the Riverland of South Australia near Barmera (data from David Baldwin, based on Gell *et al.* 2007b). The proportion of salt-tolerant taxa and diatom-inferred salinity attest to the salinisation of a wetland widely considered to be in good ecological condition.

presence is evident in sediment layers deposited earlier in the 500-year record, this sharp, sustained rise in several halophilous taxa is clear evidence for the impact of early primary industry development in the MDB. Downstream, a diatom record from a levee lake widely considered to be in 'good ecological condition', Loch Luna (Fig. 8.2), was revealed to have a natural 'baseline' condition of clear, shallow water with occasional mudflats (Gell et al. 2007b). Again, the diatoms in the sediment deposited at the commencement of river regulation showed that the commissioning of these flow control structures stimulated a rapid increase in river plankton in the wetland. The reconstructed salinity, based on a salt lake diatom transfer function (Gell 1997), revealed a sustained increase in salinity from that time, to a concentration ten times the longer term, pre-industrial average. This site, by virtue of being amongst a woodland of *Eucalyptus camaldulensis*, was protected from the early impacts of primary industry activity, despite being within kilometres of early grazing runs at Overland Corner. However, it has been impacted by the more diffuse catchment changes as a consequence of its permanent link to the river channel subsequent to regulation.

Salinity concentrations can rise in a wetland due to both the increased flux of salts but also the reduction of freshwater supply that dilutes the incoming salts and mediates the effects of evapoconcentration. Associated with the MDB on account of the building of a major inter-basin transfer and hydroelectricity facility termed the Snowy River Scheme, often considered one of the wonders of the industrialised world, is the Snowy River. Its catchment remains largely forested within national parks but, from 1966, 98% of its mean flow 'below Jindabyne' (weir) was diverted inland to supplement water for irrigation within the MDB and for HEP. At the end of the Snowy River in the floodplain, MacGregor et al. (2005) revealed a fresh and perhaps dystrophic baseline condition for Lake Curlip. The diatom-inferred salinity increased after the development of the coastal flats for cattle grazing in the late 1800s but a stepped increase was noted abruptly after the commissioning of the Snowy Scheme. Ultimately, the combined effect of local hydrological change and an inter-basin transfer for primary and power production was a near 50-fold increase in salinity from ~ 0.5 to at least $20\,\mathrm{mg\,L^{-1}}$.

An additional MDB wetland impacted by abstraction and diversion is the Coorong, a 120 km long coastal lagoon ending at the River Murray mouth that developed behind a barrier dune system after the stabilisation of sea levels ~ 7000 years ago. In 1985, it was listed as a wetland of international significance under the Ramsar protocol and its ecological character was described as a saline to hypersaline, reverse estuary (DEH 2000). This was, in fact, at odds with the case made for a brackish system made evident though ethnohistoric accounts (England 1993), but the case, and the consequent obligations under Ramsar, were influential and releases from the southeast of South Australia were limited through regulation. Through the drought of 2001–2009, the salinity of the Coorong increased to over 100 mg L^{-1} and declines in fish and migratory wader communities were well-documented. The lagoon was identified as a site deserving of River Murray environmental flows, and there were widespread calls through the media to release water under the National Water Initiative to arrest the ecological decline. Sediment-based analyses of diatoms (Fluin et al. 2007) identified salinity, but more so turbidity as the causative agent of the decline of Ruppia, the keystone autotroph in the natural system. Critically, it identified that the system was naturally sub-saline with salinities typically 5–35 mg L^{-1}, and it was a clear water, strongly tidal system that had little direct link with the River Murray over the 7000 years of its existence. Also, its decline commenced several decades before its Ramsar listing, and this was attributable to the closing of the Murray mouth due to reduced flows and reduced freshwater inputs from the hinterland. Complementary analysis of carbon and nitrogen isotopes (Krull et al. 2009) revealed that the elevated salinities, regulated to maintain a 1985 baseline, had caused a collapse in the decomposer flora leading to a net increase in sediment carbon inducing anoxia at the sediment-water interface. This led to the production of sulphides that are potentially toxic to Ruppia (Heijs et al. 2000). So in this instance, the manipulation of flows from the hinterland and the reduction of flows through the MDB, both to consequences of water manipulation to serve agricultural industries, drove a collapse in the wetland ecosystem, exacerbated by policy driven by a short-term understanding of baseline condition.

In acute cases of water management intervention, the salinisation of a wetland can be directly attributed to a cause, particularly if the site is being monitored regularly. In the case of Psyche Bend Lagoon near Mildura in the MDB the wetland was measured to be 1–2 mg L^{-1} for several seasons before it rapidly increased in salinity to over 50 mg L^{-1} within 1 year. This was attributable to a management decision to divert irrigation return flows away from the wetland with the intention of reducing salt flux to the main river channel, the principal source of water for irrigators. The catchment manager defended this as allowing the lagoon to play its role in the system implying that Psyche Bend Lagoon was naturally hypersaline. The monitoring data over this time frame (Gell et al. 2002) provided sufficient evidence, supported by evidence of dead trees and freshwater Cyperus roots around the lagoon margins, for a fresh to oligosaline baseline condition. This was denied for the purpose of pragmatic management purposes that ultimately represent an injustice to the natural system (Gell 2007). This trade-off may be a portent of the effect of the more official implementation of the derogation clause under the WFD.

The necessary corollary of the impact of water diversions on increasing wetland salinity is that the receiving waters may be freshened. This is the case for the coastal wetlands in the southeast of South Australia. Here, wetlands formed naturally between sequences of ridges that owed their origins to coastal dune fronts associated with sea level high stands of each interglacial over the last 700 000 years (Schwebel 1983). These dunes, that ran parallel to the coastline, hindered drainage of the interdune swales during wet winters. From the mid 1800s, the government commissioned a drainage scheme to allow pasture to be grown throughout the year on what were to become defined as 'seasonal wetlands'. While these wetlands inevitably dried as a consequence, those at the coast that received the drainage water became fresh. Owing to the early timing of the commencement of the diversion, these wetlands were widely recognised as naturally fresh. The fossil diatom record, however, clearly demonstrated a shallow, brackish baseline condition for several thousands of years prior to drain construction (Haynes et al. 2007). In this instance then, drainage development, intended to stimulate the local primary industry, led to the pollution of naturally brackish water wetland, not with salt, but with freshwater.

Acidity/alkalinity

One of the most striking impacts of industry on natural systems is in the release of heavy-metal pollutants to streams as acid mine drainage (AMD). Upstream and downstream monitoring is usually sufficient to demonstrate, spatially, chemical change to the receiving water and the consequent impact on aquatic biota (e.g., Ferris et al. 2002). Where the agent of change is more diffuse, it is more difficult to identify the driver. The pH of several lowland MDB rivers has been recognised as declining over recent decades. It is uncertain whether this is due to changing rates of chemical leaching across agricultural land, increases in multiple point source inputs or merely a function of improved instrumentation in recent times, also noted as a limitation of instrumental data across Europe by Bennion and Battarbee (2007). Reid et al. (2002) sought to test this by undertaking palaeolimnological analyses of wetlands along the Goulburn River, a large tributary of the Murray River in Victoria. The diatom-inferred pH showed great stability, at ~6.5, for most of the 3000 years before industrialisation, but species shifts after settlement illustrated a sustained shift to pH ~7.5. They attributed this increase in alkalinity to an increased transport of base cations through erosion, increased groundwater input due to hydrological imbalance after vegetation clearance or a reduction in humic acid input through loss of vegetation cover. These explanations point to land clearance for extensive agriculture; however, irrigation agriculture could be implicated if the regulation of the river has reduced the exchange between the river and the wetland allowing the river and the wetland to diverge chemically over time.

One of the more sinister impacts of extensive agricultural industry across the MDB is the very recent outbreak of wetland acidification in riparian contexts. Subsequent to its induced salinisation in 2000, under the shadow of an extended drought, the pH of Psyche Bend Lagoon slipped from 9.5 to 5.1 across

Figure 8.3. The red waters of Psyche Bend Lagoon attest to the impact of recent (2004) acidification of the lagoon following its sacrifice in a broader management programme (Photo: Peter Gell). See colour plate section.

several months in 2004. Return water from intensive irrigation agriculture had carried sulphate salts to the wetland which were stored in the lake's sediments. Regulation, again to ensure supply for irrigators, ensured permanency of a seasonally inundated wetland reducing the sulphate-rich sediments to a sulphidic state. Exposure to oxygen through 2004 released sulphuric acid and iron, and presumably aluminium, from the sediments into the shallow waters releasing an iron-red stain (Fig. 8.3). The remnant native fish populations were lost. This slide to ecological collapse was readily observable through wetland recent monitoring as noted above (Gell *et al.* 2002). Other sites, not quite so well-monitored, have also become acid through the same process. The preliminary, coarse resolution record from Martin's Bend Lagoon (Fig. 8.4), discovered to be acid during sediment coring, shows that the recent acid condition is indeed unprecedented. The acidophilous diatom taxa in the genus *Pinnularia* are confined to the uppermost sediments and are not found in other records across the region.

Synthesis
Multiple drivers and symptoms
Lowland systems aggregate the sediments, nutrients, metals and salts derived from the impact of humans on landscapes up-catchment as well as those derived nearby. Catchment disturbance enables the more rapid release of

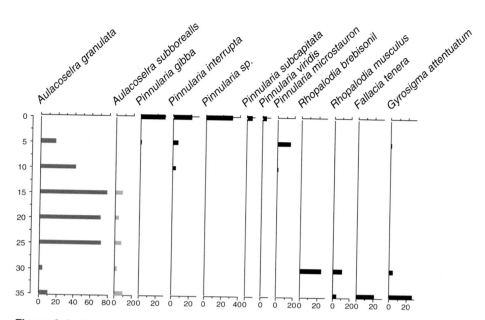

Figure 8.4. The preliminary diatom record from Martins Bend near Berri in the Riverland of South Australia. The acidophilic diatom taxa *Pinnularia* spp. have come to dominate the record due to what appears to be an unprecedented acidification event (data from Dr Jennie Fluin). The *Rhopalodia, Fallacia* and *Gyrosigma* taxa attest to a shallow, saline baseline in this case with the *Aulacoseira* rise indicative of river regulation bringing freshwater to the site on a more regular basis.

sediments and elements contained within. The development of catchments for the agriculture industry was often coincident with other developments that brought on additional direct and diffuse forms of pollution. Furthermore, the interrelationships between salinisation, erosion and nutrient release (Prosser *et al.* 2001) mean that one driver of change can induce symptoms more usually attributed to others. These factors make it difficult to attribute changes in biological and chemical indicators to particular drivers of change, and so to particular changes in land use (Gell *et al.* 2007a). The record of wetland change derived from fossil diatoms can quantitatively indicate changes in a particular water quality variable over time through the use of transfer functions. Where there are multiple drivers and stressors acting on the system coincidentally the quantitative reconstructions are confounded. In these grossly disturbed sites, several widely tolerant taxa come to dominate. Diatom species such as *Cyclotella meneghiniana* or *Staurosirella pinnata*, for example, are advantaged by increased water turbidity, salinity and nutrient concentrations. *Actinocyclus normanii*, that appears to have increased recently in abundance in both the Murray and Mississippi Rivers, is considered to be reflective of recent increased salinity along the Murray, but increased nutrients in the Mississippi. In these situations it is difficult to decipher the principal causative agent of change.

Regional integration

A better understanding of the nature and timing of wetland changes more widely may emerge from the integration of site studies into regional databases. This is occurring for lake studies across Europe with the compilation of the 'Euro-limpacs' database (Battarbee *et al.* 2007). Within Australia the records of limnological change are being integrated within the OZPACS network (Fitzsimmons *et al.* 2008) that is integrating multiple palaeoenvironmental records to understand human impact on ecosystems across the continent. It is becoming clear that, relative to the pre-industrial baseline, most wetlands across southeast Australia are impacted to a moderate to serious level and are in states unprecedented in their history. Most changes can be most easily attributed to catchment change inspired by agricultural development but direct hydrological change, through drainage schemes and inter-basin transfers, are also implicated. In many instances, the degraded nature of the wetland was not fully recognised and only revealed from sediment-based approaches. Were the acid test of 'good ecological condition' from the WFD applied to this region then no wetland would conform. Further, given the critical economic importance of the MDB to Australia's gross domestic product, it would be likely that the derogation clause would be invoked in this hypothetical scenario and managers would be directed to merely 'work towards' good condition. In the Mississippi River lakes, where degradation of condition was identified and mitigation measures put in place during the twentieth century, such improvements are already evident. Here too, as evident in the record of mercury accumulation in the lake's sediments, the path to restoration can be followed by reference to the long-term baseline available in the sediment record.

Which baseline?

Clearly, within the MDB and the Mississippi River, wetlands and rivers are severely impacted. Also, in both regions, it is clear that the changes to this impacted condition occurred early in the development of industry. In these neo-industrial landscapes, these sediment-based records of the nature and timing of change make a powerful contribution to the understanding of the challenges for restoration. It is not unreasonable for society, armed with the knowledge that a wetland was 'naturally' in better condition than anecdote or Ramsar listing would lead us to believe, to demand that action be taken to return it to its original condition even if that was 100 years ago. The longer sediment record, however, poses higher questions of the appropriate baseline to be used as a restoration target. In environments with longer settlement histories such as Denmark it is clear that even very early industry had an impact on wetland condition (Bradshaw *et al.* 2006). Given that the decision to invest in restoration is likely to be a socio-political one, it is more than likely that few would advocate the pursuit of a pre-agricultural baseline as a target for restoration. Perhaps we consider that, somewhere along this continuum, humans moved from being part of the 'natural' environment to being external agents of degradation.

   Certainly wetlands change due to drivers other than industrial development. Before industrialisation, the condition of wetlands varied with changes in climate that are both cyclical and episodic. There is clear evidence in southeast Australia of a sustained period of drying over the last few centuries (Jones *et al.* 2001), and this has led to the lowering and evapoconcentration of these crater lakes. Much of this change cannot be attributed to modern industry. More recent, accelerated climate change (IPCC 2001) is driven by industrial emissions. These changes are influencing the present water balance in rivers and lakes, and scenarios of future change are likely to shift the ecological character of wetlands further. The MDB is within a climate change hotspot (Giorgi 2006) with modelled rainfall reductions higher than most parts of the globe. The realisation of flow reduction scenarios (Jones *et al.* 2002b) on MDB wetlands will make identified natural ecological character conditions unrealistic as management targets. While catchment change appears to have been a stronger influence on wetland condition to date (Gell *et al.* 2007a), continued drying across the southern MDB will aggravate the catchment drivers, heightening the restoration challenge.

Conclusion

The use of diatoms and other complementary biological and chemical indicators archived in sediment sequences provides an understanding of the pre-impact condition of a wetland unavailable by other means. They can also aid in the identification of drivers of wetland change and flag improvements in condition. These approaches can be successfully applied to understanding the long-term changes in condition of flowing waters provided neighbouring sites of sediment accumulation exist and assumptions of the degree of exchange between channel and wetland are acknowledged.

   For restoration, there are many examples where managers have been misdirected in terms of identifying the driver of change, the means to ameliorate impact and the target condition or restoration goal. The risk of undertaking 'restorative' measures without the knowledge often available in sediment sequences is that the action itself becomes a driver of change, with the risk that it combines with other drivers to direct wetlands into unprecedented states. In these circumstances the managers lose contact with a previous baseline and can merely claim to be constructing an artificial water body and aquatic system. They then violate the core principals of frameworks or strategies intended to audit the present condition against some defensible reference or natural baseline.

Acknowledgements

This research was supported by ARC Linkage grants LP0560552 and LP0667819 and AINSE grant AINGRA05062 to P. Gell. The support of the River Murray Natural Resource Management Board and the SA Department of Water, Land and Biodiversity Conservation is appreciated. P. Gell thanks the BES and Dr Lesley Batty for the invitation to attend and support to travel to Birmingham to present this paper.

References

Balogh, S. J., Engstrom, D. R., Almendinger, J. E., Meyer, M. L. and Johnson, D. K. (1999) History of mercury loading in the upper Mississippi River reconstructed from the sediments of Lake Pepin. *Environmental Science and Technology* **33**, 3297–3302.

Barbour, M. T., Swietlik, W. F., Jackson, S. K., Courtemanch, D. L., Davies, S. P. and Yoder, C. O. (2000) Measuring the attainment of biological integrity in the USA: a critical element of ecological integrity. *Hydrobiologia* **422/423**, 453–464.

Barnett, E. (1994) A Holocene paleoenvironmental history of Lake Alexandrina, South Australia. *Journal of Paleolimnology* **12**, 259–268.

Battarbee, R. W., Jones, V. J., Flower, R. J. *et al.* (2001) Diatoms. In: *Tracking Environmental Change Using Lake Sediments. Vol. 3: Terrestrial, Algal and Siliceous Indicators* (eds. E. F. Stoermer and H. J. B. Birks), pp. 155–202. Kluwer Academic Publishers, Dordrecht, the Netherlands.

Battarbee, R. W., Morley, D., Bennion, H. and Simpson, G. L. (2007) A meta-database for regional paleolimnological studies. *PAGES News* **15**, 23–24.

Bennion, H. (1994) A diatom-phosphorus transfer function for shallow, eutrophic ponds in southeast England. *Hydrobiologia* **275/276**, 391–410.

Bennion, H. and Battarbee, R. W. (2007) The European Union Water Framework Directive: opportunities for paleolimnology. *Journal of Paleolimnology* **38**, 285–295.

Benson, L., Markham, A. and Smith, R. (2003) *The Science Behind the Living Murray Initiative*. Murray Irrigation Limited, Deniliquin, Australia.

Blumentritt, D. J., Wright, H. E. and Stefanova, V. (2009) Formation and early history of Lakes Pepin and St. Croix of the Upper Mississippi River. *Journal of Paleolimnology* **41**, 545–562.

Bradshaw, E. G., Nielsen, A. B. and Andeson, N. J. (2006) Using diatoms to assess the impacts of prehistoric, pre-industrial, and modern land-use on Danish lakes. *Regional Environmental Change* **6**, 17–24.

Chambers, J. W. and Cameron, N. G. (2001) A rod-less piston corer for lake sediments: an improved, rope-operated percussion corer. *Journal of Paleolimnology* **25**, 117–122.

Commonwealth of Australia. (2001) *Australia: State of the Environment 2001*. CSIRO, Canberra, Australia.

Davis, M. B. (1989) Retrospective studies. In: *Long Term Studies in Ecology* (ed. G. E. Likens), pp. 71–89. Springer-Verlag, New York.

Department of Environment and Heritage. (2000) *Coorong, and Lakes Alexandrina and Albert Ramsar Management Plan*. Government of South Australia.

Department of Environment and Heritage and Department of Water, Land & Biodiversity Conservation DEH and DWLBC. (2003) *Wetlands Strategy for South Australia*. DEH, South Australia.

England, R. (1993) *The Cry of the Coorong: The History of Water Flows into the Coorong: From Feast to Famine?* Kingston, South Africa. 48 pp.

Engstrom, D. R. and Swain, E. B. (1997) Recent declines in atmospheric mercury deposition in the upper midwest. *Environmental Science and Technology* **31**, 960–967.

Engstrom, D. R., Almendinger, J. E. and Wolin, J. A. (2009) Historical changes in sediment and phosphorous loading in the upper Mississippi River: mass-balance reconstructions from the sediments of Lake Pepin. *Journal of Paleolimnology* **41**, 563–588.

Farmer, J. G. (1991) The perturbation of historical pollution records in aquatic sediments. *Environmental Geochemistry and Health* **13**, 76–83.

Ferris, J., Vyverman, W., Gell, P. and Brown, P. (2002) Diatoms as biomonitors in two temporary streams affected by acid drainage from disused mines. In: *The Finniss River. A Natural Laboratory of Mining Impacts – Past, Present and Future* (eds. S. J. Markich and R. A. Jeffree), pp. 26–31. ANSTO Publications, http://hdl.handle.net/10238/327.

Fitzsimmons, K. E, Gell, P. A., Bickford, S. *et al.* (2008) The OZPACS database: a resource for understanding recent impacts on Australian ecosystems. *Quaternary Australasia* **24**, 2–6.

Fluin, J., Gell, P., Haynes, D. and Tibby, J. (2007) Paleolimnological evidence for the independent evolution of neighbouring terminal lakes, the Murray Darling Basin, Australia. *Hydrobiologia* **591**, 117–134.

Gell, P. A. (1997) The development of a diatom data base for inferring lake salinity: towards a quantitative approach for reconstructing past climates. *Australian Journal of Botany* **45**, 389–423.

Gell, P. A. (2007) River Murray wetlands: past and future. In: *Fresh Water: New Perspectives on Water in Australia* (eds. E. Potter, A. Mackinnon, S. McKenzie and J. McKay), pp. 21–30. Academy of the Social Sciences of Australia, MUP.

Gell, P., Baldwin, D., Little, F., Tibby, J. and Hancock, G. (2007b) The impact of regulation and salinisation on floodplain lakes: the lower River Murray, Australia. *Hydrobiologia* **591**, 135–146.

Gell, P., Bulpin, S., Wallbrink, P., Bickford, S. and Hancock, G. (2005b) Tareena Billabong – a palaeolimnological history of an everchanging wetland, Chowilla Floodplain, lower Murray-Darling Basin. *Marine and Freshwater Research* **56**, 441–456.

Gell, P., Fluin, J., Tibby, J. *et al* (2006) Changing fluxes of sediments and salts as recorded in lower River Murray wetlands, Australia. In: *Proceedings of the IAHS Conference, Dundee, UK, July 2006* (eds. J. Rowan, R. Duck and A. Werrity), International Association of Hydrological Sciences, **306**, 416–424.

Gell, P., Fluin, J., Tibby, J. *et al.* (2009) Anthropogenic acceleration of sediment accretion in lowland floodplain wetlands, Murray-Darling Basin, Australia. *Geomorphology* **108**, 122–126.

Gell, P., Jones, R. and MacGregor, A. (2007a) The sensitivity of wetlands and water resources to climate and catchment change, south-eastern Australia. *PAGES News* **15**, 13–15.

Gell, P. A., Sluiter, I. R. and Fluin, J. (2002) Seasonal and inter-annual variations in diatom assemblages in Murray River-connected wetlands in northwest Victoria, Australia. *Marine and Freshwater Research* **53**, 981–992.

Gell, P., Tibby, J., Fluin, J. *et al.* (2005a) Accessing limnological change and variability using fossil diatom assemblages, south-east Australia. *River Research and Applications*, **21**, 257–269.

Giorgi, F. (2006) Climate change hotspots. *Geophysical Research Letters* **33**, L08707, doi:10.1029/2006GL025734.

Goudie, A. (1993) *The Human Impact on the Natural Environment* 4th edn. Blackwell, Oxford, UK.

Grant, A. (this volume) Detecting the impacts of pollution in the aquatic environment. In: *Ecology Industrial Pollution* (eds. L. C. Batty,

and K. Hallberg), Proceedings of the Annual
Symposium of the British Ecological
Society, Birmingham, UK, April 7–8.

Grimm, E.C. (1993) TILIA Version 2.0.b.4. Illinois
State Museum, Research and Collections
Center.

Haynes, D., Gell, P., Tibby, J., Hancock, G. and
Goonan, P. (2007) Against the tide: the
freshening of naturally saline coastal lakes,
south east South Australia. *Hydrobiologia*
591, 165–183.

Heijs, S. K., Azzoni, R., Giordani, G. *et al.* (2000)
Sulphide-induced release of phosphate
from sediments of coastal lagoons and
the possible relation to the disappearance
of *Ruppia* sp. *Aquatic Microbial Ecology* **23**,
85–95.

Intergovernmental Panel On Climate Change.
(2001) *Climate Change 2001: The Scientific
Basis*. Cambridge University Press,
Cambridge, UK.

Jones, G., Hillman, T., Kingsford, R. *et al.* (2002a)
*Independent Report of the Expert Reference Panel
on Environmental Flows and Water Quality
Requirements for the River Murray System*.
CRCFE, Canberra, Australia.

Jones, R.N., McMahon, T.A. and Bowler, J.M.
(2001) Modelling historical lake levels and
recent climate change at three closed lakes,
Western Victoria, Australia (c.1840–1990).
Journal of Hydrology **246**, 159–180.

Jones, R.N., Whetton, P., Walsh, K. and Page, C.
(2002b) Future impacts of climate
variability, climate change and land use
change on water resources in the Murray
Darling Basin. CSIRO, Aspendale, Australia.

Jowsey, P.C. (1966) An improved peat sampler.
New Phytologist **65**, 245–248.

Krammer, K. and Lange-Bertalot, H. (1986)
*Susswasserflora von Mitteleuropa.
Bacillariophyceae, Teil i: Naviculaceae*. Gustav
Fischer Verlag, Stuttgart. 876 pp.

Krammer, K. and Lange-Bertalot, H. (1988)
*Susswasserflora von Mitteleuropa.
Bacillariophyceae Teil ii: Bacillariaceae,
Epithemiaceae, Surirellaceae*. Gustav Fischer
Verlag, Stuttgart. 576 pp.

Krammer, K. and Lange-Bertalot, H. (1991a)
*Susswasserflora von Mitteleuropa.
Bacillariophyceae Teil iii: Centrales,
Fragilariaceae, Eunotiaceae*. Gustav Fischer
Verlag, Stuttgart. 596 pp.

Krammer, K. and Lange-Bertalot, H. (1991b)
*Susswasserflora von Mitteleuropa.
Bacillariophyceae Teil iv: Achnanthaceae*.
Gustav Fischer Verlag, Stuttgart.
437 pp.

Krull, E., Haynes, D., Lamontagne, S., Gell, P.,
McKirdy, D., Hancock, G., McGowan, J.
and Smernik, R. (2009) Changes in the
chemistry of sedimentary organic matter
within the Coorong over space and time,
Biogeochemistry **92**, 9–25.

Langford, T.E., Shaw, P.J., Howard, S.R.,
Ferguson, A.J.D., Ottewell, D. and Eley, R.
(this volume) Ecological recovery in a river
polluted to its sources: the River Tame in
the English Midlands. In: *Ecology and
Industrial Pollution*. (eds. L.C. Batty and K.
Hallberg), Proceedings of the Annual
Symposium of the British Ecological
Society, Birmingham, UK, April 7–8.

Lawler, D.M. (2005) The importance of
high-resolution monitoring in erosion and
deposition dynamic studies: examples from
estuarine and fluvial systems. *Geomorphology*
64, 1–23.

MacGregor, A.J., Gell, P.A., Wallbrink, P.J.
and Hancock, G. (2005) Natural and
post-disturbance variability in water quality
of the lower Snowy River floodplain,
Eastern Victoria, Australia. *River Research and
Applications* **21**, 201–213.

Macumber, P.G. (1991) *Interaction Between
Groundwater and Surface Systems in Northern
Victoria*. Department of Conservation
and Environment, Melbourne, Australia.
345 pp.

Neave, M. and Rayburg, S. (2006) Salinity and
erosion: a preliminary investigation of soil
erosion on a salinized hillslope. In: *Sediment
Dynamics and the Hydromorphology of Fluvial
Systems* (eds. J.S. Rowan, R.W. Duck and
A. Werritty), pp. 531–539. International

Association of Hydrological Sciences Publishers, Wallingford, UK.

Ogden, R. W. (1996) The impacts of farming and river regulation on billabongs of the southeast Murray Basin, Australia. Unpublished PhD. Thesis, Australian National University, Canberra.

Ogden, R. W. (2000) Modern and historical variation in aquatic macrophyte cover of billabongs associated with catchment development. *Regulated Rivers: Research and Management* **16**, 487–512.

Olley, J. and Wallbrink, P. (2004) Recent trends in turbidity and suspended sediment loads in the Murrumbidgee River, NSW, Australia. International Association of Hydrological Sciences Publication No. 288, 125–129.

Prosser, I. P., Rutherfurd, I. D., Olley, J. M., Young, W. J., Wallbrink, P. J. and Moran, C. J. (2001) Large-scale patterns of erosion and sediment transport in river networks, with examples from Australia. *Marine and Freshwater Research* **52**, 81–99.

Reid, M., Fluin, J., Ogden, R., Tibby, J. and Kershaw, P. (2002) Long-term perspectives on human impacts on floodplain-river ecosystems, Murray-Darling Basin, Australia. *Verhandlungen der Internationalen Vereinigung für Theoretische und Angewandte Limnologie* **28**, 710–716.

Reid, M. A., Sayer, C. D., Kershaw, A. P. and Heijnis, H. (2007) Palaeolimnological evidence for submerged plant loss in a floodplain lake associated with accelerated catchment erosion (Murray River, Australia). *Journal of Paleolimnology* **38**, 191–208.

Reynoldson, T. B., Norris, R. H., Resh, V. H., Day, K. E. and Rosenberg, D. M. (1997) The reference condition: a comparison of multimetric and multivariate approaches to assess water quality impairment using benthic macroinvertebrates. *Journal of the North American Benthological Society* **16**, 833–852.

Roast, S., Gannicliffe, T., Ashton, D. K. *et al.* (this volume) An ecological risk assessment framework for assessing risks from contaminated land in England and Wales. In: *Ecology and Industrial Pollution* (eds. L. C. Batty and K. Hallberg), Proceedings of the Annual Symposium of the British Ecological Society, Birmingham, UK, April 7–8.

Sayer, C. D. (2001) Problems with the application of diatom-total phosphorous transfer functions: examples from a shallow English lake. *Freshwater Biology* **46**, 743–757.

Schwebel, D. A. (1983) Quaternary dune systems. In: *Natural History of the South-East* (eds. M. J. Tyler, C. R. Twidale, J. K. Ling and J. W. Holmes), pp. 15–24. Royal Society of South Australia, Adelaide, Australia.

Smol, J. P. (2008) *Pollution of Lakes and Rivers: A Paleoenvironmental Perspective.* 2nd edn. Blackwell, Oxford, UK. 383 pp.

Sonneman, J., Sincock, A., Fluin, J. *et al.* (2000) *An Illustrated Guide to Common Stream Diatom Species from Temperate Australia.* The Murray-Darling Freshwater Research Centre, Identification Guide No. 33. Wodonga, Victoria. 166 pp.

Thoms, M. C., Ogden, R. W. and Reid, M. A. (1999) Establishing the condition of lowland floodplain rivers: a palaeo-ecological approach. *Freshwater Biology* **41**, 407–423.

Tibby, J. (2004) Development of a diatom-based model for inferring total phosphorus in south-eastern Australian water storages. *Journal of Paleolimnology* **31**, 23–36.

Tibby, J. and Reid, M. (2004) A model for inferring past conductivity in low salinity waters derived from Murray River diatom plankton. *Marine and Freshwater Research* **55**, 587–607.

Triplett, L. D., Engstrom, D. R., Conley, D. J. and Schellhaass, S. M. (2008) Silica fluxes and trapping in two contrasting natural impoundments of the upper Mississippi River. *Biogeochemistry*, **87**, 217–230.

Van Dam, H., Mertens, A. and Sinkeldam, J. (1994) A coded checklist and ecological indicator values of freshwater diatoms from the Netherlands. *Netherlands Journal of Aquatic Ecology* **28**, 117–133.

Williams, A. E., Waterfall, R. J., White, K. N. and Hendry, K. (this volume) Manchester Ship Canal and Salford Quays: industrial legacy and ecological restoration. In: *Ecology and Industrial Pollution* (eds. L. C. Batty and K. Hallberg), Proceedings of the Annual Symposium of the British Ecological Society, Birmingham, UK, April 7–8.

Witkowski, A., Lange-Bertalot, H. and Metzeltin, D. (2001) *Diatom Flora of Marine Coasts 1. Iconographia Diatomologica*, Vol. 7. A.R.G. Gantner Verlag, Ruggell. 925 pp.

CHAPTER NINE

An ecological risk assessment framework for assessing risks from contaminated land in England and Wales

STEPHEN ROAST, TIM GANNICLIFFE, DANIELLE K. ASHTON, RACHEL BENSTEAD, PAUL R. BRADFORD, PAUL WHITEHOUSE AND DECLAN BARRACLOUGH

Introduction

An ecological risk assessment (ERA) may be described as 'a process that evaluates the likelihood that adverse ecological effects may occur or are occurring as a result of exposure to one or more stressors' (USEPA 1992, 1998). Arguably the most common stressors are chemical contaminants, and regulatory authorities in many countries around the world use ERA schemes to determine whether chemical contaminants are impacting on ecological systems. The USA, the Netherlands, Canada, Australia, New Zealand and the UK have all spent considerable resources developing ERAs for use under their respective regulatory regimes. But, although each country has developed its ERA schemes for its own specific regulatory needs, the approach taken is broadly similar for all ERAs.

The largest programme of work to develop ERA frameworks is that of the United States Environmental Protection Agency (USEPA), which started in the mid 1980s. Typically, the USEPA ERA follows three basic steps: problem formulation, analysis and risk characterization (USEPA 1998). The ERA process has been further adapted for use at EPA Superfund sites, to include eight specific steps, from screening level evaluations to full site investigations and risk management (USEPA 1999). Marking ten years of use in the United States, the USEPA has recently assessed how well the ERA guidelines have worked, identifying strengths and weaknesses (see Dale *et al.* 2008; Kapustka 2008; Suter & Cormier 2008).

The USEPA ERA scheme is perhaps more general than those developed in other countries, being aimed at identifying risks from a very broad range of disturbances, for example, 'habitat loss' (i.e., not just chemical contamination).

Ecology of Industrial Pollution, eds. Lesley C. Batty and Kevin B. Hallberg. Published by Cambridge University Press. © British Ecological Society 2010.

ERA schemes developed by Canada, the Netherlands and Australia, although built on the experiences of the USEPA schemes, are more targeted at identifying risks from chemical contaminants specifically. However, very few ERA schemes are used routinely for assessing ecological risks from contaminated land.

In the UK, the Environment Agency (England and Wales) has developed and recently launched (October 2008) an ERA framework to help risk assessors and regulators assess risks to ecological receptors from chemical toxicants on contaminated land. The Environment Agency framework is intended for routine use, specifically to support implementation of the Environmental Protection Act (Part 2A) (1990). However, use of the ERA framework itself is not a statutory requirement.

The aim of the Act is to identify, and remediate, land where contamination is considered to be posing unacceptable risks to human health or the environment (DETR 2000). Specifically, Part 2A refers to any ecological system, or living organism forming part of such a system, within a protected site, including:

- A site of special scientific interest (SSSI) notified under section 28 of the Wildlife and Countryside Act, 1981;
- A national nature reserve (declared under section 35 of the above Act);
- A marine nature reserve (designated under section 36 of the above Act);
- An area of special protection for birds (under section 3 of the above Act);
- Any habitat or site afforded policy protection under paragraph 6 of Planning Policy Statement 9 (PPS 9);
- Any nature reserve established under section 21 of the National Parks and Access to the Countryside Act, 1949;
- Any European site within the meaning of regulation 10 of the Conservation (Natural habitats etc.) Regulations, 1994;
- Any candidate Special Areas of Conservation or potential Special Areas of Conservation given equivalent protection.

Natural England would also like Biodiversity Action Plan (BAP) species and habitats to be included in the statutory guidance if it is revised, as they are not included at present.

Under Part 2A, land can only be determined as 'contaminated land' if there is a 'significant pollutant linkage' present (i.e., there must be evidence of a 'contaminant–pathway–receptor' relationship; CPR). Although it is the responsibility of local authorities to identify contaminated land, the Environment Agency and the Department for Environment, Food and Rural Affairs (DEFRA) aid this process by providing information and guidance on the assessment of land contamination. In this context, the Environment Agency has developed a tiered ERA. The ERA uses a tiered approach and comprises a number of different levels (tiers), each providing different chemical and/or biological information about a site (Fig. 9.1).

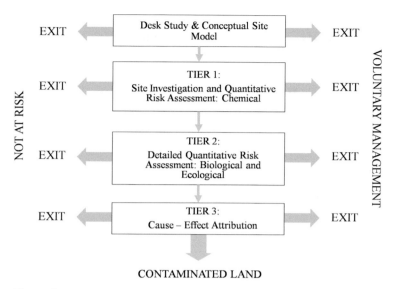

Figure 9.1. The Ecological Risk Assessment (ERA) Framework (modified from Environment Agency 2008a).

Initially, a desk study will be made drawing all the main relevant information together about a site so that a conceptual site model (CSM) can be drafted. The CSM is also updated at the end of each subsequent tier as more information becomes available about the site. At Tier 1, a site investigation is made, including quantitative chemical analysis and the use of soil screening values (SSVs). Tier 2 involves a more detailed site investigation where biological and ecological information is gathered as required by use of bioassays and/or ecological surveys. At Tier 3, all information gathered is collated to attribute cause-effect relationships. If the results at any tier demonstrate that there is no risk of significant harm or significant possibility of significant harm (SPOSH) (for example, no CPR linkage or contamination concentrations not significant) a site can exit the framework. Or, if a site is obviously contaminated with respect to the various definitions and assessment points of the framework, an appropriate person may elect to exit the framework and, in agreement with the nature conservation agency, take voluntary remedial action at his site. In contrast, if a site progresses through the entire framework, and cause-effect can be attributed to contamination of the site, the site may meet the definition of 'contaminated land' and require regulatory enforcement action, for example, through the use of a remediation notice.

This chapter presents a brief summary of the ERA framework developed by the Environment Agency, in consultation with the national conservation agencies, local authorities and industrial partners. The chapter describes the thought processes as one proceeds through the framework and describes the main processes performed at each tier.

Desk study and conceptual site model

The desk study and creation of a conceptual site model (CSM) should be the starting point for any assessment of risks to ecological receptors from land contamination (Environment Agency 2008a). The purpose is to establish whether there is a reasonable possibility of linkages existing between a potential source of contamination and ecological receptors (Fig. 9.2). It must occur early on in the ERA because it provides the background information needed to develop a CSM.

A CSM is:

A representation of the characteristics of the site in diagrammatic or written form that shows the possible relationships between contaminants, pathways and receptors

(Environment Agency 2004a).

A desk study and development of a CSM typically follows 10 steps (Environment Agency 2008b):

1. Establish the Regulatory Context. For contaminated land, the regulatory context is Part 2A of the Environmental Protection Act (1990), and so the risk assessor would refer to the statutory guidance and the list of applicable ecological receptor 'protected locations' (Table A; DEFRA 2006).
2. Collate and Assess Documentary Information. A risk assessor will need to gather as much background information about a site as possible, including identifying possible protected locations in the proximity of the potential contaminants. Depending on the CPR linkage, protected areas up to 5 km from the potential contaminants may be considered, although in most cases 1–2 km is likely to be the limit of any assessment. Information and location of protected areas is available from a number of sources, including the individual UK conservation agencies (Natural England, Countryside Council for Wales and Scottish Natural Heritage) as well as the Joint Nature Conservancy Council. Additional information is provided by the Environment Agency (2008b). Having identified the presence of one or more protected areas, information about the site will be collected including maps, existing environmental reports and records, physical information and any other generic information about the site.
3. Summarise Documentary Information. Once all the relevant information about the site has been collected, a summary report is prepared to put the information in context. The report will usually summarise the history of the site (size, land use, previous activities etc.), the chemicals present on the site (types, potential concentrations, distribution etc.) and the ecology of the site (types of habitat, key communities, conservation status etc.).
4. Identify Contaminants of Potential Concern. Once all of the information is collated, contaminants of potential concern (CoPC) can be identified

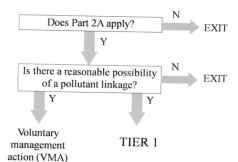

Figure 9.2. Desk study and Conceptual Site Model (CSM) (modified from Environment Agency 2008a).

and their properties described (e.g., whether they have the ability to hydrolyse or bioaccumulate/biomagnify).

5. Identify Likely Fate and Transport of Contaminants. Contaminants are prone to a number of physical processes that affect how long they persist in the environment, whether (and how) they move through the environment and their fate. Such processes include sorption, degradation, transformation, dispersion, immiscible transport, volatilisation etc. This information needs to be assessed for each CoPC so that any pathways to potential receptors of concern can be identified in the next step.

6. Identify Receptors of Potential Concern. In earlier stages, the location and ecological attributes of protected areas have been identified. Now, those ecological systems (or organisms forming part of such a system) relevant to the Part 2A legislation must be identified as receptors of potential concern (RoPC). This must be done in consultation with the nature conservation organisations.

7. Identify Pathways of Potential Concern. Having identified the CoPC and the RoPC, pathways that potentially link the two together must be identified using the information on: the physical site characteristics, the fate and transport characteristics of the CoPC and the behaviour of the RoPC collected previously.

8. Produce CSM. Having gained as much relevant information as possible about the site, contaminants and receptors, together with potential pathways by which the receptor might be exposed to a contaminant(s) a CSM is produced. A CSM is a diagrammatic or tabular representation (usually either a picture or flow-diagram) summarising the desk study information and illustrating any potential pollutant linkages to the receptors of concern (see British Standards Institution 2001; Environment Agency 2008a).

9. Identify Assessment and Measurement Endpoints. If a potential CPR linkage is identified, it must be decided how the linkage can be assessed, with particular regard to the site's protection goals/conservation objectives.

Representative assessment and measurement endpoints[1] are then decided to plan subsequent tiers of the ERA.

10. Identify Gaps and Uncertainties. Having completed the desk study and produced a CSM, a risk assessor needs to be mindful of, and document, those areas where information was lacking or where uncertainties exist. Missing or uncertain information may need further investigation or verification, but should at least be highlighted for caution when making decisions.

The desk study and production of a CSM will have gathered all of the information about a site, its contaminants and its ecological receptors. If the CSM demonstrates that there is no potential pollutant linkage the risk assessor can document this and exit the framework (Fig. 9.2). However, if there is a potential CPR linkage the risk assessor must proceed to Tier 1 which involves chemical screening. Alternatively, the appropriate person has the option to exit the ERA at this stage and take voluntary management action (Fig. 9.2).

Tier 1

In Tier 1, a simple effects assessment is performed by comparing contaminant concentrations in soil against Soil Screening Values (SSVs) to verify possible pollutant linkages and determine which contaminants may be posing a risk to receptors (Fig. 9.3) (Environment Agency 2008c).

SSVs are concentrations of chemical substances found in soils below which there are not expected to be any adverse effects on wildlife such as birds,

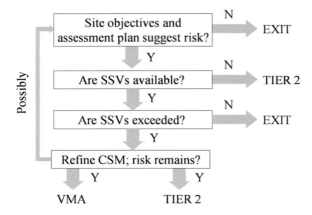

Figure 9.3. Chemical assessment and SSVs (modified from Environment Agency 2008b).

---

[1] An assessment endpoint is an explicit expression of the environmental resource that is being protected, whereas a measurement endpoint is a quantifiable indicator that relates directly to the assessment endpoint (Environment Agency 2003, 2004c).

Table 9.1. *Soil screening values derived for use in the ERA for contaminated land*

Substance	Soil screening value (mg kg^{-1})
Benzo(a)pyrene	0.15[a]
Cadmium	1.15 (0.09)[b]
Chromium	21.1
Copper	88.4 (57.8)[c]
Lead	167.9
Mercury	0.06
Nickel	25.1 (20.3)[c]
Pentachlorobenzene	0.029
Pentachlorophenol	0.6[a]
Tetrachloroethene	0.01[a]
Toluene	0.3[a]
Zinc	90.1 (72.5)[c]

Notes: [a] These SSVs are site specific, for a soil of pH 6.5, organic matter of 2% and clay content of 10%. They will change according to the site soil conditions.
[b] The second SSV for cadmium should be used where bioaccumulation is likely.
[c] These SSVs for soil are site specific. Values in brackets should be used for sandier soils with 2% organic carbon and 10% clay.

mammals, plants and soil invertebrates, or on the microbial functioning of soils. If concentrations of a chemical are found above an SSV this should prompt further investigation to examine whether there are any ecological risks. SSVs form the basis of Tier 1 of the ERA framework, and whilst it is essential that generic screening values are protective of ecosystems, they should not be so stringent that a contaminant is never screened out of the assessment; thus they act as a screening tool (Environment Agency 2008c).

To date, the Environment Agency has derived SSVs for 12 chemicals for use in the ERA framework (Table 9.1). These have been derived using a similar methodology to that used in the European Commission's Technical Guidance Documents (TGD) (EC 2003).

With SSVs available for 12 chemicals at present, there are likely to be many incidences where an SSV for the CoPC is not available. In these instances, it may be acceptable to use appropriate 'supplementary' SSVs from other acceptable regulatory organisations. Suitable supplementary SSVs include:

Soil Screening Levels (SSLs) derived by the United States Environmental Protection Agency (USEPA);

Soil Quality Guidelines (SQGs) derived by the Canadian Council of Ministers of the Environment (CCME);

Oak Ridge National Laboratory Screening Benchmarks; or

Serious Risk Concentrations (SRCs) derived by the National Institute for Public
   Health and the Environment (RIVM) in the Netherlands.

If supplementary SSVs are required for a Tier 1 assessment, their source and
concentration must be agreed with the relevant stakeholders, including the
nature conservation agency; ideally this should be done when performing the
desk study/CSM so that the ERA can be planned appropriately. In truth, it is
unlikely that SSVs from other sources will be acceptable because they will
already have been assessed in the derivation of the UK SSVs and deemed
unsuitable (Environment Agency 2008c). If a suitable value cannot be found
from any of the recommended sources, the site must automatically proceed to
Tier 2 of the ERA framework.

Although SSVs are essentially used as comparison thresholds, there may
be circumstances where simply comparing the concentration of a contaminant in the soil with its corresponding SSV is not sufficient. For example,
the contaminant in question might occur naturally at the site, and the background concentration in soil is greater than the SSV, or where physico-
chemical properties of the soil modify the contaminant to such an extent
that comparison with total soil concentration is meaningless (Environment
Agency 2008c).

Where background levels of a contaminant exist, the recommended
approach is to consider the total concentration (both background and anthropogenic) to pose a risk (Total Risk Approach, TRA) because the ecosystem is
unable to differentiate between background and anthropogenic and the contaminant exists as a combined concentration.

Where physio-chemical modification of the contaminant is likely, this can
usually be corrected using chemical kinetic equations, and expressed as the
concentration that is biologically available (i.e., can actually be taken up by the
organism) (Environment Agency 2008c). The Environment Agency is also
developing a decision tool (using Microsoft Excel) to facilitate in this process,
although it is not yet available.

If concentrations of all contaminants on a site are below their respective
SSVs, the risk assessor can document the results and exit the ERA. If the
concentration of a contaminant is greater than the SSV, or if no suitable
SSV can be found for the contaminant in question, the ERA must progress to
Tier 2.

Tier 2

By the time a site has progressed to Tier 2, it has been established that the Part 2A
regime applies to the site, that there is a contaminant present, a receptor
present and a plausible, potential exposure linkage from the contaminant to

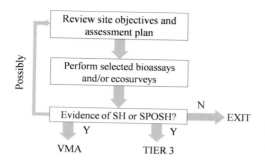

Figure 9.4. Biological and ecological assessment.

the receptor. Furthermore, it has been established that the contaminant is present at concentrations that may cause harm to the receptor. At Tier 2, detailed assessment of the effects (or potential effects) of contaminant exposure on the receptor is made, using bioassays, ecological surveys or both (Fig. 9.4). The outcome of a Tier 2 assessment is whether the risk assessor can confidently identify that significant harm (SH) or SPOSH is occurring to the ecological receptor(s).

Bioassays

Following an extensive road-testing exercise, where a number of bioassays of potential use for determining effects in industrial soils were trialled at real sites, the Environment Agency presently endorses ten biological tests for use at Tier 2 of the ERA (Environment Agency 2004b, c, 2008d, e). The tests and their potential application are presented in Table 9.2.

Although the bait lamina test can be used *in situ*, the majority of tests must be performed under controlled laboratory tests. This can be done by taking soil samples collected from the site under investigation for use in the laboratory.

The choice of which test(s) to use will be informed by the desk study and CSM. Where the receptor of concern is a plant, a combination of soil functioning and plant biology tests will probably be used (e.g., bait lamina/nitrogen mineralisation and seedling emergence, growth and vegetative vigour). In contrast, if the receptor of interest is an invertebrate a combination of soil functioning and invertebrate tests will probably be used (e.g., bait lamina and earthworm or collembolan tests). Food chain effects can also be considered, so that invertebrate tests can be used (with caution) to determine likely effects of contaminants on insectivorous animals (similarly, plant tests can be used (with caution) to determine effects on herbivorous receptors). For birds and other higher organisms, biological tests are likely to be too far removed from natural processes to be of use; in such circumstances ecological surveys are the preferred measure of SH or SPOSH.

Although some of these tests may be used as surrogates for other species (e.g., for birds), there may be receptors where bioassays are not appropriate,

Table 9.2. *Bioassays proposed for use at Tier 2 of the ERA*

Biological test	Application
Bait lamina	Measures activity of microbes in the soil; measure of general soil health
Nitrogen mineralisation	Measures the rate at which nitrogen is mineralised in a soil; provides a measure of general soil health
Earthworm reproduction	Sub-lethal test, demonstrates whether worms can survive at a given contaminant concentration, but are unable to reproduce properly; may be used as surrogate for chronic contaminant effects on a range of soil invertebrates
Earthworm neutral red retention time	Measures cellular damage (lysosomal integrity) in earthworms; may be used as surrogate for chronic contaminant effects on a range of soil invertebrates
Collembolan reproduction	Sub-lethal test, demonstrates whether springtails can survive at a given contaminant concentration, but are unable to reproduce properly; may be used as surrogate for chronic contaminant effects on a range of soil invertebrates
Plant seedling emergence	Measures the degree of successful emergence of seedlings from seed; three test species are used: tomato, cabbage and wheat, but may be used as surrogates for other plant species of interest
Plant growth	Measures growth rate of plants (tomato, cabbage and wheat) growing on contaminated soils; may be used as surrogates for other plant species of interest

and their use is therefore not appropriate. Similarly, the tests described above have physical and chemical limitations (some are pH sensitive, some are soil matrix sensitive etc.) and will not be suitable for deployment at all sites. Ideally the risk assessor will have made this decision while doing the desk study, and will agree this with the relevant natural conservation agency and other stakeholders.

Ecological surveys

It is recommended that a preliminary ecological survey, probably an extended phase 1 habitat survey (JNCC 1993), is performed as part of the desk study and development of the CSM. At Tier 2, more detailed information is required to assess whether significant harm (or SPOSH) is being caused by contaminants at a site (Environment Agency 2008a, b).

A number of information sources are available to help inform the ecosurvey, and these should be consulted when planning the survey. For sites of special scientific interest (SSSI), Natura 2000 sites, Ramsar sites and national nature reserves, as much information about a site as possible should be obtained from the nature conservation agency websites (i.e., Natural England, Countryside Council for Wales, Joint Nature Conservation Committee). These websites will also provide information on assessment endpoints; for example, 'favourable condition' for SSSIs, or 'favourable conservation status' for Natura 2000 sites. For local reserves and BAP habitats species information can be obtained from the local record centres, Local Authority ecologists or the site managers (such as the local wildlife trust or RSPB office).

There is a range of ecological survey (ecosurveys) types, and careful consideration needs to be given to what is required of the survey, which will depend on (among other things) the receptor, the exposure pathway, the contaminant's mode of action and the type of habitat. Examples of ecosurvey types (Hill *et al.* 2005) that may be undertaken include:

- size or number of individuals;
- community composition (presence of particular species/overall diversity);
- population age structure, recruitment or mortality;
- population density etc.

Whichever method is used, the ecosurvey must be adapted as appropriate to the site and type of contamination and should be specifically tailored to the features and sensitivities of the site, e.g., grassland or bat population. Considerations might include (after Environment Agency 2008b):

General issues

Contaminant pathway	How many years of data are required
Contaminant mode of action	Time of year surveys need to be made
Contaminant distribution	Are there cumulative impacts
Receptor distribution	How much detail is required
Scales of natural variation	Is expert judgement required
External influencing factors	

Practicality issues

	Safety issues (personnel health and safety)
Access to the site	
Are licenses required	

In addition to the considerations outlined above regarding surveying specific receptors, ecological functioning of the site should also be considered. Ecological functioning includes a great many factors, and some of those that might be considered in an ecosurvey at Tier 2 of the ERA are listed in Table 9.3 (Environment Agency 2008b). More information is available from IEEM (2006).

Distribution of the contaminant is another important consideration when planning an ecosurvey: the contaminant may not affect the entire site, and may move offsite via wind or water; and the habitat on the site is unlikely to be homogenous. Furthermore, offsite contamination must also be considered

Table 9.3. *A selection of factors of ecological functioning that might be considered when planning and performing an ecological survey (taken from Environment Agency* 2008b)

Aspect	Factors
Available resources	Shelter and roost sites
	Corridors for migration and dispersal
	Quantity and quality of food and water
Stochastic (random) processes	Flooding
	Drought
	Disease
	Erosion
	Climate change
Ecological processes	Population dynamics and population cycles
	Competition
	Predation
	Seasonal behaviour
Vegetation dynamics	Colonisation
	Succession
	Competition
	Nutrient cycling
Human influences	Cutting
	Mowing
	Maintenance dredging
	Pollution and contamination
	Introduction of exotics/invasive species
	Disturbance from public access and recreation, pets, and/or transport
Ecological relationships	Predator–prey relationships
Ecological role	Keystone species
Ecosystem properties	Fragility
	Stability
	Carrying capacity
	Productivity
	Connectivity
	Viable populations
	Fragmentation

because mobile species can use sites outside of the designated area. As identified in the desk study the search radius should be based on the CPR linkage(s) rather than legal or anthropogenic boundaries, but as a guide a risk assessor might consider designated sites within 5 km of the potentially contaminated land.

Clearly, a number of factors need to be considered when conducting an ecosurvey, but for a Part 2A assessment, it should be specifically focused on the regulatory objectives. To ensure that the ecosurvey is focused on the relevant points, it must draw on information collected during the desk study, and be closely linked to the CSM. Furthermore, the ecosurvey must take account of the results of Tier 1, and not rely solely on the phase 1 survey.

The Environment Agency (2008b) provides more information on various aspects of surveys for assessing ecological impacts on contaminated land, and methodologies for actually conducting the surveys are generally well-known and available in the literature (e.g., Hill *et al.* 2005). But it is essential to tailor the ecosurvey to the specific conditions present at the site being investigated so that the information supplements the other information gained during the ERA (e.g., chemical analyses at Tier 1, and bioassays – if performed – at Tier 2).

If the methods used at Tier 2 identify that significant harm or SPOSH is not occurring, the site can exit the framework, having documented the decision (probably in agreement with the relevant nature conservation agency). If the evidence suggests that significant harm or SPOSH is occurring, the site must progress to Tier 3. As in other tiers, the appropriate person can take voluntary management action at this stage, without the need to progress to Tier 3 (Fig. 9.4).

Tier 3

The objective of Tier 3 is to demonstrate whether the contaminants on site are responsible for the significant harm or SPOSH identified at Tier 2. Tier 3 aims to identify cause and effect attribution (Fig. 9.5).

By the time a site has progressed to Tier 3, it has been established that significant harm (or SPOSH) is occurring to a relevant receptor, and it is likely that the contaminants on site are at toxic concentrations and responsible for the significant harm (or SPOSH). For Part 2A, a linkage between the contaminant and the receptor must be demonstrated, but the current state of the science does not permit a mechanistic approach, such as that developed for identifying toxicants responsible for whole effluent toxicity (USEPA 1991, 1993a, b). At present, it is more appropriate to use causal criteria (such as those developed by Hill 1965), for assessing the available evidence, identifying key gaps and informing a decision about whether or not chemical contaminants can reasonably be held responsible for impacts (Environment Agency 2008d). Essentially, a list of criteria (e.g., strength of association, consistency of association, specificity, plausibility etc.) is drawn up and weighted according to perceived relevance. If the assessment demonstrates that the more heavily weighted criteria are generally satisfied, cause and effect can be attributed to the contaminant on site with reasonable confidence. Similarly, if very few of the heavily weighted criteria are satisfied it is likely that the contaminant is not causing the effects recorded. Where evidence is equivocal, more information is required and further investigative work may be necessary (Environment Agency 2008e).

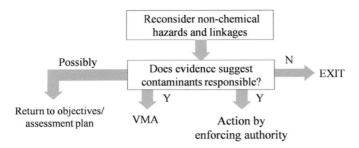

Figure 9.5. Cause-effect attribution.

Use of causal criteria only provides a guide as to whether there is a linkage between the contaminant and the effects recorded in the receptor(s). They are another 'tool' in the ERA 'toolbox' but should not be used in isolation to determine cause-effect. Instead, all evidence at Tier 3 should be reviewed thoroughly and judged by the relevant stakeholders, i.e., the risk assessor, site owner, enforcing authority and conservation agencies.

If the relevant stakeholders agree that the effects reported at Tier 2 are not attributable to the potential contaminants of concern, the risk assessor will document the decision and exit the framework. Conversely, if effects are considered attributable to the CoPC, the appropriate person may take voluntary management action or regulatory action may be required. When making decisions regarding whether the strict definitions of significant harm or significant possibility of significant harm (SPOSH) have been met, the Local Authority will consult with the nature conservation agency.

Conclusions

The Environment Agency (England and Wales), in collaboration with national nature conservation agencies, government, Local Authorities and industrial partners, has developed an ERA framework for identifying ecological risks from contaminated land.

The ERA framework has a tiered structure and is designed to assess whether there is a plausible linkage between a contaminant and a specified receptor, whether contaminants are present at concentrations that pose a risk to the receptor, whether significant harm or SPOSH is occurring and whether the contaminants on site are responsible for causing that significant harm (or SPOSH).

Ecological surveys form a substantial part of Tier 2, where detailed information is required to identify whether the specified receptors are being impacted. A wide variety of methods are available for performing ecological surveys; the type of survey used in any given risk assessment should be chosen based on the specific details of the site under investigation, and tailored to suit those details. Where specific details are required about statutory designated sites, these can be obtained from the websites of the national nature conservation agencies or the JNCC.

The ERA framework was launched in Autumn 2008.

Acknowledgements

The authors thank members of the ERA project board, including representatives from the Environment Agency, Natural England, Countryside Council for Wales, Scottish Natural Heritage, Scottish Environmental Protection Agency, Department for Environment, Food and Rural Affairs, Welsh Assembly Government, Local Authorities and industrial partners. In addition, the authors thank the numerous consultants that contributed to the various Environment Agency science reports that underpin the ERA framework. Finally, the authors thank the British Ecological Society for the invitation to present this work at the Special Symposium on Industrial Pollution.

References

British Standards Institution. (2001) *Investigation of Potentially Contaminated Sites – Code of Practice, BS10175.* ISBN 0 580 33090 7. British Standards Institution, London.

Dale, V. H., Biddinger, G. R., Newman, M. C. et al. (2008) Enhancing the ecological risk assessment process. *Integrated Environmental Assessment and Management* **4**, 306–313.

DEFRA. (2006) *Environmental Protection Act 1990: Part 2A Contaminated Land.* Circular 01/2006, September 2006. Department for Environment, Food and Rural Affairs, UK.

DETR. (2000) *Contaminated Land Environmental Protection Act 1990: Part IIA.* Circular 02/2000. Department of the Environment, Transport and the Regions, London.

EC. (2003) *Technical Guidance Document on Risk Assessment in Support of Commission Directive 93/67/EEC on Risk Assessment for New Notified Substances, Commission Regulation (EC) No. 1488/94 on Risk Assessment of Existing Substances and Directive 98/8/EC of the European Parliament and of the Council Concerning the Placing of Biocidal Products on the Market.* European Commission, Joint Research Centre, EUR 20418 EN.

Environment Agency. (2003) *Ecological Risk Assessment – A public consultation on a Framework and Methods for Assessing Harm to Ecosystems from Contaminants in Soil.* December 2003. Environment Agency, Bristol, UK.

Environment Agency. (2004a) *Model Procedures for the Management of Land Contamination. Contaminated Land Report (CLR).* 11 September 2004. Environment Agency, Bristol, UK.

Environment Agency. (2004b) *Biological Test Methods for Assessing Contaminated Land. Stage 2 – A Demonstration of the Use of a Framework for the Ecological Risk Assessment of Land Contamination.* R&D Technical Report, P5–069/TR1. Environment Agency, Bristol, UK.

Environment Agency. (2004c) *Application of Sub-lethal Ecotoxicological Tests for Measuring Harm in Terrestrial Ecosystems.* R&D Technical Report P5–063/TR2. Environment Agency, Bristol, UK.

Environment Agency. (2008a) *Guidance on Desk Studies and Conceptual Site Models for Ecological Risk Assessment.* Science Report, SC070009/SR2a. Environment Agency, Bristol, UK.

Environment Agency. (2008b) *Guidance on the use of Ecological Surveys in Ecological Risk Assessment.* Science Report, SC070009/SR2d. Environment Agency, Bristol, UK.

Environment Agency. (2008c) *Guidance on the Use of Soil Screening Values for Assessing Ecological Risks*. Science Report, SC070009/SR2b. Environment Agency, Bristol, UK.

Environment Agency. (2008d) *Standard Operational Procedures for a Selection of Biological Tests for Use in Terrestrial Ecological Risk Assessments*. Science Report, SC070009/SR3. Environment Agency, Bristol, UK.

Environment Agency. (2008e) *Guidance on Ascribing Impacts to Chemical Contamination*. Science Report, SC070009/SR2e. Environment Agency, Bristol, UK.

Hill, A. B. (1965) The environment and disease: association or causation. *Proceedings of the Royal Society of Medicine* **58**, 295–300.

Hill, D., Fasham, M., Tucker, G., Shewry, M. and Shaw, P. (2005) *Handbook of Biodiversity Methods: Survey, Evaluation and Monitoring*. Cambridge University Press, Cambridge, UK.

IEEM. (2006) *Guidelines for Ecological Impact Assessment in the United Kingdom*. Institute of Ecology and Environmental Management, Winchester, UK.

JNCC. (1993) *Handbook for Phase 1 Habitat Survey – A Technique for Environmental Audit*. Joint Nature Conservation Committee, Peterborough (originally produced by the England Field Unit, Nature Conservancy Council, reprinted JNCC).

Kapustka, L. (2008) Limitations of the current practices used to perform ecological risk assessment. *Integrated Environmental Assessment and Management* **4**, 290–298.

Suter, G. and Cormier, L. (2008) A theory of practice for environmental assessment. *Integrated Environmental Assessment and Management* **4**, 478–485.

USEPA. (1991) Summary report on issues in ecological risk assessment. US Environmental Protection Agency, Washington DC. EPA 625/3–91/018.

USEPA. (1992) Framework for ecological risk assessment. US Environmental Protection Agency, Washington DC. EPA/630/e-92/001.

USEPA. (1993a) A review of ecological assessment case studies from a risk assessment perspective. US Environmental Protection Agency, Washington DC. EPA 630/R-92/005.

USEPA. (1993b) Wildlife exposure factors handbook. US Environmental Protection Agency, Washington DC. EPA/600/R-93/187a and 187b.

USEPA. (1998) Guidelines for ecological risk assessment. US Environmental Protection Agency, Washington DC. EPA/630/R-95/002F.

USEPA. (1999) Issuance of final guidance: Ecological Risk assessment and risk management principles for Superfund. US Environmental Protection Agency, Washington DC. OSWER Directive 9825. 7–28p.

CHAPTER TEN

Diversity and evolution of micro-organisms and pathways for the degradation of environmental contaminants: a case study with the s-triazine herbicides

MICHAEL JAY SADOWSKY

Introduction

On 7 December 1854, Louis Pasteur is quoted as saying 'Dans les champs de l'observation le hasard ne favorise que les esprits préparés.' This statement, which has been often translated as 'Chance favours the prepared mind', can, after slight modification, also be applied to the interaction of micro-organisms with anthropogenic growth substrates. Namely, that chance favours the prepared bacterium! Given the strong selection pressure for growth of micro-organisms in natural environments, microbial species that have the ability to rapidly acquire new genes that allow them to utilise newly introduced anthropogenic compounds gain a selective advantage for growth over others living in the same environment. This may eventually lead to changes in microbial populations and community structure over time.

While the evolution of microbial genes, and even pathways, for the catabolism of novel compounds released into the environment was originally thought to take long periods of time (on an evolutionary scale), recent evidence indicates that microbes and their genomes are relatively plastic (Jain *et al.* 2002; Mira *et al.* 2002), and as such can evolve the ability to utilize new carbon and energy sources in a relatively short time frame, from years to tens of years (Seffernick & Wackett 2001). This phenomenon has led to paradigm shifts in the way in which we view microbial evolution and the potential impact of anthropogenic perturbations on microbial processes. While several approaches have been used to examine the evolution of bacterial genes and pathways for substrate utilization, many have been rather artificial, using idealised growth conditions, limited microbial diversity, specialised laboratory growth media, chemostats and novel substrates with structural similarities to natural

Ecology of Industrial Pollution, eds. Lesley C. Batty and Kevin B. Hallberg. Published by Cambridge University Press. © British Ecological Society 2010.

compounds. Although these approaches yield useful information for our understanding of microbial catabolic processes, these conditions rarely mimic what happens in natural systems containing a large number of micro-organisms, and their mosaic genomes (Martin 1999; Omelchenko *et al.* 2003), where growth is often limited by complex abiotic and biotic factors. Consequently, to obtain a more complete understanding of the acquisition of new metabolic potential, it is often better to examine enhancement of catabolism across microbial taxa in natural soil and water systems that have been exposed to novel new compounds over time. Moreover, catabolic expansion, defined here as the ability of micro-organisms to gain the ability to metabolise truly novel compounds, requires the examination of new chemical compounds that do not share structural or chemical properties with natural ones. In this way, the increase in catabolic ability is less likely to occur by accumulation of point mutations over time, but rather by more rapid acquisition of new enzymes whose activities are recruited into existing or newly created degradation pathways. Both Janssen *et al.* (2005) and Wackett (2004) provide excellent reviews on the evolution of new enzymes for the microbial degradation of novel and xenobiotic compounds, and the reader is pointed to these reviews for more comprehensive discussions.

Catabolic expansion vs. catabolic radiation

Several fundamentally different processes can lead to enhanced metabolic potential of micro-organisms. Micro-organisms can gain catabolic ability via slow, random, accumulation of mutations in existing genes, leading to new enzymatic functions (Wright 2000), or via acquisition of new genetic determinants via horizontal acquisition (Poelarends *et al.* 2000). Either of these processes I refer to as catabolic expansion (Fig. 10.1), the ability of a micro-organism to expand the range of substrates that can be used for growth and/or energy. While both systems have been shown to operate to expand microbial metabolism, the later processes are likely to lead to greater metabolic versatility in the short term and allow for selection of growth on radically new substrates. This is likely due to the fact that greater and faster functionality can be realised from introgression of a new gene(s) into a microbe's metabolic machinery than can be realised from mutation-induced 'tweaking' of an extant enzyme for broader substrate range. This is likely reflected in the discovery of a large number of 'foreign genes' among the genome sequences of many microbes (Berg & Kurland 2002). Catabolic expansion that leads to enhanced metabolic functionality in micro-organisms can occur either via the acquisition of a single key gene encoding for an enzyme whose product 'plugs into' existing metabolic machinery (Shapir *et al.* 2007), or via acquisition of a whole metabolic pathway encoded by a suite of genes encoding for the complete metabolism of a novel substrate (van der Meer & Sentchilo 2003). If the new pathway is plasmid borne, the later phenomenon also frequently results in catabolic radiation, defined here as the spread of the ability to degrade a novel

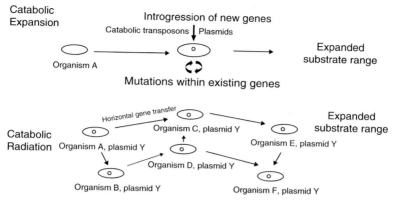

Figure 10.1. Mechanisms micro-organisms use to acquire new metabolic functions.

compound to unrelated microbial taxa living in geographical isolation. Catabolic radiation occurs chiefly by plasmid-mediated horizontal gene transfer, often coupled with transposition events, and this process is exemplified by the spread of atrazine degradation ability among phylogenetically diverse micro-organisms located across the biosphere (de Souza *et al.* 1998; Wackett *et al.* 2002).

Atrazine and the *s*-triazine herbicides

In the early 1950s, the symmetrical (*s*)-triazine compounds, simazine, atrazine, prometryn and ametryn, were synthesised by the Geigy Company in Basel, Switzerland, as selective herbicides for weed control in crops such as corn (Gast *et al.* 1955). The *s*-triazines function to disrupt photosynthetic electron flow by binding to the Q_B protein in photo system II (Sajjaphan *et al.* 2002). The *s*-triazine ring contains three symmetrical N atoms, and the herbicides differ in type and location of functional groups (i.e., Cl, SCH_3, OCH_3) at ring position 2 and other substituents (R) (Fig. 10.2).

Since the carbon atoms in the atrazine ring are at the oxidation state of CO_2, they cannot supply energy for microbial growth. The 2-chloro substituted *s*-triazine atrazine (2-chloro-4-(ethylamine)-6-(isopropylamine)-*s*-triazine), one of the most widely used herbicides in this class of compounds, contains additional N atoms in the N-ethylamino and N-isopropylamino side chains. While a few asymmetrical triazines are known to occur naturally, such as the antibiotic fervenulin (Laskin & Lechavalier 1984), the *s*-triazines were initially thought to be xenobiotic compounds (Esser *et al.* 1975), and no natural structural analogues of *s*-triazines are known to exist. This makes these compounds an ideal model to study the evolution of micro-organisms to transform these novel compounds.

Atrazine-degrading micro-organisms

While atrazine was initially considered to be a xenobiotic, over the last 40 plus years, it has been shown by many researchers that micro-organisms can

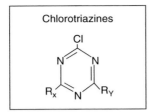

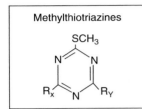

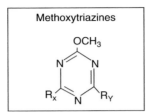

Figure 10.2. Backbone structures of three s-triazine herbicide serving as potential substrates for catabolism by micro-organisms.

partially transform atrazine by carrying out hydrolytic reactions, resulting in the cleavage of chloro, amino and alkylamino groups on the s-triazine ring (Cook 1987). In addition, while Erickson and Lee (1989) reported that some micro-organisms dealkylated the side chains of chlorinated triazines, they failed to dechlorinate atrazine, and reviews written in 1969–1975 reported that atrazine was biologically transformed in the environment by hydrolysis at carbon 2, by N-dealkylation at carbons 4 and 6 or by opening of the s-triazine ring (Knusli et al. 1969; Kaufman & Kearney 1970; Esser et al. 1975). Interestingly, while micro-organisms transforming deisopropylatrazine, deethyldeisopropylatrazine, ammeline, N-isopropylammelide and ammelide were previously reported, none were shown to degrade the chlorinated s-triazine compounds atrazine and simazine (Cook 1987). However, despite the lack of the isolation of a pure microbial culture with the ability to degrade atrazine, the half-life of atrazine in soil was found to be shorter in soils with exposure history to this compound (reviewed in Mandelbaum et al. 2008), partial mineralisation of atrazine in soils was shown, and microbial consortia were reported to degrade atrazine (Behki & Khan 1986; Mandelbaum et al. 1993). This suggested that micro-organisms were slowly evolving the ability to degrade and mineralise atrazine. In hindsight, this result may not be too surprising since melamine (the triamino substituted s-triazine), which was once considered to be xenobiotic and non-biodegradable (Scholl et al. 1937), was found nearly 40 years later to be completely biodegradable (Allan 1981; Cook et al. 1983), and has been patented as a slow-release fertiliser. However, evidence for a change in microbiologically mediated atrazine metabolism could be seen in the literature. From about 1960 to 1990, most studies reported that dealkylation of the N-alkyl substituents on the s-triazine ring of atrazine was the major route for bacterial metabolism (Shapir et al. 2007), whereas more recent reports indicated that atrazine degradation by bacteria proceeds via dechlorination to hydroxyatrazine, and not dealkylation. Despite this hopeful outlook, for the first approximately 35 years of its use, there were no reports of the isolation of a pure culture of a micro-organism that had the ability to completely mineralise atrazine to CO_2 and NH_4. A large number of studies carried out from the 1960s through the 1970s described the s-triazine herbicides in terms of their biological properties, chemistry and herbicidal properties (Knusli & Gysin 1960; Knusli et al. 1969; Kaufman & Kearney 1970; Esser et al. 1975). In 1980, Geller reported that the physical and chemical decomposition of atrazine was likely more important than microbial degradation, even though microbes with the ability to degrade atrazine are likely present in soils. This idea was widely held for some time, and the formation of hydroxyl-s-triazine derivatives in soils were thought to be due to abiotic processes. In the late 1980s, both Cook (1987) and Erickson and Lee (1989) each reported the identification of micro-organisms that degraded the side chains of chlorinated triazines. Since these microbes failed to dechlorinate atrazine before the molecule was dealkylated, it was hypothesised that the presence of both alkyl side chains inhibited dechlorination reaction. Fungi have also been reported to degrade s-triazine compounds. Sporothrix schenckii has been shown to utilise cyanuric acid, biuret and urea as sole nitrogen sources for growth (Zeyer et al. 1981), and the white rot fungi Pleurotus pulmonarius and Phanerochaete chrysosporium have been reported to degrade atrazine via N-dealkylation to deethylatrazine, deisopropylatrazine and deethyldeisopropylatrazine (Masaphy et al. 1993) and hydroxyatrazine (Lucas et al. 1993), respectively.

In 1995, Mandelbaum and co-workers reported the isolation of a pure bacterial culture, Pseudomonas sp. strain ADP, with the ability to rapidly degrade atrazine under aerobic, anoxic and non-growth conditions. Moreover, Pseudomonas ADP used atrazine as a sole source of N for growth, and the organism completely and very rapidly mineralized the s-triazine ring of atrazine. Interestingly, and within a relatively short time, several other laboratories across the United States and the world reported the isolation of pure bacterial cultures that could mineralise atrazine (Mandelbaum et al. 1995; Radosevich et al. 1995; Yanze-Kontchou & Gschwind 1995; Bouquard et al. 1997; Struthers et al. 1998). Since this time, other taxonomically unique, gram-positive and gram-negative atrazine-degrading bacteria (Table 10.1) have been isolated worldwide (Topp et al. 2000a, b; Rousseaux et al. 2001; Strong et al. 2002; Cai et al. 2003; Piutti et al. 2003; Rousseaux et al. 2003; Devers et al. 2007). These results suggest that atrazine degradation ability rapidly spread to a large number of genetically and geographically disparate soil micro-organisms in a relatively short time frame.

Table 10.1. *Some micro-organisms capable of completely or partially transforming atrazine*

Micro-organism	Reference
Agrobacterium radiobacter	Moscinski 1996
Pseudomonas ADP	Mandelbaum & Wackett 1996
Pseudomonas spp. DSM 93–99 (YAYA6)	Yanze-Kontchou & Gschwind 1995
Pseudomonas fluorescens strains LMG 10141, 10140	Vandepitte *et al.* 1994
Rhodococcus rhodochrous	Feakin *et al.* 1995
Rhodococcus B-30	Behki & Khan 1994
Rhodococcus TE1	Shao & Behki 1996
Rhodococcus corallinus strain 11	Cook & Hutter 1984
Rhodococcus N186/21	Cook & Hutter 1984
Rhodococcus corallinus	Behki *et al.* 1993
Streptomyces strain PS1/5	Shelton *et al.* 1996
Acinetobacter calcoaceticus	Mirgain *et al.* 1993
Chelatobacter heintzii	Rousseaux *et al.* 2001
Arthrobacter crystallopoietes	
Aminobacter aminovorans	
Stenotrophomonas maltophilia	
Pseudaminobacter sp. C223, C147, C195	Topp 2001
Nocardioides sp. C190	
Ralstonia basilensis M91–3	Stamper *et al.* 2002
Clavibacter michiganese	Smith *et al.* 2005
Agrobacterium tumefaciens	
Pseudomonas putida	
Sphingomonas yaniokuyae	
Nocardiodes sp.	
Rhizobium sp.	
Flavobacterium oryzihabitans	
Variovorax paradoxus	
Arthrobacter aurescens TC1	Strong *et al.* 2002
Streptomyces strain PS1/5	Shelton *et al.* 1996
Nocardioides sp. AN3, A19, N33	Satsuma 2006
Arthrobacter sp. AD1	Cai *et al.* 2003
Chelatobacter heintzii Cit1	Rousseaux *et al.* 2003
Agrobacterium tumefaciens St96–4	Devers *et al.* 2005
Bacillus sp. RK016	Korpraditskul *et al.* 1993
Nocardioides sp. SP12	Piutti *et al.* 2003
Nocardioides sp. C157	Topp *et al.* 2000a
Stenotrophomonas maltophilia	Rousseaux *et al.* 2001
Delftia acidovorans D24	Vargha *et al.* 2005
Exiguobacterium sp. BTAH1	Hu 2004

Table 10.1. (*cont.*)

Micro-organism	Reference
Rhodococcus sp. RHA1, NI86/21	Warren *et al.* 2004
Rhizobium sp.PATR	Bouquard *et al.* 1997
Arthrobacter sp. AG1	Dai *et al.* 2007
Bacillus licheniformis	Marecik *et al.* 2008
Bacillus megaterium	
Rahnella aquatilis strains	
Umbelopsis isabellina	
Volutella ciliate	
Botrytis cinerea	

Conservation of genes for atrazine degradation among disparate micro-organisms

Pioneering work done with *Pseudomonas* sp. strain ADP, and subsequently with other bacteria, indicated that atrazine mineralisation in many gram-negative bacteria proceeds via a six step degradation pathway (Martinez *et al.* 2001; Shapir *et al.* 2007). In *Pseudomonas* sp. strain ADP, atrazine degradation is initiated by a hydrolytic dechlorination reaction, catalysed via the enzyme atrazine chlorohydrolase AtzA, resulting in the formation of hydroxyatrazine (Fig. 10.3). This compound serves as the substrate for two subsequent hydrolytic deamination reactions, catalysed by AtzB (hydroxyatrazine ethylaminohydrolase) and AtzC (N-isopropylammelide isopropylamino hydrolase), resulting in the formation of cyanuric acid (Martinez *et al.* 2001; Shapir *et al.* 2007). Cyanuric acid is now regarded as a central metabolic intermediate in *s*-triazine metabolism (Cook *et al.* 1985; Fruchey *et al.* 2003), and the enzymes catabolising atrazine to cyanuric acid (AtzA, AtzB and AtzC) constitute the upper degradation pathway. All three enzymes have been purified to homogeneity (de Souza *et al.* 1996; Boundy-Mills *et al.* 1997; Shapir *et al.* 2002; Fruchey *et al.* 2003), and sequence analyses revealed that all three enzymes belong to the amidohydrolase superfamily, whose members are defined by having an $(\alpha\beta)8$ barrel structure, and conserved reaction mechanism and substrate binding ligands (Seffernick *et al.* 2002). More surprisingly, initial sequence analysis studies indicated that the *atzABC* genes were highly conserved, at nearly 99+% amino acid identity, in diverse bacterial genera isolated from across the United States and the world (de Souza *et al.* 1998; Topp *et al.* 2000a, b; Devers *et al.* 2004, 2007; Shapir *et al.* 2007). This suggests that the atrazine metabolism was rapidly spreading from bacterium to bacterium.

Not all bacteria initiate atrazine catabolism via AtzA. Results from studies done by Topp and colleagues indicate that some bacteria initiate atrazine

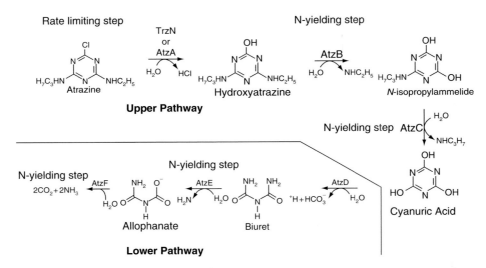

Figure 10.3. Complete atrazine degradation pathway found in *Pseudomonas* ADP and several other gram-negative bacteria. Enzymatic reactions leading to nitrogen release are indicated. The AtzA, AtzB and AtzC enzymes, hydrolyzing atrazine to cyanuric acid, constitute the upper pathway, while the enzymes AtzDEF, hydrolysing cyanuric acid to ammonia and carbon dioxide, comprise the lower pathway.

catabolism via TrzN, triazine hydrolase, an enzyme initially found in a *Norcardiodes* sp. strain (Mulbry *et al.* 2002). TrzN has been now found in a variety of gram-positive bacteria, including *A. aurescens* (Strong *et al.* 2002), where it has been purified to homogeneity (Shapir *et al.* 2006). Like AtzA, the TrzN hydrolytically displaces the chlorine from atrazine producing hydroxyatrazine. These two enzymes, however, differ in substrate range. AtzA has narrow substrate specificity and only functions to displace chlorine and fluorine leaving groups, and transforms atrazine, simazine, desethylatrazine and terbutylazine (de Souza *et al.* 1996). In contrast, TrzN, a zinc amidohydrolase, displaces cyano, azido, halide, *S*-alkyl and *O*-alkyl substituents (Shapir *et al.* 2005), allowing *A. aurescens* to metabolise a variety of *s*-triazine derivatives containing chlorine plus N-ethyl, N-propyl, N-butyl, N-*s*-butyl, N-isobutyl or N-t-butyl substituents on the *s*-triazine ring. This broad substrate range was shown to allow *A. aurescens* to use 23 *s*-triazine substrates as the sole nitrogen source, including the herbicides ametryn, atratone, cyanazine, prometryn and simazine (Strong *et al.* 2002). Moreover, atrazine substrate analogues containing fluorine, mercaptan and cyano groups in place of the chlorine substituent could also be used as growth substrates by this bacterium.

The lower atrazine degradation pathway in *Pseudomonas* ADP consists of three enzymes, AtzDEF, resulting in ring cleavage and a subsequent transformation of cyanuric acid to CO_2 and NH_4. While initial studies proposed that cyanuric acid was hydrolysed to urea, more recently we have shown that many bacteria use cyanuric acid hydrolase, AtzD, to produce biuret. Instead of AtzD, some bacteria have TrzD, a homolog of cyanuric acid hydrolase that is 44% different in amino acid sequence to AtzD (Fruchey *et al.* 2003).

The biuret produced via AtzD is subsequently deaminated by biuret hydrolase (AtzE) to generate allophanate, which in turn is the substrate for AtzF (allophanate hydrolase), which is hydrolysed to CO_2 and NH_4. This lower pathway is absent in some atrazine degrading bacteria, such as *A. aurescens*, and consequently, these bacteria excrete cyanuric acid into the growth medium (Strong *et al.* 2002; Sajjaphan *et al.* 2004). Nevertheless, these bacteria still have the ability to rapidly grow on *s*-triazines using the alkyamines released by AtzB and AtzC as carbon, nitrogen and/or energy sources (Strong *et al.* 2002).

While some studies have shown that genetically diverse bacteria contain nearly identical copies of *atzABC* (de Souza *et al.* 1998), others contain various combinations of these genes, together with *trzN*. For example, Devers *et al.* (2007) reported that 17 atrazine-degrading bacteria isolated from soils had the following combination of atrazine degradation genes: *trzN-atzBC*, *atzABC-trzD* or *atzABCDEF*. Likewise, Martin-Laurent *et al.* (2006) also reported great diversity in atrazine degradation gene content among bacterial communities isolated from French soils.

Gene diversity and distribution is driven by plasmids

Initial studies by de Souza *et al.* (1998) indicated that the atrazine degradation phenotype could be transferred from *Pseudomonas* ADP to *E. coli* by conjugation, and that this was associated with the acquisition of an ~108-kb self-transmissible plasmid in transconjugants. This suggested that atrazine degradation genes in this bacterium were localised on plasmids. Subsequent hybridization studies indicated that many gram-negative atrazine-degrading bacteria contain large molecular weight plasmids containing atrazine degradation genes (Topp *et al.* 2000b; Wackett *et al.* 2002). However, the atrazine degradation genes were located on different size plasmids, suggesting that the spread of atrazine degradation genes among disparate bacteria was not solely due to direct plasmid transfer. Moreover, this notion was further confirmed by the discovery of bacteria with various combinations of atrazine degradation genes, including *trzN-atzBC*, *atzABC-trzD* and *atzABCDEF* (Devers *et al.* 2007), and numerous reports of the selective loss of some or all atrazine degradation genes following growth under non-selective conditions.

The complete nucleotide sequencing of pADP-1, the atrazine gene-containing plasmid from *Pseudomonas* ADP, provided insights into the arrangement and composition of atrazine degradation genes in this bacterium (Martinez *et al.* 2001), and how different bacteria can contain various combinations of atrazine degradation genes. The upper pathway genes, atzA, atzB and atzC are located on different parts of the plasmid backbone, the IncPβ plasmid pR751, are not contiguous, have different mole% G + C content, appear to be constitutively expressed and are flanked by transposase TnpA-like elements from IS1071, forming nested catabolic transposon structures. In contrast, the lower degradation pathway genes, atzDEF, were located in an operon on a different portion of the plasmid under control of a LysR-like regulatory gene, atzR (García-González et al. 2005). Several lines of evidence indicate that the upper atrazine degradation pathway genes are acquired by soil bacteria via horizontal gene transfer and transposition processes, including (1) the genes are surrounded by IS elements and transposases, (2) different bacterial genera have different sized plasmids containing the degradation genes, (3) various combinations of atrazine degradation genes can be found in different micro-organisms, (4) the atrazine degradation genes can be independently lost following growth without selection pressure and (5) laboratory and soil microcosm studies indicate that atrazine degradation genes can be transferred between bacteria (de Souza et al. 1998; Devers et al. 2005). Taken together, these observations also suggest that the atrazine degradation genes in the primal degrader(s) were likely acquired from different plasmids by independent gene transfer and transposition events due to selection pressure.

Gram-positive atrazine degrading bacteria

The presence of atrazine degradation genes on plasmids is not limited to only gram-negative bacteria, and several gram-positive bacteria capable of degrading s-triazines have been reported (Giardina et al. 1982; Behki et al. 1993; Rousseaux et al. 2001; Strong et al. 2002; Cai et al. 2003; Sajjaphan et al. 2004). As before, insights into the constitution, distribution and assembly of the atrazine catabolic pathway in gram-positive bacteria came from the genome sequence analyses of Arthrobacter aurescens strain TC1. This bacterium can utilise atrazine as a sole source of nitrogen and carbon for growth, excreting cyanuric acid into the growth medium. Initial studies showed that strain TC1 contained atzB and atzC, but not atzA, and was metabolically diverse, using 22 s-triazines as growth substrates, including the herbicides simazine, terbuthylazine, propazine, cyanazine, ametryn, prometryn and terbutryn (Strong et al. 2002). We subsequently showed, by polymerase chain reaction (PCR) analysis, that strain TC1 initiated atrazine catabolism via the triazine hydrolase, encoded by trzN (Sajjaphan et al. 2004). To date, we have synthesised about 64 different s-triazine compounds, and these have been shown by functional analyses to be substrates for the TrzN, AtzB and AtzC. However, based on substrate specificity studies of all three atrazine degradation genes (de Souza et al. 1996; Boundy-Mills et al. 1997;

Sadowsky *et al.* 1998; Seffernick *et al.* 2002; Shapir *et al.* 2002; Fruchey *et al.* 2003), *A. aurescens* TC1 can theoretically metabolize 560 *s*-triazine compounds (Shapir *et al.* 2007).

Sequence analysis of a 160-kb genomic DNA region of strain TC1 cloned in bacterial artificial chromosomes (BACs), and PCR and hybridisation studies, established that this bacterium contained the *trzN*, *atzB* and *atzC* genes, but not the lower pathway, localised on a ∼380-kb plasmid, pTC1 (Sajjaphan *et al.* 2004). Like the arrangement of atrazine catabolizing genes on pADP-1, those on pTC1 are not contiguous and not organised in an operon-like structure. To get a better understanding of the genomic context of the atrazine catabolism genes relative to the rest of the genome, the complete genome of *A. aurescens* TC 1 was sequenced (Mongodin *et al.* 2006). The genome of this bacterium consists of a 4.59-Mbp circular chromosome, and two circular plasmids, pTC1 and pTC2, which are 408 kb and 300 kb, respectively. As predicted by hybridisation studies, sequence analysis confirmed that the *trzN*, *atzB* and *atzC* genes are localised on pTC1. Genomic analyses indicated that the combination of three, plasmid-encoded, atrazine degradation genes, along with a variety of chromosome-borne amine-catabolizing enzymes, and plasmid localised Ipu pathway gene clusters, make this bacterium very metabolically diverse. The later pathway is likely involved in the catabolism of ethyl- and propyl-amines liberated from *s*-triazine compounds by the action of AtzB and AtzC, resulting in TC1's ability to grow on the large range of *s*-triazines as C and N sources (Shapir *et al.* 2007). Interestingly, genome sequence analysis indicated that *A. aurescens* strain TC1 contains six copies of *trzN* located within six identical direct tandem repeats of about 16 kb (Mongodin *et al.* 2006). This increase in the copy number of *trzN* may enhance catabolism, via gene dosage effects, or provide a competitive advantage to strain TC1 versus other atrazine-degrading micro-organisms, such as *Pseudomonas* ADP, which contains a single triazine hydrolase gene. This may be reflected in the growth rate differential between these two organisms, strain TC1 grows much faster on atrazine than *Pseudomonas* ADP. Moreover, multiple copies of the catabolism initiating enzyme may protect the bacterium from complete loss of atrazine degradation potential in growth conditions lacking adequate selection pressure.

Regulation of atrazine catabolism in *Pseudomonas* ADP reflects the recent evolution and assembly of the degradation pathway

Several lines of evidence support the contention that the atrazine degradation pathway in *Pseudomonas* ADP, and other gram-negative and -positive bacteria, was recently assembled. This include reports that (1) the mole% G + C data showing that while *atzA*, *atzB* and *trzN* have mole% G + C contents of 58.3, 64.1 and 63.1%, respectively, within the range of mole% G + C contents found in total *Pseudomonas* sp. DNA (58 to 70%) and *Arthrobacter* (59 to 66%), the *atzC* gene has a 39.5 mole% G + C content, well outside the range for total *Pseudomonas* and *Arthrobacter* DNA (Sadowsky *et al.* 1998); (2) individual atrazine degradation genes can be lost from cultures grown without selection pressure (Topp *et al.* 2000b; Martinez *et al.* 2001; Sajjaphan *et al.* 2004); (3) atrazine degradation genes reside on different size plasmids in different bacteria (Topp *et al.* 2000b; Martinez *et al.* 2001; Wackett *et al.* 2002); (4) atrazine degradation genes are flanked by insertion sequence-like elements and transposases (Martinez *et al.* 2001, Devers *et al.* 2007); and (5) triazine-degrading bacteria can have various combinations of atrazine degradation genes, some with complete and incomplete degradation pathways (Strong *et al.* 2002; Devers *et al.* 2007; Shapir *et al.* 2007).

In *Pseudomonas* strain ADP, sequence, Northern hybridization and RT-PCR analyses indicate that the *atzA*, *atzB* and *atzC* genes are constitutively expressed (Martinez *et al.* 2001; Devers *et al.* 2004). However, atrazine degradation in this bacterium appears to be influenced by inorganic nitrogen sources, such as nitrate and ammonia, likely under the control of a genome-wide system influencing general nitrogen metabolism in response to decreased nitrogen availability (García-González *et al.* 2003, 2005). In *Pseudomonas* ADP, ammonia repression of atrazine degradation has been shown to be relieved by growth of cells in the presence of L-methionine sulfoximine (MSX), an inhibitor of glutamine synthetase, and in a nitrate assimilation mutant (García-González *et al.* 2003). This, along with the above observations, is consistent with the recent acquisition of these atrazine catabolism genes, likely in a time scale of tens of years. In contrast, the *atzDEF* genes, which are located in an operon-like structure on pADP-1 (Martinez *et al.* 2001) are regulated by a LysR-like regulatory element, *atzR*, located upstream, and divergently transcribed, from *atzDEF* (Martinez *et al.* 2001; García-González *et al.* 2005). The regulation of cyanuric acid degradation in *Pseudomonas* ADP is apparently complex and involves AtzR activity, the presence of cyanuric acid and Ntr-mediated signal transduction in response to nitrogen limitation. Nevertheless, the presence of these genes in an operon suggests that they were acquired/recruited and assembled together much earlier, in an evolutionary sense, than the other atrazine catabolism genes. Accordingly, it is not surprising that many soil bacteria have been shown to have the ability to use cyanuric acid as a growth substrate (Fruchey *et al.* 2003). Recent studies have shown that expression of *atzR* is induced by nitrogen limitations, transcription of *atzDEF* is driven from a sigma70-type promoter, and *atzD* expression is induced by cyanuric acid (García-González *et al.* 2005). Taken together, these results strongly suggest (1) that pADP-1 independently acquired the *atzA*, *atzB*, *atzC* and the *atzDEF* genes from different sources, (2) that gene acquisition was likely a recent event, and (3) that the LysR-controlled *atzDEF* operon likely evolved much earlier than did the acquisition and assembly of the upper pathway genes on pADP-1.

Despite the spread of atrazine degradation genes to disparate genera of soil bacteria, atrazine still remains largely effective in weed control. This is likely due to suppression of the atrazine degradation phenotype by inorganic nitrogen sources. Several studies have convincingly shown that the addition of exogenous nitrogen inhibits atrazine catabolism by indigenous soil populations and pure cultures of atrazine-degrading bacteria, including *Pseudomonas* sp. strain ADP (Entry *et al.* 1993; Alvey & Crowley 1995; Struthers *et al.* 1998; Gebendinger & Radosevich 1999; Abdelhafid *et al.* 2000; García-González *et al.* 2003). Thus, despite some limited reports that atrazine is losing effectiveness for weed control in some soils (Krutz *et al.* 2007), elevated soil nitrogen levels likely repress atrazine degradation by indigenous soil bacteria in agricultural systems, and thus this herbicide can remain efficacious in the presence of a large number of degrading micro-organisms.

Enhanced atrazine degradation at the field and molecular levels

Over the past several years, there have been numerous reports of the apparent reduced half-life of atrazine in soils (Pussemier *et al.* 1997; Vanderheyden *et al.* 1997; Shaner & Henry 2007). It is widely accepted that microbial populations adapt to and are selected for by various C and N sources, and this also appears to be the case for atrazine-degrading micro-organisms since accelerated atrazine degradation in soils has been widely reported (Barriuso & Houot 1996; Ostrofsky *et al.* 1997; Vanderheyden *et al.* 1997; Houot *et al.* 2000). While some variation in reported half-lives may be attributable to abiotic soil characteristics at the sites under study, it is more likely due to the rapid spread of atrazine degradation genes, greater numbers of atrazine-degrading bacteria and changes in expression patterns in various soil microbial populations. Increased substrate availability is conventionally thought to increase soil population levels of atrazine-degrading micro-organisms (Shapir & Mandelbaum 1997; Rhine *et al.* 2003), resulting in enhanced degradation capacity. This may occur at the very local level, in soil particles, where atrazine concentrations are likely much greater than the average amounts found in bulk soils following liquid extraction. In addition, these bacteria can also respond to atrazine availability by altering gene dosage or expression levels of atrazine degradation genes. For example, *Arthrobacter* strain TC1 was shown to have six copies of *trzN* located within six identical direct tandem repeats of about 16 kb (Mongodin *et al.* 2006); *Nocardioides* sp. strain SP12 contains at least two copies of the *trzN, atzB* and *atzC* genes (Devers *et al.* 2007); and growth of *Pseudomonas* ADP for 320 generations in a liquid medium containing atrazine as the sole N source resulted in a more rapidly degrading evolved population containing a tandem duplication of *atzB* (Devers *et al.* 2008). In addition, RT-qPCR analyses indicated that expression levels of *atzA* and *atzB* significantly increased in response to atrazine addition in *Chelatobacteria heintzii* (Devers *et al.* 2004).

These results indicate that atrazine-catabolizing micro-organisms can respond to increased substrate availability in several ways, resulting in enhanced degradation capacity.

Evolution of atrazine degradation genes

While it is relatively easy to envision how strong selection pressure for growth, horizontal gene transfer and transposition can lead to the rapid assembly of complete catabolic pathways in bacteria, it is often much more difficult to determine where the degradation genes arose from. Conventional wisdom dictates that these genes likely arose from mutational events impacting extant genes with initially different functions. Although initial degradation studies indicated that major atrazine metabolites detected in soils and water were due to dealkylation reactions catalysed by a cytochrome P450 monooxygenase from *Rhodococcus* strains (Nagy *et al.* 1995), the reactions carried out by this enzyme were subsequently shown to be non-specific. In contrast, the reactions carried out by AtzA, AtzB and AtzC are very specific, and there are constraints on the substrates catalysed by these enzymes, all must contain the *s*-triazine ring and specific R groups. For example, AtzA catalysed hydrolysis of the chloride group in atrazine requires that at least one of the two nitrogen side chains contain an alkyl group (Seffernick & Wackettt 2001). Since there has been no evidence that natural compounds contain the *s*-triazine ring, it is likely that atrazine degradation ability arose subsequent to the release of this compound in the environment. Moreover, since many bacteria that have the ability to degrade atrazine contain identical enzymes, and that atrazine degradation is initiated via AtzA or TrzN, it suggests that atrazine dechlorination was the rate limiting step in further catabolism of this substrate. Sequence analysis indicates that all three upper pathway enzymes are members of the amidohydrolase superfamily (Sadowsky *et al.* 1998), all members have a conserved reaction mechanism whereby one to two divalent metals, coordinated by the enzymes, activate water for nucleophilic attack on their substrate (Seffernick *et al.* 2001). Superfamily members include cytosine deaminase, urease and adenine deaminase, where reactions result in the hydrolytic displacement of amino groups from purine and pyrimidine rings, and the *s*-triazine ring-containing compounds in many ways resemble the pyrimidine ring-containing compounds. This suggests that the atrazine catabolic enzymes may have evolved from those involved in intermediary metabolism for the displacement of amino groups from purine and pyrimidine compounds. Further evidence for the rapid evolution of AtzA comes from comparison to an enzyme involved in the catabolism of the related *s*-triazines, 2,4,6-triamino-1,3,5-triaizine and melamine, TriA. The AtzA from *Pseudomonas* sp. strain ADP is 98% identical at the amino acid level to TriA, melamine deaminase (Seffernick *et al.* 2001). It is striking to note that the nine amino acid differences between these two enzymes originate from nine nucleotide changes, and this results in different enzymatic activities: TriA is a deaminase and AtzA a chlorohydrolase. DNA shuffling studies done using both enzymes indicated that (1) leaving group specificity was likely imparted by residue 328, (2) each parent enzyme is fairly well optimised for their respective reactions and (3) a few amino acid changes increased TriA activity for displacement of methylthioether and methoxy substituents on the s-triazine ring, leaving groups displaced by TrzN (Raillard *et al.* 2001; Seffernick & Wackett 2001). While the original enzyme which evolved into one hydrolysing atrazine remains unknown, taken together, these results strongly suggest that atrazine chlorohydrolase and melamine deaminase recently diverged from a common ancestral amidohydrolase superfamily enzyme. These enzymes include those essential for microbial metabolism, such as cytosine deaminase and adenosine deaminase. Like AtzA, these two superfamily member enzymes contain a mononuclear active site metal, which is required for catalysis (Seffernick *et al.* 2002).

Conclusions

The metabolic versatility of micro-organisms is truly astounding! Micro-organisms often respond to anthropogenic inputs of chemicals by evolving the ability to use these compounds as growth substrates. However, these compounds create a powerful selection pressure for the evolution of enzymes with new activities and the assembly of these enzymes into new or existing metabolic pathways. While it was previously thought that such processes occurred over long evolutionary time scales, rapid microbial growth rates and gene introgression, facilitated by a plastic microbial genome, has resulted in more rapid evolution of metabolic functionality than previously thought possible. Such is the case for microbial acquisition of the ability to degrade atrazine, a novel compound that has been on the planet for only about 50 years. As with most halogenated compounds, the ability to use atrazine as a growth substrate was due to the evolution of a suitable dechlorinating enzyme, atrazine chlorohydrolase (AtzA) or the broad triazine hydrolase TrzN. Both of these metalloenyzmes are members of the amidohydrolase superfamily and have commonalities both in structure and function. Since substrate hydrolysis by AtzA and TrzN fails to release carbon or nitrogen from their respective preferred substrates, atrazine and ametryn, respectively, subsequent hydrolytic steps are required to produce metabolically useful compounds. Genes encoding for these enzymatic steps were likely initially present in separate bacteria, which lived in a consortium capable of growing on these substrates. While syntrophic interactions like this are pervasive among micro-organisms in most ecosystems, they are likely metabolically and energetically inefficient, and selection pressure from competitive micro-organisms living in the same environment likely led to the eventual recruitment of the six enzymatic steps required for complete atrazine mineralisation into a single bacterium.

The ability to rapidly build this new pathway in a single micro-organism was likely driven by independent and multiple plasmid transfer and transposition events, the genetic machinery facilitating metabolic diversity. The analysis of the complete genomes of many micro-organisms has repeatedly shown that microbial genomes are plastic, mosaic and composite, and these properties play a prominent role in the ability of micro-organisms to rapidly respond to changing environments and the input of novel compounds in to the environment.

References

Abdelhafid, R., Houot, S. and Barriuso, E. (2000) How increasing availabilities of carbon and nitrogen affect atrazine behaviour in soils. *Biology and Fertility of Soils* **30**, 333–340.

Allan, G. G. (1981) Use of technical melamine crystallized from water as plant fertilizer. Patent 3 034 062, Germany.

Alvey, S. and Crowley, D. E. (1995) Influence of organic amendments on biodegradation of atrazine as a nitrogen source. *Journal of Environmental Quality* **24**, 1156–1162.

Barriuso, E. and Houot, S. (1996) Rapid mineralisation of s-triazine ring of atrazine in soils in relation to soil management. *Soil Biology and Biochemistry* **28**, 1341–1348.

Behki, R. M. and Khan, S. U. (1986) Degradation of atrazine by *Pseudomonas*: N-dealkylation and dehalogenation of atrazine and its metabolites. *Journal of Agricultural Food and Chemistry* **34**, 746–749.

Behki, R. M. and Khan, S. U. (1994) Degradation of trazine, propazine and simazine by *Rhodococcus* strain B-30. *Journal of Agricultural Food and Chemistry* **42**, 1237–1241.

Behki, R., Topp, E., Dick, W. and Mantelli, M. (1993) Metabolism of the herbicide atrazine by *Rhodococcus* strains. *Applied and Environmental Microbiology,* **59**, 1955–1959.

Berg, O. G. and Kurland, C. G. (2002) Evolution of microbial genomes: sequence acquisition and loss. *Molecular Biology and Evolution* **19**, 2265–2276.

Boundy-Mills, K. L., de Souza, M. L., Wackett, L. P., Mandelbaum, R. and Sadowsky, M. J. (1997) The *atzB* gene of *Pseudomonas* sp. strain ADP encodes hydroxyatrazine ethylaminohydrolase, the second step of a novel atrazine degradation pathway. *Applied and Environmental Microbiology* **63**, 916–923.

Bouquard, C., Ouazzani, J., Prome, J., Michel-Briand, Y. and Plesiat, P. (1997) Dechlorination of atrazine by a *Rhizobium* sp. isolate. *Applied and Environmental Microbiology* **63**, 862–866.

Cai, B., Han, Y., Liu, B., Ren, Y. and Jiang, S. (2003) Isolation and characterization of an atrazine-degrading bacterium from industrial wastewater in China. *Letters in Applied Microbiology* **36**, 272–276.

Cook, A. M. (1987) Biodegradation of s-triazine xenobiotics. *FEMS Microbiology Reviews* **46**, 93–116.

Cook, A. M. and Hutter, R. (1984) Deethylsimazine: bacterial dechlorination, deamination, and complete degradation. *Journal of Agricultural Food and Chemistry* **32**, 581–585.

Cook, A. M., Beilstein, P., Grossenbacher, H. and Hutter, R. (1985) Ring cleavage and degradative pathway of cyanuric acid in bacteria. *Biochemical Journal* **231**, 25–30.

Cook, A. M., Grossenbacher, H. and Hutter, R. (1983) Isolation and cultivation of microbes with biodegradative potential. *Experientia* **39**, 1191–1198.

Dai, X., Jiang, J., Gu, L., Pan, R. and Li, S. (2007) Study on the atrazine-degrading genes in *Arthrobacter* sp. AG1. *Chinese Journal of Biotechnology* **23**, 789–793.

de Souza, M. L., Sadowsky, M. J. and Wackett, L. P. (1996) Atrazine chlorohydrolase from *Pseudomonas* sp. strain ADP: gene sequence, enzyme purification, and protein characterization. *Journal of Bacteriology* **178**, 4894–4900.

de Souza, M. L., Seffernick, J., Sadowsky, M. J. and Wackett, L. P. (1998) The atrazine catabolism genes *atzABC* are widespread and highly conserved. *Journal of Bacteriology* **180**, 1951–1954.

Devers, M., El Azhari, N., Kolic, N. U., Rouard, N. and Martin-Laurent, F. (2007) Detection and organization of atrazine-degrading genetic potential of seventeen bacterial isolates belonging to divergent taxa indicate a recent common origin of their catabolic functions. *FEMS Microbiology Letters* **273**, 78–86.

Devers, M., Henry, S., Hartmann, A. and Martin-Laurent, F. (2005) Horizontal gene transfer of atrazine-degrading genes (*atz*) from *Agrobacterium tumefaciens* St96-4 pADP1:Tn5 to bacteria of maize-cultivated soil. *Pest Management Science* **61**, 870–880.

Devers, M., Rouard, N. and Martin-Laurent, F. (2008) Fitness drift of an atrazine-degrading population under atrazine selection pressure. *Environmental Microbiology* **10**, 676–684.

Devers, M., Soulas, G. and Martin-Laurent, F. (2004) Real-time reverse transcription PCR analysis of expression of atrazine catabolism genes in two bacterial strains isolated from soil. *Journal of Microbiological Methods* **56**, 3–15.

Entry, J. A., Mattson, K. G. and Emmingham, W. H. (1993) The influence of nitrogen on atrazine and 2,4-dichlorophenoxyacetic acid mineralization in grassland soils. *Biology and Fertility of Soils* **16**, 179–182.

Erickson, L. E. and Lee, H. K. (1989) Degradation of atrazine and related s-triazines. *Critical Reviews in Environmental Control* **19**, 1–13.

Esser, H. O., Dupuis, G., Ebert, E., Marco, G. J. and Vogel, C. (1975) S-triazines. In: *Herbicides: Chemistry, Degradation and Mode of Action*, 2nd edn (eds. P. C. Kearney and D. D. Kaufman). Marcel Dekker, New York.

Feakin, S. J., Gubbins, B., McGhee, I., Shaw, L. J. and Burns, R. G. (1995) Inoculation of granular activated carbon with s-triazine-degrading bacteria for water treatment at pilot-scale. *Water Research* **29**, 1681–1688.

Fruchey, I., Shapir, N., Sadowsky, M. J. and Wackett, L. P. (2003) On the origins of cyanuric acid hydrolase: purification, substrates and prevalence of *AtzD* from *Pseudomonas* sp. strain ADP. *Applied and Environmental Microbiology* **69**, 3653–3657.

García-González, V., Govantes, F., Porrua, O. and Santero, E. (2005) Regulation of the *Pseudomonas* sp. strain ADP cyanuric acid degradation operon. *Journal of Bacteriology* **187**, 155–167.

García-González, V., Govantes, F., Shaw, L. J., Burns, R. G. and Santero, E. (2003) Nitrogen control of atrazine utilization in *Pseudomonas* sp. strain ADP. *Applied and Environmental Microbiology* **69**, 87–93.

Gast, A., Knusli, E. and Gysin, H. (1955) Chlorazine as a phytotoxic agent. *Experientia* **11**, 107.

Gebendinger, N. and Radosevich, M. (1999) Inhibition of atrazine degradation by cyanazine and exogenous nitrogen in bacterial isolate M91-3. *Applied Microbiology and Biotechnology* **51**, 375–381.

Geller, A. (1980) Studies on the degradation of atrazine by bacterial communities enriched from various biotopes. *Archives of Environmental Contamination and Toxicology* **9**, 289–305.

Giardina, M. C., Giardi, M. T. and Filacchioni, G. (1982) Atrazine metabolism by *Nocardia*: elucidation of initial pathway and synthesis of potential metabolites. *Agricultural Biology and Biochemistry* **46**, 1439–1445.

Houot, S., Topp, E., Yassir, A. and Soulas, G. (2000) Dependence of accelerated degradation of atrazine on soil pH in French and Canadian soils. *Soil Biology and Biochemistry* **32**, 615–625.

Hu, J., Dai, X. and Li, S. (2004) Effects of atrazine and its degrader *Exiguobaterium* sp. BTAH1 on soil microbial community. *Chinese Journal of Applied Ecology* **16**, 1518–1522.

Jain, R., Rivera, M., Moore J. and Lake, J. A. (2002) Horizontal gene transfer in microbial genome evolution. *Theoretical Population Biology* **61**, 489–495.

Janssen, D. B., Dinkla, I. J. T., Poelarends, G. J. and Terpstra, P. (2005) Bacterial degradation of xenobiotic compounds: evolution and distribution of novel enzyme activities. *Environmental Microbiology* **7**, 1868–1882.

Kaufman, D. D. and Kearney, P. C. (1970) Microbial degradation of triazine herbicides. In: *Residue Reviews 32: The Triazine Herbicides* (eds. F. A. Gunther and J. D. Gunther), pp. 235–265. Springer-Verlag, New York.

Knusli, E. and Gysin, H. (1960) Chemistry and herbicidal properties of triazine derivatives. *Advances in Pest Control Research* **3**, 289–357.

Knusli, E., Berrer, D., Dupuis, G. and Esser, H. O. (1969) S-triazines. In: *Degradation of Herbicides* (eds. P. C. Kearney and D. D. Kaufman), pp. 51–78. Marcel Dekker, New York.

Korpraditskul, R., Katayama, A. and Kuwatsuka, S. (1993) Degradation of atrazine by soil bacteria in the stationary phase. *Journal of Pesticide Science* **18**, 293–298.

Krutz, L. J., Zablotowicz, R. M., Reddy, K. N., Koger, C. H. and Weaver, M. A. (2007) Enhanced degradation of atrazine under field conditions correlates with a loss of weed control in the glasshouse. *Pest Management Science* **63**, 23–31.

Laskin, A. I. and Lechavalier, H. A. (1984) In: *CRC Handbook of Microbiology*, 2nd edn (ed. A. I. Laskin), pp. 557. CRC Press, Boca Raton.

Lucas, A. D., Bekheit, H. K. M., Goodrow, M. H. *et al.* (1993) Development of antibodies against hydroxyatrazine and hydroxysimazine – application to environmental samples. *Journal of Agricultural Food and Chemistry* **41**, 1523–1529.

Mandelbaum, R. T. and Wackett, L. P. (1996) *Pseudomonas* strain for degradation of s-traizines in soil and water. US Patent 5 508 193.

Mandelbaum, R. T., Allan, D. L. and Wackett, L. P. (1995) Isolation and characterization of a *Pseudomonas* sp. that mineralizes the s-triazine herbicide atrazine. *Applied and Environmental Microbiology* **61**, 1451–1457.

Mandelbaum, R. T., Sadowsky, M. J. and Wackett, L. P. (2008) Microbial degradation of s-triazine herbicides. In: *The Triazine Herbicides* (eds. H. LeBaron, J. McFarland and O. Burnside). Elsevier, Amsterdam.

Mandelbaum, R. T., Wackett, L. P. and Allan, D. L. (1993) Mineralization of the s-triazine ring of atrazine by stable bacterial mixed cultures. *Applied Environmental Microbiology* **59**, 1695–1701.

Marecik, R., Kroliczak, P., Czaczyk, K., Bialas, W., Olejnik, A. and Cyplik, P. (2008) Atrazine degradation by aerobic micro-organisms isolated from the rhizosphere of sweet flag (*Acorus calamus* L.). *Biodegradation* **19**, 293–301.

Martin, W. (1999) Mosaic bacterial chromosomes: a challenge en route to a tree of genomes. *Bioessays* **21**, 99–104.

Martin-Laurent, F., Barres, B., Wagschal, I. *et al.* (2006) Impact of the maize rhizosphere on the genetic structure, the diversity and the atrazine-degrading gene composition of the cultivable atrazine-degrading communities. *Plant and Soil* **282**, 99–115.

Martinez, B., Tomkins, J., Wackett, L. P., Wing, R. and Sadowsky, M. J. (2001) Complete nucleotide sequence and organization of the atrazine catabolic plasmid pADP-1 from *Pseudomonas* sp. strain ADP. *Journal of Bacteriology* **183**, 5684–5697.

Masaphy, S., Levanon, D., Vaya, J. and Henis, Y. (1993) Isolation and characterization of a novel atrazine metabolite produced by the fungus *Pleurotus pulmonarius*, 2-chloro-4-ethylamino-6-(1-hydroxyisopropyl)amino-1,3,5-triazine. *Applied and Environmental Microbiology* **59**, 4342–4346.

Mira, A., Klasson, L. and Andersson, S. D. (2002) Microbial genome evolution: sources of variability. *Current Opinions in Microbiology* **5**, 506–512.

Mirgain, I., Green, G. A. and Monteil, H. (1993) Degradation of atrazine in laboratory microcosms: isolation and identification of the degrading bacteria. *Environmental Toxicology and Chemistry* **12**, 1627–1634.

Mongodin, E. F., Shapir, N., Daugherty, S. C. *et al.* (2006) Secrets of soil survival revealed by the genome sequence of *Arthrobacter aurescens* TC1. *PLoS Genetics* **2**, 214.

Moscinski, J. K. (1996) Bioaugmentation and biostimulation technologies to bioremediate soils contaminated with herbicide mixtures. Unpublished MSc Thesis, Department of Microbiology, Iowa State University, Ames, IA, pp. 1–158.

Mulbry, W., Zhu, H., Nour, S. and Topp, E. (2002) The triazine hydrolase gene *trzN* from *Nocardioides* sp. strain C190: cloning and construction of gene-specific primers. *FEMS Microbiology Letters* **206**, 75–79.

Nagy, I., Verheijen, S., De Schrijver, A. *et al.* (1995) Characterization of the *Rhodococcus* sp. NI86/21 gene encoding alcohol: N,N′-dimethyl-4-nitrosoaniline oxidoreductase inducible by atrazine and thiocarbamate herbicides. *Archives of Microbiology* **163**, 439–446.

Omelchenko, M. V., Makarova, K. S., Wolf, Y. I., Rogozin, I. B. and Koonin, E. V. (2003) Evolution of mosaic operons by horizontal gene transfer and gene displacement in situ. *Genome Biology* **4**, R55.

Ostrofsky, E. B., Traina, S. J. and Tuovinen, O. H. (1997) Variation in atrazine mineralization rates in relation to agricultural management practices. *Journal of Environmental Quality* **26**, 647–657.

Piutti, S., Semon, E., Landry, D. *et al.* (2003) Isolation and characterization of *Nocardioides* sp. SP12, an atrazine-degrading bacterial strain possessing the gene *trzN* from bulk- and maize rhizosphere soil. *FEMS Microbiology Letters* **221**, 111–117.

Poelarends, G. J., Kulakov, L. A., Larkin, M. J., van Hylckama Vlieg, J. E. and Janssen, D. B. (2000) The role of horizontal gene transfer and gene integration in the evolution of 1,3-dichloropropene- and 1,2-dibromoethane-degradative pathways. *Journal of Bacteriology* **182**, 2191–2199.

Pussemier, L., Goux, S., Vanderheyden, V., Debongnie, P., Tresinie, I. and Foucart, G. (1997) Rapid dissipation of atrazine in soils taken from various maize fields. *Weed Research* **37**, 171–179.

Radosevich, M., Traina, S. J., Hao, Y. L. and Tuovinen, O. H. (1995) Degradation and mineralization of atrazine by a soil bacterial isolate. *Applied and Environmental Microbiology* **61**, 297–302.

Raillard, S. A., Krebber, A., Chen, Y. *et al.* (2001) Novel enzyme activities and functional plasticity revealed by recombining homologous enzymes. *Chemistry and Biology* **8**, 891–898.

Rhine, E. D., Fuhrmann, J. J. and Radosevich, M. (2003) Microbial community responses to atrazine exposure and nutrient availability: linking degradation capacity to community structure. *Microbial Ecology* **46**, 145–160.

Rousseaux, S., Hartmann, A., Lagacherie, B., Piutti, S., Andreux, F. and Soulas, G. (2003) Inoculation of an atrazine-degrading strain, *Chelatobacter heintzii* Cit1, in four different soils: effects of different inoculum densities. *Chemosphere* **51**, 569–576.

Rousseaux, S., Hartmann, A. and Soulas, G. (2001) Isolation and characterisation of new Gram-negative and Gram-positive atrazine degrading bacteria from different French soils. *FEMS Microbiology Ecology* **36**, 211–222.

Sadowsky, M. J., Tong, Z., de Souza, M. and Wackett, L. P. (1998) AtzC is a new member of the amidohydrolase protein superfamily and is homologous to other atrazine-metabolizing enzymes. *Journal of Bacteriology* **180**, 152–158.

Sajjaphan, K., Shapir, N., Judd, A. K., Wackett, L. P. and Sadowsky, M. J. (2002) A novel *psbal* gene from a naturally-occurring atrazine-resistant cyanobacterial isolate. *Applied and Environmental Microbiology* **168**, 1358–1366.

Sajjaphan, K., Shapir, N., Wackett, L. P. *et al.* (2004) *Arthrobacter aurescens* TC1 atrazine catabolism genes *trzN, atzB,* and *atzC* are linked on a 160-kilobase region and are functional in *Escherichia coli. Applied and Environmental Microbiology* **70**, 4402–4407.

Satsuma, K. (2006) Characterisation of new strains of atrazine-degrading *Nocardioides* sp. isolated from Japanese riverbed sediment using naturally derived river ecosystem. *Pest Management Science* **62**, 340–349.

Scholl, W., Davis, R. O. E., Brown, B. E. and Reid, F. R. (1937) Melamine: a possible plant food value. *Chemtech* **29**, 202–205.

Seffernick, J. L. and Wackett, L. P. (2001) Rapid evolution of bacterial catabolic enzymes: a case study with atrazine chlorohydrolase. *Biochemistry* **40**, 12747–12753.

Seffernick, J. L., de Souza, M. L., Sadowsky, M. J. and Wackett, L. P. (2001) Melamine deaminase and atrazine chlorohydrolase: 98% identical but functionally different. *Journal of Bacteriology* **183**, 2405–2410.

Seffernick, J. L., McTavish, H., Osborne, J. P., de Souza, M. L., Sadowsky, M. J. and Wackett, L. P. (2002) Atrazine chlorohydrolase from *Pseudomonas* sp. strain ADP is a metalloenzyme. *Biochemistry* **41**, 14430–14437.

Shaner, D. L. and Henry, W. B. (2007) Field history and dissipation of atrazine and metolachlor in Colorado. *Journal of Environmental Quality* **36**, 128–134.

Shao, Z. Q. and Behki, R. (1996) Characterization of the expression of the *thcB* gene, coding for a pesticide-degrading cytochrome P-450 in *Rhodococcus* strains. *Applied and Environmental Microbiology* **62**, 403–407.

Shapir, N. and Mandelbaum, N. T. (1997) Atrazine degradation in subsurface soil by indigenous and introduced micro-organisms. *Journal of Agricultural Food and Chemistry* **45**, 4481–4486.

Shapir, N., Mongodin, E. F., Sadowsky, M. J., Daugherty, S. C., Nelson, K. E. and Wackett, L. P. (2007) Evolution of catabolic pathways: genomic insights into microbial s-triazine metabolism. *Journal of Bacteriology* **189**, 674–682.

Shapir, N., Osborne, J. P., Johnson, G., Sadowsky, M. J. and Wackett, L. P. (2002) Purification, substrate range and metal center of AtzC: the N-isopropylammelide hydrolase involved in bacterial atrazine metabolism. *Journal of Bacteriology* **184**, 5376–5384.

Shapir, N., Pedersen, C., Gil, O. *et al.* (2006) TrzN from *Arthrobacter aurescens* TC1 is a zinc amidohydrolase. *Journal of Bacteriology* **188**, 5859–5864.

Shapir, N., Rosendahl, C., Johnson, G., Andreina, M., Sadowsky, M. J. and Wackett, L. P. (2005) Substrate specificity and colorimetric assay for recombinant TrzN derived from *Arthrobacter aurescens* TC1. *Applied and Environmental Microbiology* **71**, 2214–2220.

Shelton, D. R., Khader, S., Karns, J. S. and Pogell, B. M. (1996) Metabolism of twelve herbicides by *Streptomyces. Biodegradation* **7**, 129–136.

Smith, D., Alvey, S. and Crowley, D. E. (2005) Cooperative catabolic pathways within an atrazine-degrading enrichment culture isolated from soil. *FEMS Microbiology Ecology* **53**, 265–275.

Stamper, D. M., Radosevich, M., Hallberg, K. B., Traina, S. J. and Tuovinen, O. H. (2002) *Ralstonia basilensis* M91–3, a denitrifying soil bacterium capable of using s-triazines as nitrogen sources. *Canadian Journal of Microbiology* **48**, 1089–1098.

Strong, L. C., Rosendahl, C., Johnson, G., Sadowsky, M. J. and Wackett, L. P. (2002) *Arthrobacter aurescens* TC1 metabolizes diverse s-triazine ring compounds. *Applied and Environmental Microbiology* **68**, 5973–5980.

Struthers, J. K., Jayachandran, K. and Moorman, T. B. (1998) Biodegradation of atrazine by *Agrobacterium radiobacter* J14a and use of this strain in bioremediation of contaminated soil. *Applied and Environmental Microbiology* **64**, 3368–3375.

Topp, E. (2001) A comparison of three atrazine-degrading bacteria for soil bioremediation. *Biology and Fertility of Soils* **33**, 529–534.

Topp, E., Mulbry, W. M., Zhu, H., Nour, S. M. and Cuppels, D. (2000a) Characterization of s-triazine herbicide metabolism by a *Norcardiodes* sp. isolated from agricultural soils. *Applied and Environmental Microbiology* **66**, 3134–3141.

Topp, E., Zhu, H., Nour, S. M., Houot, S., Lewis, M. and Cuppels, D. (2000b) Characterization of an atrazine-degrading *Pseudaminobacter* sp. isolated from Canadian and French agricultural soils. *Applied and Environmental Microbiology* **66**, 2773–2782.

Vandepitte, V., Wierinck, I., De Vos, P., De Poorter, M. P., Houwen, F. and Verstraete, W. (1994) N-dealkylation of atrazine by hydrogenotrophic fluorescent *Pseudomonads. Water, Air, and Soil Pollution* **78**, 335–341.

Vanderheyden, V., Debongnie, P. and Pussemier, L. (1997) Accelerated degradation and mineralization of atrazine in surface and subsurface soil materials. *Pesticide Science* **49**, 237–242.

van der Meer, J. R. and Sentchilo, V. (2003) Genomic islands and the evolution of catabolic pathways in bacteria. *Current Opinions in Biotechnology* **14**, 248–254.

Vargha, M., Takáts, Z. and Márialigeti, K. (2005) Degradation of atrazine in a laboratory scale model system with Danube river sediment. *Water Research* **39**, 1560–1568.

Wackett, L. P. (2004) Evolution of enzymes for the metabolism of new chemical inputs into the environment. *Journal of Biological Chemistry* **279**, 41259–41262.

Wackett, L. P., Sadowsky, M. J., Martinez, B. and Shapir, N. (2002) Biodegradation of atrazine and related triazine compounds: from enzymes to field studies. *Applied Microbiology and Biotechnology* **58**, 39–45.

Warren, R., Hsiao, W., Kudo, H. *et al.* (2004) Functional characterization of a catabolic plasmid from polychlorinated-biphenyl-degrading *Rhodococcus* sp. strain RHA1. *Journal of Bacteriology* **186**, 7783–7795.

Wright, B. E. (2000) A biochemical mechanism for nonrandom mutations and evolution. *Journal of Bacteriology* **182**, 2993–3001.

Yanze-Kontchou, C. and Gschwind, N. (1995) Mineralization of the herbicide atrazine in soil inoculated with a *Pseudomonas* strain. *Journal of Agricultural Food and Chemistry* **43**, 2291–2294.

Zeyer, J., Bodmer, J. and Hutter, R. (1981) Microbial degradation of ammeline. *Zentralblatt für Bakteriologie Mikrobiologie und Hygiene* **3**, 289–298.

CHAPTER ELEVEN

The microbial ecology of land and water contaminated with radioactive waste: towards the development of bioremediation options for the nuclear industry

ANDREA GEISSLER, SONJA SELENSKA-POBELL,
KATHERINE MORRIS, IAN T.BURKE, FRANCIS
R.LIVENS AND JONATHAN R.LLOYD

Introduction

The release of radionuclides from nuclear and mining sites and their subsequent mobility in the environment is a subject of intense public concern and has promoted much recent research into the environmental fate of radioactive waste (Lloyd & Renshaw 2005b). Naturally occurring radionuclides can input significant quantities of radioactivity into the environment while both natural and artificial/manmade radionuclides have also been released as a consequence of nuclear weapons testing in the 1950s and 1960s, and via accidental release, e.g., from Chernobyl in 1986. The major burden of anthropogenic environmental radioactivity, however, is from the nuclear facilities themselves and includes the continuing controlled discharge of process effluents produced by industrial activities allied to the generation of nuclear power.

Wastes containing radionuclides are produced at the many steps in the nuclear fuel cycle, and vary considerably from low level, high-volume radioactive effluents produced during uranium mining to the intensely radioactive plant, fuel and liquid wastes produced from reactor operation and fuel reprocessing (Lloyd & Renshaw 2005b). The stewardship of these contaminated waste-streams needs a much deeper understanding of the biological and chemical factors controlling the mobility of radionuclides in the environment. Indeed, this is highly relevant on a global stage as anthropogenic radionuclides have been dispersed to the environment both by accident and as part of a controlled/monitored release, e.g., in effluents. Micro-organisms have adapted

Ecology of Industrial Pollution, eds. Lesley C. Batty and Kevin B. Hallberg. Published by Cambridge University Press. © British Ecological Society 2010.

to live even in radioactively contaminated environments (Selenska-Pobell 2002; Lloyd 2003; Fredrickson *et al.* 2004; Ruggiero *et al.* 2005; Akob *et al.* 2007), and they can affect radionuclide speciation via a number of mechanisms which are potentially useful for scalable, cost-effective bioremediation of sediments and waters impacted by nuclear waste (Keith-Roach & Livens 2002; Lloyd 2003; Suzuki *et al.* 2003; Lloyd & Renshaw 2005a; Brodie *et al.* 2006; Fomina *et al.* 2007).

The aim of this chapter is to give an overview of known interactions of micro-organisms with key radionuclides, focusing on potential roles in controlling radionuclide mobility in the subsurface. We will then discuss the influence of microbial processes on the immobilisation of radionuclides described in recent laboratory studies from our groups, focusing on two priority contaminants: technetium and uranium. The former studies address the impact of bioreduction strategies on solubility and include work on stimulated bioreduction (achieved via added organic electron donor) on technetium solubility, while the latter studies demonstrate the immobilisation of uranium in sediments from a uranium mining waste pile without the addition of a carbon source. Finally, due to the widespread use of nitric acid in the nuclear sector, the multiple influences of nitrate and nitrate-reducing bacteria on the solubility of radionuclides will be discussed, especially where the activities of these organisms impact on the biogeochemistry of uranium and technetium.

Microbial interactions with radionuclides

The environmental fate of a radionuclide is governed by the interplay between the background matrix of the radioactive material, the often complex chemistry of the radionuclide in question and a broad range of chemical factors associated with the environment that has been impacted by the radioactive material in question (Lloyd & Renshaw 2005b). In addition, microbial activity will have a profound effect on the solubility of radionuclides via a complex range of often overlapping mechanisms including biosorption, bioaccumulation, biotransformation, biomineralisation and microbially enhanced chemisorption of heavy metals (Fig. 11.1). For a more extensive discussion of these microbe–radionuclide interactions, the reader is directed to the following reviews (Lloyd & Lovley 2001; Keith-Roach & Livens 2002; Pedersen 2005; Renshaw *et al.* 2007). A brief synopsis, focusing on interactions with subsurface micro-organisms, is given below.

Biosorption describes the metabolism-independent sorption of heavy metals and radionuclides to biomass and encompasses both adsorption and absorption. Multiple studies over the past three decades have confirmed that a wide range of micro-organisms are capable of efficient biosorption of radionuclides (Francis *et al.* 2004; Merroun *et al.* 2005; Ohnuki *et al.* 2005). Both living and dead biomass is capable of biosorption, with ligands involved in metal binding including carboxyl, amine, hydroxyl, phosphate and sulphhydryl groups. Most

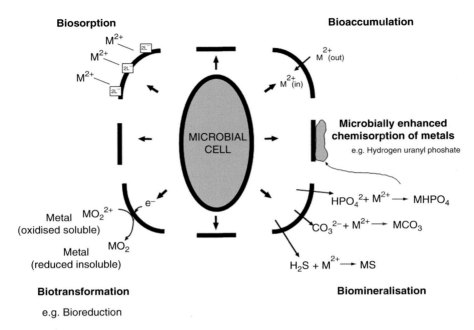

Figure 11.1. Mechanisms of metal/radionuclide-microbe interactions.

studies have investigated sorption of U(VI) (Francis *et al.* 2004; Merroun *et al.* 2005), with far less known about the sorption of other radionuclides, such as Pu (Ohnuki *et al.* 2007). For a more detailed overview of this research area, the reader is referred to Lloyd and Macaskie (2000).

Bioaccumulation, which can be defined as energy-dependent metal uptake, has been demonstrated for most physiologically important metal ions and some radionuclides which can enter the cell as chemical 'surrogates' using these transport systems. There have been just a few investigations of the intracellular accumulation of actinides, and almost all of these have concentrated on uranium (Francis *et al.* 2004, Suzuki & Banfield 2004). It has been suggested that intracellular uptake of uranium is metabolism-independent, and results from increased cell-membrane permeability caused, for example, by uranium toxicity (Suzuki & Banfield 1999). In other studies, bioaccumulation has been proposed as a mechanism promoting uranium tolerance in *Arthrobacter* spp., although the final fate of the uranium taken up into the cell remains to be confirmed (Suzuki & Banfield 2004; Martinez *et al.* 2006; Geissler 2007). In contrast, the intracellular uptake of Pu(IV) in *Microbacterium flavescens* was reported to be an active, metabolism-dependent transport process (John *et al.* 2001).

Micro-organisms can also catalyse the direct transformation of toxic metals and metalloids to less soluble or more volatile forms via two distinct enzymatic mechanisms. Bioreduction can result in precipitation of some metals/metalloids, where the reduced form is less soluble than the high oxidation form, as is the case for enzymatic U(VI) reduction to U(IV) (Lovley 1993), while biomethylation can yield highly volatile derivatives. Both mechanisms can result in a decrease in the concentration of soluble metals in contaminated water. While the microbial enzymatic reduction of radionuclides including U(VI), Tc(VII), Np(V) and Pu(IV) has been demonstrated (Lovley 1993; Lloyd & Macaskie 1996; Lloyd et al. 2000a; Boukhalfa et al. 2007), biomethylation of radionuclides has received little attention.

Lovley and co-workers were the first to demonstrate the dissimilatory reduction of U(VI) by the Fe(III)-reducing bacteria *Geobacter metallireducens* GS-15 and *Shewanella oneidensis* (formerly *Alteromonas putrefacies*), a process by which energy is conserved for anaerobic growth of these organisms (Lovley et al. 1991; Lovley 1993). Other bacteria able to reduce U(VI) (but without conserving energy), include the sulphate-reducing bacteria *Desulfovibrio desulfuricans* (Lovley & Phillips 1992) and *Desulfosporosinus* sp. (Suzuki et al. 2004), in addition to *Clostridium* sp. ATCC 53464 (Francis et al. 1994), *Salmonella subterranean* strain FRC1 (Shelobolina et al. 2004) and *Anaeromyxobacter dehalogenans* strain 2CP-C (Wu et al. 2006).

The mechanism of enzymatic reduction of uranium has been characterised most comprehensively in the sulphate-reducing bacterium *Desulfovibrio vulgaris*, which contains a periplasmic cytochrome c3, identified as the terminal reductase for the reduction of U(VI) (Lovley et al. 1993). Similar mechanisms may be important in *Geobacter* spp. (Lloyd et al. 2002), but the terminal reductase for U(VI) in this organism remains to be identified unequivocally. Interestingly, it was shown recently that *Geobacter sulfurreducens* cannot reduce NpO_2^+, even though it reduces UO_2^{2+} efficiently (Renshaw et al. 2005). The authors suggested that the enzyme system responsible for uranium reduction is capable of transferring one electron to an actinyl ion, and the instability of the resulting U(V) then generates U(IV) via disproportionation. The reduction of Np(V) is not possible, however, because it appears the enzyme is specific for hexavalent actinides, and cannot transfer an electron to NpO_2^+ (Renshaw et al. 2005). In contrast, it has been known for some time that *S. oneidensis* is able to reduce Np(V) (Lloyd et al. 2000b), and more recent studies have confirmed that cell suspensions of *S. oneidensis* are able to enzymatically reduce unchelated Np(V) to insoluble Np(IV)(s), but cell suspensions of *G. metallireducens* are unable to reduce Np(V) (Icopini et al. 2007), suggesting the factors controlling enzymatic reduction of Np(V) are complex.

These two organisms are also able to reduce highly soluble Tc(VII) to insoluble lower-valence species enzymatically (Lloyd & Macaskie 1996). However, in the subsurface, Fe(III)-reducing bacteria can also indirectly reduce and precipitate Tc(VII) via biogenic Fe(II), and this mechanism is especially efficient when the Fe(II) is associated with mineral phases, e.g., when biogenic magnetite is present. This was first studied in G. *sulfurreducens,* which used hydrogen as the electron donor for enzymatic reduction of Tc(VII) (presumably via a NiFe hydrogenase identified in other bacteria as the terminal reductase for Tc(VII)) (Lloyd *et al.* 1997; Marshall *et al.* 2008). When acetate was supplied as the electron donor, however, enzymatic reduction of Tc(VII) was not possible, but Tc(VII) reduction was very efficient when Fe(III) oxides were provided with the acetate and reduced to biogenic Fe(II)-bearing magnetite, which could then act as an electron shuttle to Tc(VII) and cause reduction. Thus, Fe(II)-bearing biomagnetite was shown to be an excellent reductant for Tc(VII).

Finally, the mechanism of plutonium reduction by G. *metallireducens* and S. *oneidensis* MR-1 has also been studied recently, and it was demonstrated that both organisms produce very little Pu(III) enzymatically from $Pu(IV)(OH)_4$, whereas in the presence of ethylenediaminetetraacetic acid (EDTA), most of the $Pu(IV)(OH)_4(am)$ was reduced to Pu(III) (Boukhalfa *et al.* 2007). Inefficient enzymatic production of Pu(III) from $Pu(IV)(OH)_4$ was also identified in G. *sulfurreducens*, even though the organism is very adept at reducing a broad range of extracellular Fe(III) and Mn(IV) minerals (Renshaw *et al.* 2008).

Biomineralisation is the process by which metals and radionuclides can be precipitated with microbially generated ligands, e.g., phosphate, sulphide or carbonate. In these examples, the microbial ligands accumulate to high concentrations around the cell, and the cell surface provides a nucleation site for precipitation, resulting in efficient removal of the radionuclide from solution (Renshaw *et al.* 2007). The process of biomineralisation can be induced by secretion of inorganic compounds, such as orthophosphate groups, which can directly bind U(VI) in insoluble polycrystalline uranyl hydrogen phosphate or in meta-autunite-like mineral phases (Macaskie *et al.* 1992; Merroun *et al.* 2006; Beazley *et al.* 2007; Jroundi *et al.* 2007). In addition to direct precipitation by microbially generated ligands, actinide ions can also be removed from solution by chemisorption to biogenic minerals ('microbially enhanced chemisorption') (Macaskie *et al.* 1994).

Biogeochemistry of technetium reduction in sediments

Technetium is a significant, long-lived (^{99}Tc half-life $= 2.15 \times 10^5$ years) radio-active contaminant from nuclear fuel cycle operations. It is highly mobile in its oxidised form (as $Tc(VII)O_4^-$) but is scavenged to sediments in its reduced forms (predominantly poorly soluble Tc(IV)). As part of a long-term collaboration between our groups in Manchester and Leeds, we have been studying the biogeochemical behaviour of Tc and its potential environmental mobility, to better inform bioremediation approaches and safety case assessments.

Initial experiments used microcosms constructed from Tc-free Humber Estuary surface sediments (with their indigenous microbial populations), which were spiked with low levels (<5 μM) of TcO_4^-, and technetium solubility was then monitored as anoxia developed in the microcosms (Burke *et al.* 2005). As expected, the microcosms progressed through a cascade of terminal electron accepting processes during which time over 99% of the Tc was removed from the pore waters. In sterile controls, Tc remained in solution (presumably as Tc(VII)), indicating that removal to sediments was biologically mediated (Burke *et al.* 2005). A detailed analysis of geochemical indicators demonstrated that Tc removal occurred as Fe(III)-reducing conditions developed after the consumption of most of the nitrate and accumulation of Mn(II) in pore waters.

Pure culture microcosms were established by inoculating sterilised mixtures of sediment slurry and river water with cultures of either a nitrate-reducing bacterium (*Pseudomonas stutzeri*), an Fe(III)-reducing bacterium (*Shewanella* sp.) or a sulphate-reducing bacterium (*Desulfovibrio desulfuricans* sp. Essex), all with the addition of an appropriate electron donor. The generation of Fe(II) and the concomitant removal of Tc occurred only in the presence of *Shewanella* and *Desulfovibrio* spp., which suggested that Tc removal is linked to Fe(II) ingrowth in these sediments (Burke *et al.* 2005). The 16S rRNA gene analysis confirmed the presence of organisms related to known nitrate- (*Rhodobacter capsulatus*), sulphur/metal- (*Pelobacter* sp.) and sulphate-reducing bacteria (*Desulfovibrio senezii*) in the sediments. Additionally, *Geobacteraceae*-specific primers were used to detect Fe(III)-reducing bacteria from this phylogenetic group. Thus, there was a complex range of Fe(III)- and sulphate-reducing bacteria present in the sediments that could have been responsible for production of Fe(II), or potentially sulphide, and the subsequent indirect reduction of Tc(VII) mediated by these reduction products. X-ray absorption spectroscopic analysis from progressive anoxia samples spiked with $1000\,\mu M\,TcO_4^-$ confirmed that TcO_4^- removal was due to reduction to hydrous $Tc(IV)O_2$ in Fe(III)- and sulphate-reducing estuarine sediments (Burke *et al.* 2005).

In addition to working with estuarine sediments that initially contained no background radioactivity, we have also worked extensively with sediments from (or representative of) several 'nuclear' sites. For example, microcosm experiments containing soil samples representative of the UKAEA site at Dounreay have been performed with unamended sediments, carbonate buffered sediments and microcosms amended with EDTA, a complexing ligand used in nuclear fuel cycle operations (Begg *et al.* 2007). During the development of anoxia mediated by indigenous microbial populations, $Tc(VII)O_4^-$ was again removed from solution, during periods of microbial Fe(III) reduction when Fe(II) was growing into the microcosms. A pivotal role for biogenic Fe(II) in Tc(VII) reduction and precipitation (Fig. 11.2) was confirmed in microcosms which had been prereduced to the point that Fe(III) reduction dominated, and then sterilised by autoclaving. In these sterile Fe(III)-reducing sediments, the Tc-spike was removed from solution to below the liquid scintillation counting detection limit (>98%) over 21 days.

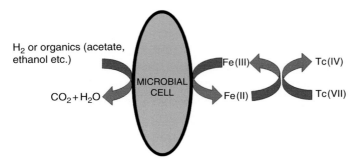

Figure 11.2. Indirect reduction of Tc(VII) mediated by Fe(II).

Similar trends were also noted in sediments collected from the Drigg low-level radioactive waste storage site (Wilkins *et al.* 2007), confirming a generic indirect mechanism for Tc(VII) reduction that is not strongly site specific. Interestingly, in both studies, the reduced insoluble Tc was surprisingly resistant to remobilisation by strong oxidising agents such as nitrate, consistent with data from other parallel studies (Burke *et al.* 2006).

Finally, the interplay between nitrate and the reduction of Tc(VII) was explored in more detail in microcosms prepared from sediments from the US Department of Energy Field Research Center (FRC) in Oak Ridge, Tennessee, USA (McBeth *et al.* 2007). Here the impact of 0, 10 and 100 mM added nitrate on the progression of a range of terminal electron accepting processes and ^{99}Tc immobilisation was assessed. In the nitrate unamended and 10 mM nitrate amended systems, bioreduction proceeded and extractable Fe(II) ingrowth and concomitant Tc(VII) removal was observed. Interestingly, the relatively low (10 mM) addition of nitrate seemed to augment the development of bioreducing conditions. In contrast, in the 100 mM nitrate amended system, Fe(II) ingrowth was limited and no Tc(VII) removal occurred, suggesting strong inhibition of microbial metal reduction at high, but nuclear site relevant concentrations of NO_3^-.

Bacterial community changes induced by uranyl or sodium nitrate treatments and the fate of the added U(VI)

The fate and transport of uranium are governed by the contrasting chemistry of U(IV) and U(VI). U(VI) generally forms soluble, and thus mobile, complexes with carbonate and hydroxide, while U(IV) precipitates as the highly insoluble mineral uraninite. Many studies have focused on *in situ* bioremediative stimulation of native U(VI)-reducing bacteria by the addition of different organic electron donors for aqueous U(VI) reduction such as acetate, lactate, glucose and ethanol to uranium contaminated waters and sediments (Holmes *et al.* 2002; Anderson *et al.* 2003; Suzuki *et al.* 2003; North *et al.* 2004; Brodie *et al.* 2006; Nyman *et al.* 2006).

Although not the focus of such intensive recent research, the fate of uranium in complex natural systems without the addition of organic substances is also of great environmental importance, in order to predict the potential risks of uranium migration within piles, tailings and depository sites and to prevent their spread via groundwater flow. Thus, different microcosm experiments were performed in order to investigate the influence of uranyl or sodium nitrate on the natural bacterial community of a uranium mining waste site in Germany under acidic and oligotrophic conditions (Geissler & Selenska-Pobell 2005; Geissler 2007; Selenska-Pobell *et al.* 2008).

One of the most important observations obtained from the analyses of the uranyl and sodium nitrate treated subsamples was the extremely high diversity of the indigenous bacterial community and the strong changes in community structure noted by increasing the uranyl or sodium nitrate concentrations, as well as a strict dependence on aeration conditions (Geissler & Selenska-Pobell 2005; Geissler 2007; Selenska-Pobell *et al.* 2008). After longer incubations, even with higher U(VI) concentrations, uranium sensitive populations were established under aerobic as well as under anaerobic conditions. This indicated that U(VI) was no longer bioavailable in these long-term experiments (Geissler & Selenska-Pobell 2005; Geissler 2007; Selenska-Pobell *et al.* 2008). Surprisingly, no U(VI) reduction was observed under anaerobic oligotrophic conditions even after longer incubations, when all the available nitrate was depleted (Geissler 2007; Selenska-Pobell *et al.* 2008). Time-Resolved Laser-induced Fluorescence Spectroscopic (TRLFS) analysis demonstrated that, in the uranyl nitrate treated sample incubated for 14 weeks under reducing conditions, most of the added U(VI) was bound by phosphate phases of biotic origin. U(VI) added to this sample was bound in mixed organic and inorganic phosphate compounds, suggesting that, at this site, U(VI)-phosphate chemistry may have a major role in controlling uranium fate.

Recent publications have also shown that representatives of *Rahnella* spp. recovered from uranium contaminated samples are able to immobilise U(VI) via the secretion of orthophosphate (Martinez *et al.* 2007) leading to the precipitation of meta-autunite-like mineral phases under laboratory conditions (Beazley *et al.* 2007). Moreover, it was demonstrated that various bacteria, phylogenetically unrelated to *Rahnella* spp., are able to protect themselves against potentially toxic U(VI) at acidic pH by secretion of inorganic phosphate groups which are involved in precipitation of uranyl hydrophosphate-like phases (Macaskie *et al.* 1992) or of meta-autunite-like compounds (Merroun *et al.* 2006; Jroundi *et al.* 2007; Martinez *et al.* 2007) when studied in laboratory conditions.

The results of the TRLFS spectra of the uranyl nitrate treated soil samples differed significantly from those of the meta-autunite-like phases produced by different bacterial strains in the presence of U(VI) observed in fully defined and relatively simple microbiological media under laboratory conditions

(Merroun *et al.* 2003; Beazley *et al.* 2007). The organic uranyl phosphate complexes detected in the field-based study are the product of interactions with phosphorylated biopolymers supplied by both dead or live bacteria which, along with the orthophosphate released by components of microbial populations such as *Rahnella* spp., can contribute to U(VI) immobilisation.

Mössbauer spectroscopic analyses revealed increased Fe(III) reduction during incubation from 4 to 14 weeks in the uranyl nitrate treated samples incubated under anaerobic conditions (Geissler 2007; Selenska-Pobell *et al.* 2008). It is suggested that after the reduction of the added nitrate, microbially mediated Fe(III) reduction took place even without bioaugmentation via addition of external carbon sources. However, no previously characterised dissimilatory Fe(III) reducers, for example, *Geobacter, Shewanella* or *Geothrix* spp., were identified in the uranyl nitrate treated microcosms. It is possible, therefore, that the *Rahnella* spp. or other components of the microbial community detected in the microcosms were able to respire Fe(III). These could include the betaproteobacterial populations detected, by analogy to other Fe(III)-reducing betaproteobacterial species such as *Ferribacterium limneticum* (Cummings *et al.* 1999) or *Rhodoferax ferrireducens* (Finneran *et al.* 2003).

Multiple influences of nitrate and nitrate-reducing bacteria on radionuclide solubility during bioremediation

As noted already, nitrate is a common co-contaminant at sites where nuclear materials have been processed or stored. In these settings, nitrate and nitrate-reducing bacteria have the potential to influence strongly the environmental behaviour of not only uranium but also of other radionuclides. Here we review some important aspects of these influences on U and Tc solubility, highlighted in recent publications.

Due to anthropogenic contamination, e.g., by nitric acid, subsurface nitrate concentrations can often be very high (exceeding $100\,mM\,NO_3^-$; McBeth *et al.* 2007), resulting in a dramatic inhibition of metal reduction (Finneran *et al.* 2002; Senko *et al.* 2002; Istok *et al.* 2004). Typical *in situ* strategies involving the stimulation of metal-reducing bacteria are generally hindered by two factors: the low pH of the environment and the requirement that the nitrate must first be removed via denitrification prior to uranium being utilised as an electron acceptor (Madden *et al.* 2007). Istok and co-workers (2004) have also demonstrated that the addition of electron donors such as ethanol, glucose or acetate can stimulate nitrate reduction, which was apparently followed by a simultaneous bioreduction of U(VI) and Tc(VII) (Istok *et al.* 2004). In contrast, Shelobolina and co-workers (2003) reported that the biological reduction of nitrate, stimulated by acetate amendment, can cause a rise in pH and reduced solubility of UO_2^{2+}. Here uranium removal was via hydrolysis and precipitation rather than through bioreduction (Shelobolina *et al.* 2003). The impact of nitrate on

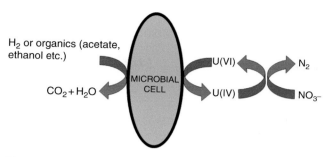

H₂ or organics (acetate, ethanol etc.)

$CO_2 + H_2O$

MICROBIAL CELL

U(VI)

U(IV)

N₂

NO_3^-

Figure 11.3. Reduction of U(VI) and reoxidation of U(IV) by nitrate.

uranium removal is clearly very complex, and probably site/sediment specific. Indeed, a recent study has even demonstrated that the stimulation of a natural microbial community to immobilise U through bioreduction is possible without the removal of nitrate (Madden *et al.* 2007). Incubations with uranium-contaminated sediments demonstrated nearly complete reduction of U(VI) with very little loss of nitrate from pH 5.7 to 6.2 using methanol or glycerol as a carbon source. The majority of the micro-organisms stimulated by these enrichment conditions consisted of low G + C gram-positive bacteria most closely related to *Clostridium* and *Clostridium*-like organisms (Madden *et al.* 2007).

Finally, it is also obvious from several recent papers that nitrate can not only inhibit U(VI) reduction (as noted in a majority of studies), but it can also have a dramatic impact on pre-reduced sediments containing insoluble U(IV). When added to such systems, nitrate (and oxygen also) can reoxidise the U(IV) (Fig. 11.3), remobilising the uranium as U(VI) (Moon *et al.* 2007). In the case of technetium, reoxidation behaviour is more complex, with remobilisation dependent on the nature of the oxidant (Burke *et al.* 2006; Morris *et al.* 2008).

The abundance of nitrate in the subsurface environments discussed provides a selective pressure that favours denitrifiers among the soil micro-organisms that can tolerate an acidic, nutrient-starved environment. In agreement with the high nitrate concentrations observed in sediments from the Oak Ridge Field Research Center (ORFRC), 16S rRNA gene sequences related to those of nitrate-reducing bacteria, such as members of the Proteobacteria (including the genera *Sphingomonas, Acidovorax, Acinetobacter, Alcaligenes* and *Ralstonia*), showed a high relative abundance in the total and metabolically active fractions of the microbial community (Akob *et al.* 2007). In a parallel study, the microbial community structure from ethanol-biostimulated sediments of a high-nitrate (>130 mM), low-pH, uranium-contaminated site at the ORFRC, taken at a time point when denitrification was most likely, showed an abundance of betaproteobacterial clones in biostimulated sediments (Spain *et al.* 2007). Multiple lines of evidence suggest the dominance of *Castellaniella* species in these biostimulated sediments and their role in nitrate removal *in situ* (Spain *et al.* 2007).

In a recent study from our laboratories, *Bacillus*-like and *Herbaspirillum*-like bacteria were abundant during denitrification in acetate biostimulated microcosms prepared from sediments representative of the Sellafield nuclear site in North West England and supplemented with 10 mM nitrate and a 5 µM technetium spike (Law *et al.* 2008). Their abundance was demonstrated by 16S rRNA and functional *narG* (alpha subunit of the membrane bound nitrate reductase) gene analyses of samples taken after 30-day incubation when nitrate depletion was most pronounced but before any metal reduction or Tc(VII) had been observed (Law *et al.* 2008). Similar *narG* gene sequences of *Bacillus* spp. were abundant in sodium nitrate treated samples from uranium mining waste piles in Haberland that had been incubated for 4 weeks under anaerobic conditions (Geissler 2007). The dominance of *Bacillus* spp. in this sample was also demonstrated by 16S rRNA gene analysis (Selenska-Pobell *et al.* 2008). This suggests that in radionuclide contaminated sediments *Bacillus* spp. are also involved in the reduction of nitrate, in agreement with the detection of distinct *Bacillus* spp. in a broad range of heavy-metal and radionuclide contaminated samples (Selenska-Pobell *et al.* 1999; Martinez *et al.* 2006). This is of interest, given their ability to sorb large amounts of uranium and other heavy metals (Selenska-Pobell *et al.* 1999).

Conclusions

It is important that we improve our understanding of the mechanisms underpinning the biogeochemistry and mobility of radionuclides in the environment, to underpin safety case assessments and bioremediation efforts. The studies presented in this chapter are examples where microbiological and geochemical analyses have been combined in the laboratory to obtain a better understanding of the biogeochemical cycle of priority radionuclides. These approaches, applied to field-scale investigations, and operating at environmentally relevant concentrations of radionuclides, are potentially challenging but crucial to drive this area forward and realise the potential of *in situ* bioremediation at nuclear facilities in Europe.

There are also specific questions raised by the studies described in this review. For example, the impact of competing 'direct' enzymatic and 'indirect' (Fe(II)/sulphide-mediated) reductive processes needs clarifying for some radionuclides, most notably Tc(VII). In the case of uranium, more research is clearly needed to understand both the precise mechanisms of U(VI) reduction (and long-term stability of U(IV) phases under field conditions) and also the potential roles that bacteria can play in the *in situ* precipitation of other insoluble phases such as U(VI) phosphate. The tools of molecular ecology can play a role in these investigations, but it is necessary to expand our studies not only to look at the microbial community by targeting long-lived DNA, but also to target more transient mRNA to track the metabolism of the 'active' components of the bacterial community using functional gene probes. The availability of genome sequences and genetic systems for additional micro-organisms (other than *Geobacter, Shewanella and Desulfovibrio* spp.) will also help better understand their role in the biogeochemical cycling of radionuclides.

Acknowledgements

The authors acknowledge the financial support of NERC (Grants NE/D00473X/1 and NE/D005361/1) and access to beam time at Daresbury SRS via STFC funding.

References

Akob, D. M., Mills, H. J. and Kostka, J. E. (2007) Metabolically active microbial communities in uranium-contaminated subsurface sediments. *FEMS Microbiology Ecology* **59**, 95–107.

Anderson, R. T., Vrionis, H. A., Ortiz-Bernad, I. et al. (2003) Stimulating the *in situ* activity of *Geobacter* species to remove uranium from the groundwater of a uranium-contaminated aquifer. *Applied and Environmental Microbiology* **69**, 5884–5891.

Beazley, M. J., Martinez, R. J., Sobecky, P. A., Webb, S. M. and Taillefert, M. (2007) Uranium biomineralization as a result of bacterial phosphatase activity: insights from bacterial isolates from a contaminated subsurface. *Environmental Science and Technology* **41**, 5701–5707.

Begg, J. D., Burke, I. T. and Morris, K. (2007) The behaviour of technetium during microbial reduction in amended soils from Dounreay, UK. *The Science of the Total Environment* **373**, 297–304.

Boukhalfa, H., Icopini, G. A., Reilly, S. D. and Neu, M. P. (2007) Plutonium(IV) reduction by the metal-reducing bacteria *Geobacter metallireducens* GS15 and *Shewanella oneidensis* MR1. *Applied and Environmental Microbiology* **73**, 5897–5903.

Brodie, E. L., Desantis, T. Z., Joyner, D. C. et al. (2006) Application of a high-density oligonucleotide microarray approach to study bacterial population dynamics during uranium reduction and reoxidation. *Applied and Environmental Microbiology* **72**, 6288–6298.

Burke, I. T., Boothman, C., Lloyd, J. R. et al. (2006) Reoxidation behavior of technetium, iron, and sulfur in estuarine sediments. *Environmental Science and Technology* **40**, 3529–3535.

Burke, I. T., Boothman, C., Lloyd, J. R., Mortimer, R. J., Livens, F. R. and Morris, K. (2005) Effects of progressive anoxia on the solubility of technetium in sediments. *Environmental Science and Technology* **39**, 4109–4116.

Cummings, D. E., Caccavo, F., Spring, S. and Rosenzweig, R. F. (1999) *Ferribacterium limneticum*, gen. nov., sp. nov., an Fe(III)-reducing micro-organism isolated from mining-impacted freshwater lake sediments. *Archives of Microbiology* **171**, 183–188.

Finneran, K. T., Housewright, M. E. and Lovley, D. R. (2002) Multiple influences of nitrate on uranium solubility during bioremediation of uranium-contaminated subsurface sediments. *Environmental Microbiology* **4**, 510–516.

Finneran, K. T., Johnsen, C. V. and Lovley, D. R. (2003) *Rhodoferax ferrireducens* sp. nov., a psychrotolerant, facultatively anaerobic bacterium that oxidizes acetate with the reduction of Fe(III). *International Journal of Systematic and Evolutionary Microbiology* **53**, 669–673.

Fomina, M., Charnock, J. M., Hillier, S., Alavarez, R. and Gadd, G. M. (2007) Fungal transformations of uranium oxides. *Environmental Microbiology* **9**, 1696–1710.

Francis, A. J., Dodge, C. J., Lu, F. L., Halada, G. P. and Clayton, C. R. (1994) XPS and XANES studies of uranium reduction by

Clostridium sp. *Environmental Science and Technology* **28**, 636–639.

Francis, A. J., Gillow, J. B., Dodge, C. J., Harris, R., Beveridge, T. J. and Papenguth, H. W. (2004) Uranium association with halophilic and non-halophilic bacteria and archaea. *Radiochimica Acta* **92**, 481–488.

Fredrickson, J. K., Zachara, J. M., Balkwill, D. L. *et al.* (2004) Geomicrobiology of high-level nuclear waste-contaminated vadose sediments at the Hanford Site, Washington State. *Applied and Environmental Microbiology* **70**, 4230–4241.

Geissler, A. (2007) *Prokaryotic Micro-organisms in Uranium Mining Waste Piles and their Interactions with Uranium and Other Heavy Metals*. TU Bergakademi Freiberg. Freiberg, Germany.

Geissler, A. and Selenska-Pobell, S. (2005) Addition of U(VI) to a uranium mining waste sample and resulting changes in the indigenous bacterial community. *Geobiology* **3**, 275–285.

Holmes, D. E., Finneran, K. T., O'Neil, R. A. and Lovley, D. R. (2002) Enrichment of members of the family *Geobacteraceae* associated with stimulation of dissimilatory metal reduction in uranium-contaminated aquifer sediments. *Applied and Environmental Microbiology* **68**, 2300–2306.

Icopini, G. A., Boukhalfa, H. and Neu, M. P. (2007) Biological reduction of Np(V) and Np(V) citrate by metal-reducing bacteria. *Environmental Science and Technology* **41**, 2764–2769.

Istok, J. D., Senko, J. M., Krumholz, L. R. *et al.* (2004) In situ bioreduction of technetium and uranium in a nitrate-contaminated aquifer. *Environmental Science and Technology* **38**, 468–475.

John, S. G., Ruggiero, C. E., Hersman, L. E., Tung, C. S. and Neu, M. P. (2001) Siderophore mediated plutonium accumulation by *Microbacterium flavescens* (JG-9). *Environmental Science and Technology* **35**, 2942–2948.

Jroundi, F., Merroun, M. L., Arias, J. M., Rossberg, A., Selenska-Pobell, S. and

Gonzalez-Munoz, M. T. (2007) Spectroscopic and microscopic characterization of uranium biomineralization in *Myxococcus xanthus*. *Geomicrobiology Journal* **24**, 441–449.

Keith-Roach, M. J. and Livens, F. R. (eds.) (2002) *Interactions of Micro-organisms with Radionuclides*. Elsevier, London.

Law, G. T. W., Geissler, A., Lloyd, J. R. *et al.* (2008) Manuscript in preparation.

Lloyd, J. R. (2003) Microbial reduction of metals and radionuclides. *FEMS Microbiology Reviews* **27**, 411–425.

Lloyd, J. R. and Lovley, D. R. (2001) Microbial detoxification of metals and radionuclides. *Current Opinion in Biotechnology* **12**, 248–253.

Lloyd, J. R. and Macaskie, L. E. (1996) A novel PhosphorImager-based technique for monitoring the microbial reduction of technetium. *Applied and Environmental Microbiology* **62**, 578–582.

Lloyd, J. R. and Macaskie, L. E. (2000) Bioremediation of radioactive metals. In: *Environmental Microbe-Metal Interactions* (ed. D. R. Lovley). ASM Press, Washington, DC.

Lloyd, J. R. and Renshaw, J. C. (2005a) Bioremediation of radioactive waste: radionuclide-microbe interactions in laboratory and field-scale studies. *Current Opinion in Biotechnology* **16**, 254–260.

Lloyd, J. R. and Renshaw, J. C. (2005b) Microbial transformations of radionuclides: fundamental mechanisms and biogeochemical implications. *Metal Ions in Biological Systems* **44**, 205–240.

Lloyd, J. R., Chesnes, J., Glasauer, S., Bunker, D. J., Livens, F. R. and Lovley, D. R. (2002) Reduction of actinides and fission products by Fe(III)-reducing bacteria. *Geomicrobiology Journal* **19**, 103–120.

Lloyd, J. R., Cole, J. A. and Macaskie, L. E. (1997) Reduction and removal of heptavalent Tc from solution by *Escherichia coli*. *Journal of Bacteriology* **179**, 2014–2021.

Lloyd, J. R., Sole, V. A., Van Praagh, C. V. G. and Lovley, D. R. (2000a) Direct and Fe(II)-mediated reduction of technetium by

Fe(III)-reducing bacteria. *Applied and Environmental Microbiology* **66**, 3743–3749.

Lloyd, J. R., Yong, P. and Macaskie, L. E. (2000b) Biological reduction and removal of Np(V) by two micro-organisms. *Environmental Science and Technology* **34**, 1297–1301.

Lovley, D. R. (1993) Dissimilatory metal reduction. *Annual Review of Microbiology* **47**, 263–290.

Lovley, D. R. and Phillips, E. J. P. (1992) Reduction of uranium by *Desulfovibrio desulfuricans*. *Applied and Environmental Microbiology* **58**, 850–856.

Lovley, D. R., Phillips, E. J. P., Gorby, Y. A. and Landa, E. R. (1991) Microbial reduction of uranium. *Nature* **350**, 413–416.

Lovley, D. R., Widman, P. K., Woodward, J. C. and Phillips, E. J. P. (1993) Reduction of uranium by cytochrome c3 of *Desulfovibrio vulgaris*. *Applied and Environmental Microbiology* **59**, 3572–3576.

Macaskie, L. E., Empson, R. M., Cheetham, A. K., Grey, C. P. and Skarnulis, A. J. (1992) Uranium bioaccumulation by a *Citrobacter* sp. as a result of enzymatically mediated growth of polycrystalline HUO2PO4. *Science* **257**, 782–784.

Macaskie, L. E., Jeong, B. C. and Tolley, M. R. (1994) Enzymically accelerated biomineralization of heavy metals – application to the removal of americium and plutonium from aqueous flows. *FEMS Microbiology Reviews* **14**, 351–367.

Madden, A. S., Smith, A. C., Balkwill, D. L., Fagan, L. A. and Phelps, T. J. (2007) Microbial uranium immobilization independent of nitrate reduction. *Environmental Microbiology* **9**, 2321–2330.

Marshall, M. J., Plymale, A. E., Kennedy, D. W. *et al.* (2008) Hydrogenase- and outer membrane c-type cytochrome-facilitated reduction of technetium(VII) by *Shewanella oneidensis* MR-1. *Environmental Microbiology* **10**, 125–136.

Martinez, R. J., Beazley, M. J., Taillefert, M., Arakaki, A. K., Skolnick, J. and Sobecky, P. A. (2007) Aerobic uranium (VI) bioprecipitation by metal-resistant bacteria isolated from radionuclide- and metal-contaminated subsurface soils. *Environmental Microbiology* **9**, 3122–3133.

Martinez, R. J., Wang, Y. L., Raimondo, M. A., Coombs, J. M., Barkay, T. and Sobecky, P. A. (2006) Horizontal gene transfer of P-IB-type ATPases among bacteria isolated from radionuclide- and metal-contaminated subsurface soils. *Applied and Environmental Microbiology* **72**, 3111–3118.

McBeth, J. M., Lear, G., Lloyd, J. R., Livens, F. R., Morris, K. and Burke, I. T. (2007) Technetium reduction and reoxidation in aquifer sediments. *Geomicrobiology Journal* **24**, 189–197.

Merroun, M., Nedelkova, M., Rossberg, A., Hennig, C. and Selenska-Pobell, S. (2006) Interaction mechanisms of bacterial strains isolated from extreme habitats with uranium. *Radiochimica Acta* **94**, 723–729.

Merroun, M. L., Geipel, G., Nicolai, R., Heise, K. H. and Selenska-Pobell, S. (2003) Complexation of uranium (VI) by three eco-types of *Acidithiobacillus ferrooxidans* studied using time-resolved laser-induced fluorescence spectroscopy and infrared spectroscopy. *Biometals* **16**, 331–339.

Merroun, M. L., Raff, J., Rossberg, A., Hennig, C., Reich, T. and Selenska-Pobell, S. (2005) Complexation of uranium by cells and S-layer sheets of *Bacillus sphaericus* JG-A12. *Applied and Environmental Microbiology* **71**, 5532–5543.

Moon, H. S., Komlos, J. and Jaffe, P. R. (2007) Uranium reoxidation in previously bioreduced sediment by dissolved oxygen and nitrate. *Environmental Science and Technology* **41**, 4587–4592.

Morris, K., Livens, F. R., Charnock, J. M. *et al.* (2008) An X-ray absorption study of the fate of technetium in reduced and reoxidised sediments and mineral phases. *Applied Geochemistry* **23**, 603–617.

North, N. N., Dollhopf, S. L., Petrie, L., Istok, J. D., Balkwill, D. L. and Kostka, J. E. (2004) Change in bacterial community structure during *in situ* biostimulation of subsurface sediment cocontaminated with uranium and nitrate. *Applied and Environmental Microbiology* **70**, 4911–4920.

Nyman, J. L., Marsh, T. L., Ginder-Vogel, M. A., Gentile, M., Fendorf, S. and Criddle, C. (2006) Heterogeneous response to biostimulation for U(VI) reduction in replicated sediment microcosms. *Biodegradation* **17**, 303–316.

Ohnuki, T., Yoshida, T., Ozaki, T. *et al.* (2005) Interactions of uranium with bacteria and kaolinite clay. *Chemical Geology* **220**, 237–243.

Ohnuki, T., Yoshida, T., Ozaki, T. *et al.* (2007) Chemical speciation and association of plutonium with bacteria, kaolinite clay, and their mixture. *Environmental Science and Technology* **41**, 3134–3139.

Pedersen, K. (2005) Micro-organisms and their influence on radionuclide migration in igneous rock environments. *Journal of Nuclear and Radiochemical Sciences* **6**, 11–15.

Renshaw, J. C., Butchins, L. J., Livens, F. R., May, I., Charnock, J. M. and Lloyd, J. R. (2005) Bioreduction of uranium: environmental implications of a pentavalent intermediate. *Environmental Science and Technology* **39**, 5657–5660.

Renshaw, J. C., Law, N., Geissler, A., Livens, F. R. and Lloyd, J. R. (2008) Impact of the Fe(III)-reducing bacteria *Geobacter sulfurreducens* and *Shewanella oneidensis* on the speciation of plutonium. Short communication. *Biogeochemistry* submitted.

Renshaw, J. C., Lloyd, J. R. and Livens, F. R. (2007) Microbial interactions with actinides and long-lived fission products. *Comptes Rendus Chimie* **10**, 1067–1077.

Ruggiero, C. E., Boukhalfa, H., Forsythe, J. H., Lack, J. G., Hersman, L. E. and Neu, M. P. (2005) Actinide and metal toxicity to prospective bioremediation bacteria. *Environmental Microbiology* **7**, 88–97.

Selenska-Pobell, S. (2002) Diversity and activity of bacteria in uranium waste piles. In: *Interactions of Micro-organisms with*

Radionuclides (eds. M. J. Keith-Roach and F. R. Livens). 225: Elsevier Sciences Ltd, Oxford, UK.

Selenska-Pobell, S., Geissler, A., Merroun, M., Flemming, K., Geipel, G. and Reuther, H. (2008) Biogeochemical changes induced by uranyl nitrate in a uranium mining waste pile. In: *Uranium, Mining and Hydrogeology* (eds. B. J. Merkel and A. Hasche-Berger), pp. 743–752. Springer Verlag, New York.

Selenska-Pobell, S., Panak, P., Miteva, V., Boudakov, I., Bernhard, G. and Nitsche, H. (1999) Selective accumulation of heavy metals by three indigenous *Bacillus* strains, *B. cereus*, *B. megaterium* and *B. sphaericus*, from drain waters of a uranium waste pile. *FEMS Microbiology Ecology* **29**, 59–67.

Senko, J. M., Istok, J. D., Suflita, J. M. and Krumholz, L. R. (2002) In-situ evidence for uranium immobilization and remobilization. *Environmental Science and Technology* **36**, 1491–1496.

Shelobolina, E. S., O'Neill, K., Finneran, K. T., Hayes, L. A. and Lovley, D. R. (2003) Potential for *in situ* bioremediation of a low-pH, high-nitrate uranium-contaminated groundwater. *Soil and Sediment Contamination* **12**, 865–884.

Shelobolina, E. S., Sullivan, S. A., O'Neill, K. R., Nevin, K. P. and Lovley, D. R. (2004) Isolation, characterization, and U(VI)-reducing potential of a facultatively anaerobic, acid-resistant bacterium from low-pH, nitrate- and U(VI)-contaminated subsurface sediment and description of *Salmonella subterranea* sp. nov. *Applied and Environmental Microbiology* **70**, 2959–2965.

Spain, A. M., Peacock, A. D., Istok, J. D. *et al.* (2007) Identification and isolation of a *Castellaniella* species important during biostimulation of an acidic nitrate- and uranium-contaminated aquifer. *Applied and Environmental Microbiology* **73**, 4892–4904.

Suzuki, Y. and Banfield, J. (1999) Geomicrobiology of uranium. *Reviews in Mineralogy and Geochemistry* **38**, 393–432.

Suzuki, Y. and Banfield, J. F. (2004) Resistance to, and accumulation of, uranium by bacteria from a uranium-contaminated site. *Geomicrobiology Journal* **21**, 113–121.

Suzuki, Y., Kelly, S. D., Kemner, K. A. and Banfield, J. F. (2003) Microbial populations stimulated for hexavalent uranium reduction in uranium mine sediment. *Applied and Environmental Microbiology* **69**, 1337–1346.

Suzuki, Y., Kelly, S. D., Kemner, K. M. and Banfield, J. F. (2004) Enzymatic U(VI) reduction by *Desulfosporosinus* species. *Radiochimica Acta* **92**, 11–16.

Wilkins, M. J., Livens, F. R., Vaughan, D. J., Beadle, I. and Lloyd, J. R. (2007) The influence of microbial redox cycling on radionuclide mobility in the subsurface at a low-level radioactive waste storage site. *Geobiology* **5**, 293–301.

Wu, Q., Sanford, R. A. and Loffler, F. E. (2006) Uranium(VI) reduction by *Anaeromyxobacter dehalogenans* strain 2CP-C. *Applied and Environmental Microbiology* **72**, 3608–3614.

CHAPTER TWELVE

The microbial ecology of remediating industrially contaminated land: sorting out the bugs in the system

KEN KILLHAM

Introduction

Understanding and manipulating the microbial ecology of contaminated land are increasingly critical steps towards meeting the challenge of remediating the considerable legacy of industrial pollution in the UK and worldwide.

Understanding the microbial ecology of contaminated land is key to its remediation in two ways. First, the microbial community in the soil, occasionally enhanced through inoculation, can be exploited to bioremediate (through either *in situ* or *ex situ* approaches) contaminated sites (Alexander 1999; Atlas & Philp 2005). This is an increasingly attractive, environmentally sustainable option as excavation and landfill of contaminated site waste become both increasingly expensive (both through landfill tax increases, ever-increasing transport costs and the introduction of the aggregrate levy on material brought in as fill), fewer landfill sites are available (in Scotland, for example, there are no hazardous landfills and material must be transported great distances, further adding to disposal costs) and environmental regulators rightly press for more sustainable approaches to site clean-up. Second, the indigenous microbial communities of sites are often impacted by the contamination, particularly where toxic contaminants such as free phase solvents and available heavy metals are present, and part of the remediation challenge is to restore soil/aquifer biological function. Because of the wide range of microbial functions carried out in the soil, in particular, this restoration may be linked, for example, to carbon and nutrient (N, P and S) cycling or to the wide range of plant–microbe interactions on which ecosystem health depends.

In this chapter, two aspects of the microbial ecology of remediating industrially contaminated soils are considered:

Key soil bacteria, such as those involved in the biodegradation of organic contaminants and in key soil functions, have been selected for construction

Ecology of Industrial Pollution, eds. Lesley C. Batty and Kevin B. Hallberg. Published by Cambridge University Press. © British Ecological Society 2010.

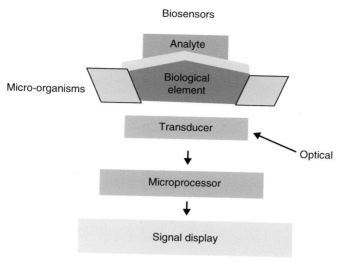

Figure 12.1. Microbial biosensors represent the biological element that, when exposed to an analyte (a contaminated soil sample or soil extract), provide a diagnostic signal via an appropriate transducer (in the case of light emitting biosensors, this transducer is optical). In the case of biosensors constructed for diagnosis of contamination, the diagnostic signal may reflect the general toxicity of site contamination or the bioavailable fraction of a specific contaminant or group of contaminants.

of biosensors (Fig. 12.1) to assess the extent and impact of contamination, to support the bioremediation process and to establish restoration of soil health.

The support of bioremediation using biosensors involves both determining the potential for bioremediation and identifying possible constraints to bioremediation so that they can be alleviated through engineered interventions, as part of bioremediative management support. In particular, *lux*-marked bacterial biosensors, expressing bioluminescence reporter genes either under the control of a strong, general, constitutive cell promoter (in the case of toxicity biosensors) or under the control of a contaminant catabolic or resistance promoter (in the case of contaminant specific biosensors), have proved extremely useful in this regard. This is because their inherent flexibility, in terms of host selection and construction, offers strong environmental relevance and both the ability to screen for overall toxicity (due to contaminants or some other soil factor) or for the availability of a specific site contaminant.

A key series of degradation assessment methods for supporting microbially driven remediation (bioremediation) of target contaminants includes determination of the presence of key catabolic genes (Fig. 12.2) and mineralisation of ^{14}C labelled contaminants.

Although the identification of key catabolic genes for contaminant degradation provides useful information, it does not necessarily mean that the contaminants

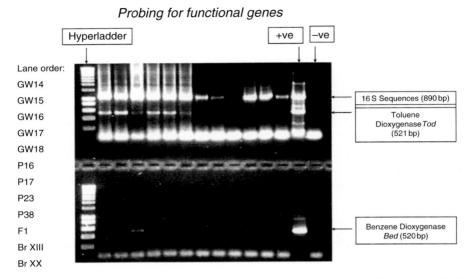

Figure 12.2. Molecular probes for functional genes such as toluene dioxygenase (*Tod*) and benzene dioxygenase (*Bed*) which are critical to the bioremediation of BTEX contaminated sites (mainly by aerobic bacteria) can now be included as part of a routine screen to assess the potential of an environment sample/site to degrade target contaminants (Brockman 1995). For contaminated soils, DNA and RNA can be readily extracted and probed for the above catabolic genes. Aquifer material is more difficult to obtain and groundwater contains very low population densities of microbial degraders. A useful approach is to lower inert, high surface area 'tokens' into boreholes so that they become colonised by representative degraders. DNA/RNA can then be extracted readily from the tokens and potential contaminant degrader activity of aquifers assessed.

will be degraded by the microbes possessing these genes at the site. More direct evidence for assessing this likelihood comes from the mineralisation of ^{14}C labelled contaminants. The latter are usually added as spikes to soil, sediment or groundwater. The resulting ^{14}CO$_2$ is trapped and the ^{14}C counted to determine the fraction of the contaminant pool degraded (Reid *et al.* 2001).

Assessment/remediation of two contaminated site case studies is presented to exemplify applications of the above techniques to characterise microbial ecology and to demonstrate large/field-scale successes where exploiting an understanding of microbial ecology has enabled effective bioremediation to an environmentally relevant and acceptable endpoint. This heralds an encouraging future for more sustainable approaches to managing industrially contaminated land.

Case study 1

The first case study involves a substantial (8.5 ha) contaminated petrochemical plant in the Rhinelands of Germany (Fig. 12.3). The site had been a paint

Aerial Photo 1997

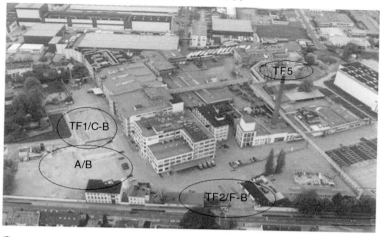

 Project Rheingold

KCL 18

Figure 12.3. Aerial photograph of the Rhinelands case study site, a large BTEX contaminated paint factory site in Germany, with co-contamination from chlorinated hydrocarbons and metals.

manufacturing plant for more than 100 years, was characterised by deep (30 m plus) Rhine sediments with groundwater flowing towards the Rhine and was found in the 1990s to be contaminated predominantly by monoaromatic hydrocarbons (benzene, ethyl-benzene, toluene and xylene – the 'BTEX'). These compounds were used in the manufacture of paints and were spilled into the soil and groundwater over decades of storage and use. Also present were some chlorinated hydrocarbons (e.g., trichloroethylene, TCE) and heavy metals such as lead, copper and zinc. The metals were used to pigment the paint and were primarily introduced as metal oxides. The chlorinated solvents were used as paint thinners and, like the non-chlorinated aromatics, were introduced to the environment over many years. The BTEX contaminants are partly soluble in groundwater and were transported across the site, while the remainder of the BTEX compounds float on the groundwater, being less dense than water. The latter are examples of Light, Non Aqueous Phase Liquids (LNAPLs). In contrast, the chlorinated solvents are denser than the water and so sink deeper into the system. They are examples of Dense, Non Aqueous Phase Liquids (DNAPLs).

The extent of the contamination in the deep sediments and the associated groundwater and the presence of high copy numbers of BTEX catabolic genes led to the conclusion that the best way to tackle the site was to exploit the indigenous microbial community in the sediment to attenuate the main organic contaminants, in a process referred to as *in situ* bioremediation.

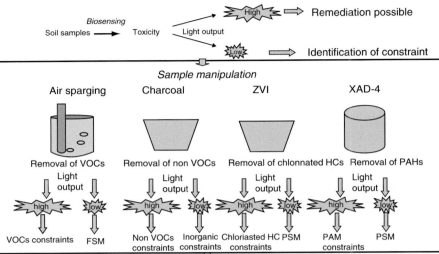

Figure 12.4. A sample manipulation protocol coupled to a biosensor assay to identify constraints to microbial degradation in the soils, sediments and groundwater of a BTEX contaminated site. An example of the kind of results obtained from this type of screening is provided in Fig. 12.5 (adapted from Sousa *et al.,* 1998).

To do this, and since the main risk contaminants were the BTEX which require aerobic conditions for their rapid degradation (Fritsche & Hofrichter 2005), it was first necessary to identify key constraints to the aerobic BTEX mineralising bacterial population and investigate the scope to alleviate these constraints. Biosensor technology was deployed to understand the bioavailability and toxicity of the contaminants to the indigenous microbial population and to identify key constraints to BTEX mineralisation resulting from other specific, bioavailability contaminants (Paton *et al.* 1995). Bacterial biosensors are now routinely used for screening of contaminated soil (Killham & Paton 2003) as a complementary tool to standard, chemical analysis. It is the capacity of microbial biosensors to address contaminant bioavailability that so effectively complements chemical analysis, as well as to screen (using toxicity biosensors) all toxic contaminants and to identify otherwise unknown problems that the highly specified, analytical approach may miss (Killham & Paton 2003). The use of biosensors to assess contaminant bioavailability involves integration of the many physical, chemical and biological factors that influence this availability.

   A scheme of sample manipulation coupled to a biosensor assay was developed (Fig. 12.4) so that the remediative potential of this part of the soil

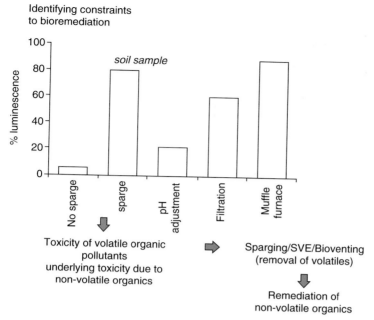

Figure 12.5. An example of the identification of a constraint to bioremediation using the sample manipulation/biosensor protocol illustrated in Fig. 12.1. In this case, most of the toxicity of a soil sample was removed by air sparging, suggesting that volatile organics were responsible for much of the toxicity.

or sediment bacterial community could be fully realised. This proved to be a highly successful approach (Sousa *et al.* 1998a, b), and a number of constraints were successfully identified and alleviated (Fig. 12.5), so that the bulk of the site could be bioremediated over a 3- to 5-year window. The efficacy of this approach was confirmed by ^{14}C mineralisation studies using radiolabelled BTEX as well as by identifying enhanced contaminant remediation at the site.

Scale-up of biosensor-based assessment of contaminated sites from the small sample scale (i.e., extraction of a few grams of soil) associated with most biosensor-based toxicity testing is currently achieved by using geostatistical tools such as block krieging which facilitate interpolation of spatial data assuming the sampling coverage/density of the area is adequate (Standing *et al.* 2007). Killham and Staddon (2002) highlight the strengths of geostatistics for quantifying soil health in this way. Figure 12.6 shows examples of this type of plot, for the study site, using the 'Surfer' software package (Golden Software Inc, Colorado, USA) in which the contours on the figure join points of equal soil toxicity. The use of a simple colour scale which grades from non-toxic through to highly toxic (in terms of biosensor response) is qualitative but enables such maps to be readily interpreted and form a valuable

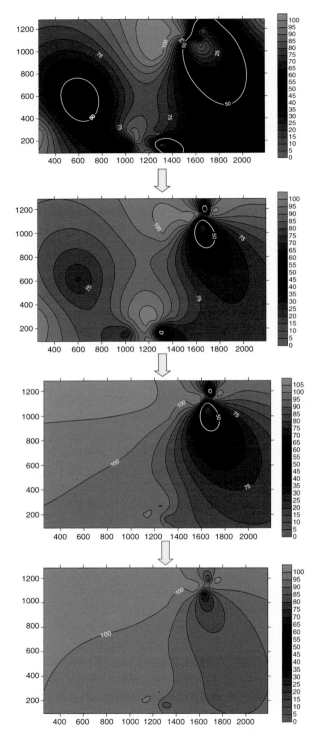

Figure 12.6. Tox-map (of the study site) constructed time course of bioremediation, leading from a highly toxic to a virtually non-toxic site. Each step was linked to use of

tool in screening sites for toxic pollutants and identifying and quantifying hotspots where there are likely to be serious toxicity constraints to intrinsic bioremediation of normally degradable, organic pollutants (Killham & Paton 2003).

Figure 12.6 therefore represents a time course of bioremediation at the site as various constraints were alleviated and the bioremediation proceeded progressively further. At the start of the time course, the tox-map highlights that there were three particularly toxic areas of the site (designated by increasingly bright red) which are delineated by a yellow boundary representing site material that is too toxic for degradation by micro-organisms (predominantly BTEX degrading bacteria) and hence for any intrinsic bioremediation to occur.

Removal of the co-contaminants on this site was particularly challenging. The chlorinated solvents generally require anaerobic conditions for the reductive dechlorination of these hydrocarbons (Bouwer 1994). It was therefore necessary to only air-sparge sufficiently (often in controlled pulses) to encourage aerobic BTEX degradation but not to damage or kill the anaerobic microbial community able to degrade the chlorinated solvents. The metals represent an even greater challenge as they cannot be removed by microbial bioremediation, but they can be converted into a non-available form, and it was this latter strategy that was adopted (through urea hydrolysis which raised the pH and precipitated out the metals). This at least temporarily removed the toxic constraint imposed on mineralisation of the target organic contaminants by the metals, and enabled the hydrocarbon contamination to be successfully bioremediated.

Case study 2

The second case study involves a large hydrocarbon contaminated site in Aberdeen. It was originally a railway siding and train fuelling depot and so most of the contamination was diesel from trains. The contamination was quite extensive, and had migrated to considerable depth in the alluvial, sandy

Caption for Figure 12.6. (*cont.*)
the manipulations coupled to microbial biosensor testing detailed in Fig. 12.1 to progressively identify constraints to bioremediation and alleviate them through site engineering. Figure 12.6 therefore represents a time course of bioremediation at the site as various constraints were alleviated and the bioremediation proceeded progressively further. At the start of the time course, the tox-map highlights that there were three particularly toxic areas of the site (designated by increasingly bright red) which are delineated by a yellow boundary representing site material that is too toxic for degradation by micro-organisms (predominantly BTEX degrading bacteria) and hence for any intrinsic bioremediation to occur. See colour plate section.

soils. The amount of soil excavated in exceedance of regulatory limits was 27 000 tonnes. Because of this considerable amount of soil and the presence of diesel hydrocarbons as the primary contaminant, it was decided to avoid unnecessary landfill and follow the more sustainable option of *ex situ* bioremediation using windrows, by particularly exploiting the activity of aerobic, microbial degraders. In the past, the primary strategy for such sites was to excavate and landfill the contaminated soil. Increasingly, however, bioremediation offers a cost-effective and sustainable alternative to the traditional 'dig and dump' approach (Alexander 1999; Atlas & Philp 2005).

Windrowing of hydrocarbon contaminated soil is, along with biopiling, an *ex situ* bioremediation technique which relies on the action of aerobic microorganisms (mainly bacteria) to break down organic compounds (Semple *et al.* 2001; Khan *et al.* 2004; Li *et al.* 2004). The methods involve the excavation and piling of contaminated soils into piles, usually to a height of 2–4 m, in order to enhance aerobic microbial activity through aeration (if this is forced, then biopile is the correct term; if aeration is achieved by turning with passive diffusion of oxygen, then windrow is the appropriate term), the addition of nutrients and the control of moisture and pH (Jørgensen *et al.* 2000; Khan *et al.* 2004). Windrows and biopiles have been effectively used to remediate a wide range of contaminants such as petroleum hydrocarbons, pesticides, PAHs and sewage sludge (Semple *et al.* 2001; Thassitou & Arvanitoyannis 2001; Khan *et al.* 2004).

The non-engineered windrows/biopiles rely mainly on wind-induced pressure gradients (i.e., natural airflow) which are non-uniform and particularly weak in the central part of the pile, which may lead to a local O_2 deficiency (Eweis *et al.* 1998; Li *et al.* 2004). Engineered windrows and biopiles are often covered and lined with waterproof plastic to control water infiltration, runoff and volatilisation as well as to enhance solar heating (Fahnestock *et al.* 1998; Khan *et al.* 2004). Furthermore, an impermeable membrane or clay layer may be used as a basis to reduce the risk of pollutant leaching into uncontaminated soil.

The distribution of soil characteristics such as texture, permeability, water content and bulk density/porosity which are critical to the activity of the hydrocarbon degrading microbial population are often non-uniform, and therefore turning the contaminated soil may be required to promote optimal biodegradation conditions, including improved aeration (Khan *et al.* 2004).

Windrows and biopiles containing organic matter (e.g., wood waste, sewage sludge and food waste) are usually referred to as 'composting' systems (Vidali 2001). In this type of biopile/windrow, the degradation of organic matter results in an increase in the biopile temperature leading to changes in the microbial community structure during the course of bioremediation (Semple *et al.* 2001; Thassitou & Arvanitoyannis 2001). The degradation process in the composting method is initiated by mesophilic bacteria which grow optimally between 20 and 40 degrees (Thassitou & Arvanitoyannis 2001). However, the increase in the biodegradation rate results in heat production leading to temperature increases of up to 65° C (Semple *et al.* 2001; Thassitou & Arvanitoyannis 2001). The increase in the pile temperature results in an increase in thermophilic bacteria and a decline in the mesophilic microbial population (Semple *et al.* 2001; Thassitou & Arvanitoyannis 2001).

The advantages of windrow/biopile strategies include their cost efficiency (approximately £15–35/tonne) (Semple *et al.* 2001), they are easy to design and they can also be applied on site (Khan *et al.* 2004). In addition, the area and time required for biopile treatments are less than those required for land farming. Furthermore, vapour emissions can be controlled using a closed system, and they can be designed to fit a range of products and site conditions (Khan *et al.* 2004). The limitations include space requirements and the evaporation of volatile compounds which are required to be treated prior to discharge to the atmosphere (Mueller *et al.* 1996; Khan *et al.* 2004). Contamination at the former railway marshalling yard in Aberdeen required intensive management such as covering during heavy rainfall to not only prevent slumping, but also to ensure a soil matric potential and aeration/oxygen diffusion commensurate with optimal hydrocarbon degradation activity.

On the study site, an algorithm (1) was used which, coupled to optimal management of water, pH, nutrients and aeration, enabled optimisation of the hydrocarbon degradation by the soil microbial community.

Bioremediation factor α (f) [TPH] (f) availability (f) degraders (f) constraints (1)

The terms in the algorithm represent a combination of contamination factors which drive the bioremediation. 'TPH' is the total petroleum hydrocarbon concentration in the soil and 'availability' refers to the fraction of these hydrocarbons that is available for bioremediation. This is measured using hydrocarbon catabolic bacterial biosensors where *lux* genes are downstream of hydrocarbon catabolic promoters and so the induction of light output from the sensors indicates hydrocarbon bioavailability (Killham & Paton 2003) and microbial factors ('degraders' refers to the population density of hydrocarbon degraders in the soil measured by most probable number techniques, and 'constraints' refers to the toxicity imposed on the microbial community by the chemical toxicity of the contaminated soil environment and is measured using the *lux*-based toxicity biosensors previously described).

Using the algorithm to predict the likely success of bioremediation, the windrows of contaminated soil were carefully managed and the hydrocarbon contamination degraded down to compliance concentrations over 6–8 weeks (Fig. 12.7). The windrows were monitored during this period to ensure high rates of hydrocarbon degrader activity (in the sandy soils under remediation,

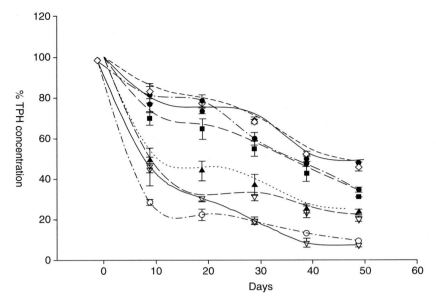

Figure 12.7. Monitoring of hydrocarbon degradation rates (as a percentage of original hydrocarbon concentration) across the different windrows of the site. Each symbol represents a different windrow on the site, and the data for each point are the mean of 5 observations.

this was simply achieved by measuring oxygen uptake and carbon dioxide evolution by the soil, as well as monitoring reduction in hydrocarbon concentration). When the O_2 consumption/CO_2 evolution rates decreased markedly, the windrows were turned mechanically to introduce air and oxygenate the microbial degraders throughout the anatomy of the windrow.

Conclusions

In this chapter, understanding of the potential of key species of the soil microbial community responsible for certain metabolic pathways has been exploited to bioremediate a range of industrial site contaminants. This also involved using modern, molecular methods to identify and alleviate constraints to their bioremediative activity. Some of these constraints are due to the site contaminants themselves and some are due to the inherent physicochemical nature of the microbial habitat. The importance of understanding and exploiting the microbial ecology of contaminated land has therefore been highlighted, and, by drawing on two contrasting case studies (one employing *in situ* bioremediation and the other *ex situ* bioremediation), it has been demonstrated that sustainable approaches to remediation of many contaminated sites, particularly where hydrocarbons are the dominant contaminants, can be highly successful over attractive time frames. By using the algorithm approach where bioremediation success and rate can be predicted at the onset based on measuring key properties of the contaminated soil and its hydrocarbon-degrading microbial community, a great deal more certainty can be introduced into an approach that has previously been associated with considerable uncertainty. Furthermore, by monitoring the activity of the contaminant degrader microbial community during the bioremediation process, conditions can be maintained near optimal for the process and progress monitored to ensure confident achievement of clean-up targets for regulatory sign-off. Finally, exploiting the microbial ecology of industrially contaminated land in sustainable remediation strategies will be increasingly required by government regulators to ensure that we no longer rely so heavily on landfill in the future.

References

Alexander, M. (1999) *Biodegradation and Bioremediation*. Academic Press, New York.

Atlas, R. M. and Philp, J. C. (eds.) (2005) *Bioremediation: Applied Microbial Solutions for Real-world Environmental Cleanup*. American Society for Microbiology Press, Washington, DC.

Bouwer, E. J. (1994) Bioremediation of chlorinated solvents using alternate electron acceptors. In: *Handbook of Bioremediation* (eds. R. D. Norris, R. E. Hinchee, R. Brown *et al.*), pp. 149–175. Lewis Publishers, Boca Raton, FL.

Eweis, J. F., Ergas, S. J., Chang, D. P. and Schroeder, E. D. (1998) *Bioremediation Principles*. International Edition. Malaysia.

Fahnestock, F. M., Wickramanayake, G. B., Kratzke, R. J. and Major, W. R. (1998) *Biopile Design, Operation, and Maintenance Handbook for Treating Hydrocarbon-Contaminated Soils*. Battelle Press, Columbus, OH.

Fritsche, W. and Hofrichter, M. (2005) Aerobic degradation of recalcitrant organic compounds by micro-organisms. In: *Environmental Biotechnology* (eds. H. J. Jördening and J. Winter). Wiley, New York.

Jørgensen, K. S., Puustinen, J. and Suortti, A. M. (2000) Bioremediation of petroleum hydrocarbon-contaminated soil by composting in biopiles. *Environmental Pollution* **107**, 245–254.

Khan, F. I., Husain, T. and Hejazi, R. (2004) An overview and analysis of site remediation technologies. *Journal of Environmental Management* **71**, 95–122.

Killham, K. and Paton, G. I. (2003) Intelligent site assessment: a role for ecotoxicology. In: *Bioremediation: A Critical Review* (eds. I. Singleton, M. G. Milner and I. M. Head). Horizon Press, London.

Killham, K. and Staddon, W. (2002) Bioindicators and sensors of soil health and the application of geostatistics. In: *Enzymes in the Environment* (eds. R. G. Burns and R. P. Dick), pp. 391–406. Dekker, New York.

Li, L., Cunningham, C. J., Pas, V., Philp, J. C., Barry, D. A. and Anderson, P. (2004) Field trial of a new aeration system for enhancing biodegradation in a biopile. *Waste Management* **24**, 127–137.

Mueller, J. G., Cerniglia, C. E. and Pritchard, P. H. (1996) Bioremediation of environments contaminated by polycyclic aromatic hydrocarbons. In: *Bioremediation: Principles and Applications* (eds. R. L. Crawford and D. L. Crawford), pp. 125–194. Cambridge University Press, Cambridge, UK.

Paton, G. I., Campbell, C. D., Cresser, M. S., Glover, L. A., Rattray, E. A. S. and Killham, K. (1995) *Bioluminescence-based Ecotoxicity Testing of Soil and Water*. OECD special publication on Bioremediation, Tokyo 94, OECD Press, pp. 547–552.

Reid, B.J., MacLeod, C.J. A., Lee, P.H., Morriss, A.W.J., Stokes, J.D. and Semple, K.T. (2001) A simple 14C-respirometric method for assessing microbial catabolic potential and organic contaminant bioavailability. *FEMS Microbiology Letters* **196**, 141–146.

Semple, K.T., Reid, B.J. and Fermor, T.R. (2001) Impact of composting strategies on the treatment of soils contaminated with organic pollutants: review. *Environmental Pollution*, **112**, 269–283.

Sousa, S., Duffy, C., Weitz, H., Glover, L.A., Henkler, R. and Killham, K. (1998a) Use of a *lux* bacterial biosensor to identify constraints to remediation of contaminated environmental samples. *Environmental Toxicology and Chemistry* **17**, 1039–1045.

Sousa, S., Weitz, H., Duffy, C. *et al.* (1998b) Contribution of *lux* bacterial biosensors to remediation of BTEX contaminated land. In: *Consoil'98, Sixth International Conference on Contaminated Soil.* pp. 839–840. Thomas Telford, London.

Standing, D., Baggs, E.M., Wattenbach, M., Smith, P. and Killham, K. (2007) Meeting the challenge of scale-up in the plant-soil-microbe system. *Biology and Fertility of Soils* **44**, 245–257.

Thassitou, P.K. and Arvanitoyannis, I.S. (2001) Bioremediation: a novel approach to food waste management. *Trends in Food and Technology* **12**, 185–196.

Vidali M. (2001) Bioremediation. An overview. *Pure and Applied Chemistry* **73**, 1163–1172.

CHAPTER THIRTEEN

Ecological recovery in a river polluted to its sources: the River Tame in the English Midlands

TERRY E. L. LANGFORD, PETER. J. SHAW, SHELLEY R. HOWARD, ALASTAIR J. D. FERGUSON, DAVID OTTEWELL AND ROWLAND ELEY

Introduction

In many industrialised regions particularly in Britain, rivers have been impounded for use by mills, polluted by multiple point sources and channelised to the very source over many centuries (e.g., Bracegirdle 1973; Lester 1975; Harkness 1982; Holland & Harding 1984; Haslam 1991). Since the 1960s, the ecological recovery of such historically polluted and disturbed rivers in Britain has been remarkable. Long reaches of once black, foetid, fishless watercourses, some almost completely devoid of macroscopic biota, have been transformed into clear streams and rivers with diverse floras and faunas and prolific fish populations. This transformation is perceived to have been the result of a number of factors, including law, public pressure, new technologies, new infrastructure and changes in the economy and industry. Even so, ecological recovery is still poorly advanced in some rivers and the reasons for this have not been explained in any detail. This short chapter uses sets of long-term chemical and biological data from three sites on a Midland river in a preliminary analysis of the possible reasons for the variable rates of ecological recovery and the relationship between the long-term chemical and biological changes in the river. It is part of a series of longer term studies of the problems associated with ecological recovery of polluted rivers (e.g., Langford et al. 2009).

Ecological recovery of any ecosystem from a disturbance has been defined as 'the return to an ecosystem which closely resembles unstressed surrounding areas' (Gore 1985). Gore and Milner (1990) state that 'as normally used,... recovery... implies that the system is moving toward a condition that existed before it was disturbed'. In a river recovering from pollution, the ultimate

Ecology of Industrial Pollution, eds. Lesley C. Batty and Kevin B. Hallberg. Published by Cambridge University Press. © British Ecological Society 2010.

composition of the flora and fauna will depend mainly on the physico-chemical condition of the water unless the structure of the channel is extremely modified physically, for example, as a concrete channel or culvert with no natural or semi-natural structure (Langford & Frissell 2009). Ecological recovery from gross pollution can therefore be expected to be a progress toward the unpolluted state, with a flora and fauna intolerant of the polluted conditions. Such recovery is also dependent upon recolonisation and subsequent succession by species from relevant sources (Yount & Niemi 1990; Milner 1996). The rate of recovery is dependent upon the severity of the disturbance, the proximity of a potential source of species, the mechanism of recolonisation and the mobility of the relevant species. Ultimate recovery to the unpolluted ecological condition will depend on the availability of the original species assemblage, though knowledge of this is unlikely in streams polluted to their sources over hundreds of years (e.g., Holland & Harding 1984). Excellent predictive systems of the potential composition of unpolluted communities have been developed based on reference conditions. In the UK, the River Invertebrate Prediction and Classification System (RIVPACS) (e.g., Wright *et al.* 2000 and Chapter 6) has been developed over some 30 years based on data from over 600 sites. However, the sites used for the predictions are on cleaner rivers which may or may not have been subjected to marked historical chemical and physical disturbances and such reference sites are often viewed with caution.

Where gross pollution occurs to the absolute sources of a river system, ecological classification may be misleading once chemical quality has improved, in that the absence of potential colonisers does not allow ecological recovery to follow chemical recovery closely (Langford *et al.* 2009). This could be important for industry or municipal authorities in that ecological classification of the river may suggest that further treatment of discharges entailing further costs are necessary, whereas the real problem is in the biological recovery processes and not the chemical quality of the water. Limitations on recolonisation may also stem from some extreme physical factors such as absence of natural margins or substrates in culverted or heavily engineered channels (Davenport *et al.* 2004).

In this present account, long-term data originate from three sites on the River Tame, but further studies are using more extensive data from the catchment. The main source of material is a 50-year sequence of biological data from the 1950s to the present day originating from records of the Trent River Board and its successors (Woodiwiss 1964). Other material has been extracted from published data (e.g., Hawkes 1956, 1962, 1975) and from local archives. These extensive datasets were used to develop the Trent Biotic Index (Woodiwiss 1964), one of the earliest biological indices for monitoring ecological effects of river pollution, variations of which are still in use in Europe (e.g., De Pauw & Vanhooren 1983; Metcalf-Smith 1996; Sweeting 1996).

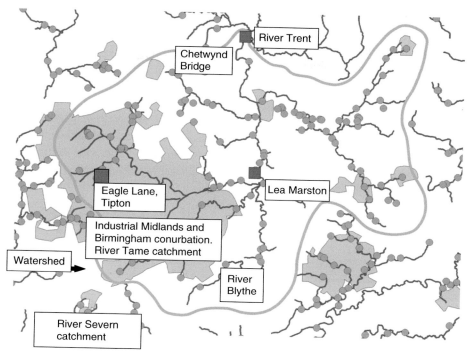

Figure 13.1. River Tame system showing relative positions of the three sampling sites used for this analysis. Green spots indicate EA biological sampling stations. Red squares, the sites used in this analysis. See colour plate section.

The River Tame and its tributaries
The landscape and drainage

The Tame system drains industrial South Staffordshire, the Birmingham conurbation, parts of Warwickshire and Leicestershire in the English Midlands. The River Tame itself (Fig. 13.1) has most of its sources in the Middle and Upper Coal measures of the West Midlands (Wills 1950). The upper part of the catchment, known as the Black Country, probably because of the effects of industrial smoke in the nineteenth century, comprises urban and industrial development which reaches to the extreme upper ridge of the Tame watershed dividing the Trent and Severn river basins. The river and many of its upper tributaries have been grossly polluted by industrial and domestic sewage effluents for well over 150 years (Spicer 1950; Lester 1975; Harkness 1982). Although Birmingham was initially the location of small industries and mills responsible for pollution of the middle river, it was the growth of mining and industry in the Black Country that eventually caused the major pollution to the sources.

The Tame system, upstream of its confluence with the River Trent comprises streams and rivers with varied histories of pollution. The most polluted were the Tame itself, the River Rea through Birmingham which drained the industrial urban areas and streams such as the Hockley Brook which received highly toxic wastes from metal and metal treatment industries (Hawkes 1956). The River Blythe, running through Warwickshire, in contrast, drains a mainly rural area and has contained a diverse flora and fauna throughout its known history. The Blythe is a Site of Special Scientific Interest (SSSI) and contains a high diversity of plants, fish and macroinvertebrates (Box & Walker 1994). Rivers such as the Anker, the Cole and the Sence are intermediate with histories of both good and poor water quality, depending upon the reach and the period (Hawkes 1956; Skerry & Green 1986; Martin 1993, 1997; Farrimond & Martin 1996).

In the Black Country coal measures, the Tame originates as two main streams, the Oldbury arm and the Wolverhampton arm which join at Bescot near Walsall. The Oldbury arm also comprises a number of very small streams, for example, Tipton Brook, Lea Brook and Hobnail Brook. Feeder streams for the Wolverhampton arm include Darlaston Brook, Ford Brook and Waddens Brook. All of these were also polluted to their sources over many years. Gradients are generally low, and though there were 70 mill sites operating in the thirteenth century, water was often scarce and flows poor (Dilworth 1976). This lack of water for powering forge hammers delayed the development of the metal industries in the Black Country until the advent of steam power removed the necessity for water-driven mills. The soils of the Black Country are essentially acidic with low phosphates, and historically, liming was necessary to make them agriculturally productive. The soils of the Blythe and Cole are more alkaline and productive without liming (Twyman 1950).

The landscape and vegetation of the upper Tame valley in the first century AD comprised marsh, heathland and dense woodland. The Domesday Book records a thinly spread, poor human population in the eleventh century, in contrast with the denser populations of the areas of richer soils, for example, in the Blythe catchment. On the coal-measures, mainly in South Staffordshire, population densities were about 1.9–2 individuals per square mile compared with 9–13 individuals per square mile on the richer lands (Kinvig 1950). The low number of watermills also indicated impoverished lands in South Staffordshire compared with Warwickshire. The changes caused by the discovery of large coal seams, ironstone and clay were dramatic. Wise (1950) notes that 'this originally impoverished and underpopulated area became the industrial heart of England and the scene of the first large-scale industrial revolution in the world'. By the end of the nineteenth century, the middle and upper Tame catchment included some of the most densely populated and industrialised regions in the world with a landscape comprising spoil heaps, mines, mills, houses, furnaces, forges and factories. The area was typically covered by the pall of smoke which eventually gave it its name. Lester (1975) noted that in the 1970s, the Tame catchment from Lea Marston upwards was 47% urbanised

Figure 13.2. Black Country landscape in the 1960s. Mine spoil heaps and pit pools such as these were common in the seventeenth century and stocked with coarse fish to provide food. The large block in the background is the remains of a slag-heap from a blast-furnace demolished in the 1940s. Today the area is all housing. (© T. E. L. Langford.) See colour plate section.

with 33% 'man-made impenetrable area' which affected both runoff rates and surface water quality. In the 1991 census there were 1.7 million people in the Tame catchment of which 80% lived in the Black Country and Birmingham conurbations (NRA 1996).

The relatively few watermills which did exist in the upper Tame area were used for various processes from at least the eleventh century until the late nineteenth and early twentieth century, and included fulling mills, tanneries, paper-mills and forge-hammer mills, mills for powering bellows for blast-furnaces, blade mills and slitting mills for cutting iron into bars or strips (Dilworth 1976). Small-scale metal industries were common by the Middle Ages (Pelham 1950), and by 1540 there were many smithies, lorimers and nailmakers in both Birmingham and the Black Country. Although all the industries were on a relatively small-scale individually, noxious effluents were associated with tanneries, fulling mills, abbatoirs and metal working causing localised river pollution (Raistrick 1973; Wohl 1983; Haslam 1991). The disturbed and industrialised state of the land by the mid seventeenth century is illustrated by the plethora of mine pools and flooded clay pits in the Black Country which contained mixed fish populations, often introduced by residents to augment their diet (Plot 1686) (Fig. 13.2).

By 1900, with the rapid growth of coal mining, the iron, steel and associated industries, the River Tame had become the drainage and waste disposal system for the whole West Midlands and Birmingham conurbation and industrial region. The economy of this region continued to expand until the 1950s as a result of two world wars and subsequent regeneration (Bird 1970, 1973, 1974; Hillier 1976; Parsons 1986). The regional economy began to decline in the 1960s and the industrial pattern changed with consequent effects on the river, as we will see. 'Between 1971 and 1985 Birmingham lost … 29% of all employment and manufacturing employment virtually halved' (Spencer *et al.* 1986). At the same time local gross domestic product (GDP) fell by more than 10% when compared with national GDP.

Pollution and degradation of the river

In the seventeenth, eighteenth and early nineteenth centuries, exchanges of fishing rights were important transactions along the Tame, but by the middle of the nineteenth century pollution of the river was a major concern for fishery owners and remained so for over 150 years (Harkness 1982). Before 1700, the Tame was a noted salmonid fishery and supported a diverse fish community in its middle and lower reaches. Little is known about the fish of the smaller feeders and uppermost reaches, but the underlying geology suggests that the water was acidic and probably supported mainly trout and sea-trout. Further down the river, mills paid part of their rents in eels and salmonids until the early nineteenth century. The biological degradation of the whole river was not noted significantly until the middle of the nineteenth century (NRA 1996). Indeed from 1831to 1866, the river in its middle reaches was used as potable water but this was discontinued in 1872 (Harkness 1982). Eventually, potable water was supplied from reservoirs in Wales and much of the base flow-water in the river was chemically different from the original, pre-pollution water. Despite the increasing domestic, surface and industrial wastes through the nineteenth century, the development of gas-works producing gas from coal and discharging highly toxic effluents appears to have been the final factor in the almost complete chemical and biological degradation of the river (Spicer 1937, 1950). By 1918, there were no fish in the river from the source to the confluence with the River Trent (NRA 1996) and most of the invertebrate fauna had disappeared from much of the upper and middle Tame. Below Birmingham, the river, which had been a well used trout fishery for 40 km before 1800, contained little evidence of fish or invertebrate communities (Spicer 1950). The worst polluted streams were the Black Country feeder streams (i.e., Oldbury and Wolverhampton arms of the Tame), the Rea, Cole and Hockley Brook all of which carried metallic as well as gasworks effluents and sewage wastes (Fig. 13.3). In the 1940s and 1950s, the reach of the Tame from near Perry Barr to Minworth was devoid of invertebrates for some 15 km (Hawkes 1956)

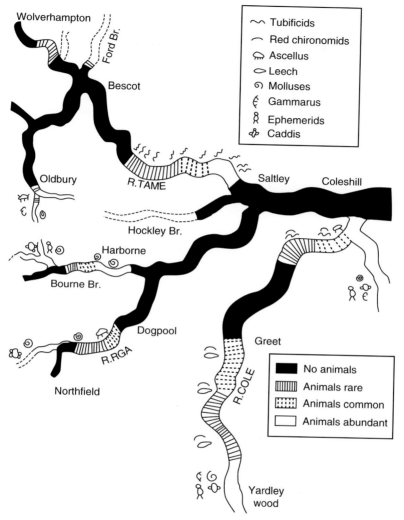

Figure 13.3. Sketch map of the Upper River Tame showing the extent of pollution in 1945 and indicating the invertebrates recorded (from Butcher 1946).

as a result of toxic effluents. '... it is only after dilution with Birmingham's sewage effluents and the Rivers Blythe and Bourne that tubificids re-appear and self-purification is re-established' (Hawkes 1956). The Blythe was then, as now, relatively clean with a mixed fishery and diverse invertebrate fauna (Hawkes 1975) and the Anker was moderately polluted but retained a fish population.

   The process of cleaning up the river began in earnest in 1923 (Spicer 1937, 1950). Indeed Spicer (1937) quoted a Trent Fishery Board Report of 1936, which suggested that in the River Trent catchment, including the Tame, 'practically all the sources of major pollution are under some control'. Given the recorded water quality conditions at the time, this was certainly untrue for the Tame and the onset of the Second World War in 1939 exacerbated the already polluted state. Little effect of any pollution control was observed until legislation in the 1960s began to place quality and quantity controls on all domestic and industrial point discharges (Lester 1975). Further initiatives included treatment lakes in the lower river and rerouting of major sewers which took effluents from the upper reaches to treatment works downstream. This development also allowed the closure of many ineffective smaller treatment works. Also in the 1960s, gas production from coal was replaced by natural gas and the grossly polluting gas-works effluents ceased. These plus economic decline in the region led, as we will see, to a considerable reduction of the pollution load in the river.

Sites and methodology

Three sites on the main River Tame were selected for this initial study, one on the lower reaches, one in the middle reaches and one in the upper reaches within the Black Country region.

Site 1. Chetwynd Bridge (OS Ref. SK 187138) is located 1.6 km upstream of the confluence with the River Trent. The mean flow of the river is about 2000 million litres per day. The nearest tributary upstream is the River Anker which is mostly of medium to good quality. This site integrates practically the whole of activities that affect water quality in the Tame basin.

Site 2. Lea Marston (OS Ref 206934) is located 32 km upstream of the confluence with the Trent. The main tributaries upstream are the Blythe, the Cole and the Rea. The Blythe is a clean river and the others, once polluted, have now cleaned to a great extent.

Site 3. Eagle Lane, Tipton (OS Ref. SO 979931) is located some 65.5 km upstream of the confluence with the Trent and about 5 km downstream of the Oldbury source. The feeders are all the once grossly polluted streams of the Black Country and the coal measures, and there is no known 'natural' or unpolluted reach upstream of this site.

The main datasets originate from biological and chemical surveys begun in the 1950s (Woodiwiss 1964) and continuing to the present day (Environment Agency Registers BIOSYS and WIMS). Raw chemical data from surveys before 1980 were not available; therefore, summarised data were extracted from various annual and special reports dating back to the 1930s. There have been changes to chemical or biological techniques over the years, but the data treatments and methods of representation have been standardised here as much as possible for comparability.

For this initial study, we have carried out a univariate analysis using ammonia concentrations (mg L^{-1} as N) as the primary chemical indicator of pollution as it can reflect both industrial and non-industrial categories (Klein 1962). There are also data on the toxicity of ammonia to many aquatic organisms (e.g., Alabaster & Lloyd 1980). Relationships between selected chemical indicators were analysed using Pearson correlation analysis. Biological recovery is indicated by taxon richness (Ntaxa), using identification to the BMWP 'family' level (Wright *et al.* 2000). As the weighted indicators of ecological quality, e.g., ASPT and BMWP (Wright *et al.* 2000), are based on the tolerance of selected taxa to chemical determinands such as ammonia and Biochemical Oxygen Demand (BOD), using these would not give as clear an indication of the biological processes of recolonisation and succession as taxon richness. Successional changes in the invertebrate fauna are summarised by chronological analysis of the raw data. 'Clean-water' taxa include all those which score 8–10 in the BMWP pollution scoring system (Wright *et al.* 2000). The presence and absence of fish is also indicated from survey results supplied by the Environment Agency.

Results
Chronological changes
Chemistry and taxon richness

The responses of all three sites are similar chronologically in that, from the mid 1960s to the present, there has been a reduction in average annual ammonia concentrations from in excess of $15 \, mg \, L^{-1}$ to less than $2 \, mg \, L^{-1}$. At Chetwynd Bridge (Fig. 13.4) and Lea Marston (Fig. 13.4), average concentrations were below $1 \, mg \, L^{-1}$ from the mid 1990s. At Eagle Lane (Fig. 13.4), ammonia concentrations from the 1990s were slightly higher than at the other two sites, averaging between 1 and $2 \, mg \, L^{-1}$. Also, at Eagle Lane, there was a sharp fall in ammonia to an average of *c.* $5 \, mg \, L^{-1}$ in the late 1960s (coinciding with the closure of two large coal-based gasworks), but a subsequent increase to between 5 and $10 \, mg \, L^{-1}$ (reasons unknown to date) which was maintained until the mid 1980s when a second decline began. BOD values did not increase at Eagle Lane in the late 1960s, though as with the ammonia concentrations, there was a period from 1970 to 1990 when there was no decline (Environment Agency WIMS database). Increases in taxon richness also followed similar upward trends though the increases began in 1975–1977 at Chetwynd Bridge and Lea Marston but as late as the early 1990s at Eagle Lane. The increases in taxon richness coincided with periods when the ammonia concentrations fell below $5 \, mg \, L^{-1}$ with accelerated increases below $2.5 \, mg \, L^{-1}$. The $2.5 \, mg \, L^{-1}$ threshold is shown as dotted lines on Fig. 13.4.

The changes in ammonia concentrations were significantly correlated with BOD (F = 0.965, $p < 0.001$, n = 24) and, as a result of improved oxidation, negatively with nitrate (NO_3) concentrations (F = −0.756, $p < 0.01$, n = 24). Correlations between other determinands were not investigated for this univariate analysis, though further work will include multivariate analyses using other determinands. The variation in ammonia levels was greater before 1990 than in more recent years at all three sites with maxima declining from about $10 \, mg \, L^{-1}$ to $2 \, mg \, L^{-1}$ or less.

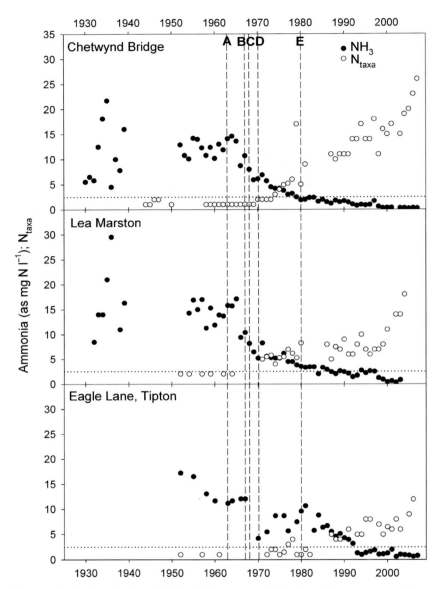

Figure 13.4. Chemical and biological recovery of the River Tame at three sites.
(Annual average ammonia (asN) concentrations in relation to average number
of taxa (BMWP families) taken from a long-term dataset (>50 years). Major relevant
events are shown as vertical dotted lines. Data from Southampton University
Historic River Data Archive (acknowledgements to Mr F. S. Woodiwiss and the
Environment Agency.) A, 1961 Rivers (Prevention of Pollution) Act implemented;
B, Midlands recession starts; C, coal gas production ceases; D, Black Country Sewer
completed; E, Tame treatment lakes fully open.

Taxon recolonisation and successions

Recovery of rivers from catastrophic disturbances is a function of both habitat conditions such as water chemistry and to a lesser extent channel geomorphology and the proximity of species for recolonisation (Milner 1996; Langford & Frissell 2009). Despite chemical controlling factors, the fauna of streams can be heavily dependent on the fauna of neighbouring streams (Sanderson *et al.* 2005). Four potential recolonisation mechanisms have been identified for river invertebrates, namely downstream migration (drift), upstream migration, aerial migration (mainly by insects) and upward migration from deeper substrates (Williams & Hynes 1976; Williams 1981; Milner 1996). Migrations from upstream and downstream are the only natural recolonisation mechanisms for fish, though often fish are introduced by anglers or fishery managers once conditions are judged satisfactory. Plants might use the first three mechanisms, with wind or birds as an aerial agent.

In many rivers, severe pollution or disturbance to the source has destroyed the fauna and therefore the major upstream (and downstream) sources of recolonisation. Further, many feeders and tributaries are similarly polluted so there may be no nearby sources of aerial colonisers other than the pollution-tolerant taxa such as chironomids. The Tame tributaries in the Black Country are classic examples (Harkness 1982).

The general patterns of taxonomic recolonisation and succession in rivers recovering from pollution have been recorded in many places over many years (e.g., Richardson 1921; Butcher 1946; Hawkes 1956, 1975; Hynes 1960; Milner 1996) but the variations in relationships between absolute chemical and biological parameters are less well-known. The onset of taxonomic succession in the Tame was similar at the three sites with predictable increases in taxon richness as ammonia concentrations fell, indicating general chemical and biological improvement (Figs. 13.5, 13.6). The presence of specific taxa is still difficult to relate to absolute values of chemical determinands because of the use of annual averages and restricted numbers of biological collections, but there is clearly a basis for predicting communities from ranges of chemical quality. At both Chetwynd Bridge and Eagle Lane, average ammonia concentrations exceeding $5\,mg\,L^{-1}$ were associated with Oligochaeta (sludge worms), Asellidae (hog-lice), Chironomidae (bloodworms and fly-larvae) and Erpobdellidae (leeches) (Figs. 13.5 and 13.7). However, at Lea Marston, average ammonia concentrations around $5\,mg\,L^{-1}$ were associated with a greater number of taxa, including those above plus moderately tolerant groups such as Planariidae (flatworms), Corixidae (water-boatmen), Baetidae (mayflies) and Gammaridae (freshwater shrimps) (Fig. 13.6). At Chetwynd Bridge clean-water taxa (Wright *et al.* 2000) such as free-living caddis (Psychomyiidae), damsel flies (Calopterygidae) and cased caddis (Molannidae, Leptoceridae) colonised as ammonia concentrations fell below an average of around $2.5\,mg\,L^{-1}$. At Lea Marston,

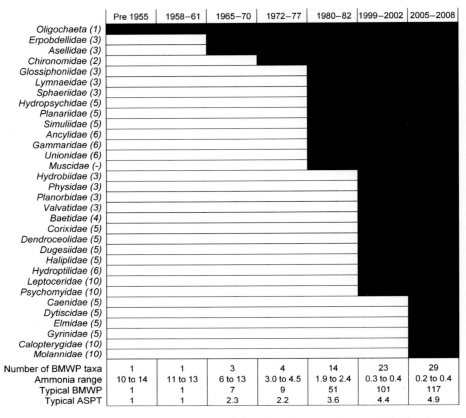

	Pre 1955	1958–61	1965–70	1972–77	1980–82	1999–2002	2005–2008
Number of BMWP taxa	1	1	3	4	14	23	29
Ammonia range	10 to 14	11 to 13	6 to 13	3.0 to 4.5	1.9 to 2.4	0.3 to 0.4	0.2 to 0.4
Typical BMWP	1	1	7	9	51	101	117
Typical ASPT	1	1	2.3	2.2	3.6	4.4	4.9

Figure 13.5. Colonisation and succession of the River Tame at Chetwynd Bridge by macroinvertebrate taxa. (Identification is to BMWP families/groups, see Wright *et al.* 2000.) All taxa recorded over the relevant year are shown.

Oligochaeta, Chironomidae, Asellidae and Erpobdellidae were all, predictably, recorded at ammonia concentrations around 5–10 mg L^{-1} but taxon richness was much lower by 2006 than at Chetwynd Bridge and no clean-water taxa were present at concentrations below 2.5 mg L^{-1}, though moderately tolerant taxa such as Gammaridae, Ancylidae (limpets) and flatworms (e.g., Planariidae) had recolonised. At Eagle Lane, moderately tolerant taxa such as Gammaridae and Planariidae were present at the lower concentrations of ammonia (1–2 mg L^{-1}), but no clean-water taxa were recorded even to 2006. Thus, the extent of succession differed at each site with the most improvement being made at Chetwynd Bridge and the least at Eagle Lane, despite indications of similar chemical improvements. The numbers of taxa observed per sample compared with the numbers predicted by RIVPACS (Wright *et al.* 2000) were 26/30 (2007) for Chetwynd Bridge, 18/32 (2005) for Lea Marston and 12/29

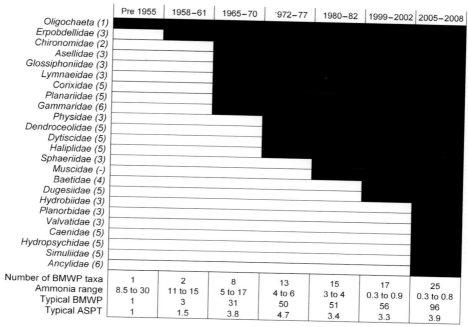

Figure 13.6. Colonisation and succession of the River Tame at Lea Marston by macroinvertebrate taxa. (Identification is to BMWP families/groups, see Wright *et al.* 2000.) All taxa recorded in the relevant year are shown.

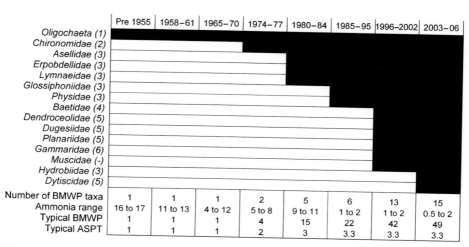

Figure 13.7. Colonisation and succession of the River Tame at Eagle Lane, Tipton, by macroinvertebrate taxa. (Identification is to BMWP families/groups, see Wright *et al.* 2000.) All taxa recorded in the relevant year are shown.

Table 13.1. *List of fish species recorded at two sites on the Tame (data supplied by the Environment Agency)*

Common names	Species	Chetwynd and Elford		Lea Marston	
		1992	2005	1992	2005
Chub	*Leuciscus cephalus*	X	X	X	
Dace	*Leuciscus leuciscus*	X	X	X	
Bleak	*Alburnus alburnus*	X	X		
Gudgeon	*Gobio gobio*	X	X	X	X
Rudd	*Scardinius erythrophthalmus*				X
Roach	*Rutilus rutilus*	X	X	X	X
Perch	*Perca fluviatilis*	X	X	X	X
Pike	*Esox lucius*	X	X	X	X
Eel	*Anguilla anguilla*	X		X	
Rainbow trout	*Onchorhynchus mykiss*			X	
Brown trout	*Salmo trutta*				X
Tench	*Tinca tinca*				
Common Bream	*Abramis brama*				
Common Carp	*Cyprinus carpio*			X	
Barbel	*Barbus barbus*		X		
Minnow	*Phoxinus phoxinus*		X		
Stickleback	*Gasterosteus aculeatus*		X		
Bullhead	*Cottus gobio*		X		
Stoneloach	*Nemacheilus barbatulus*		X		X
Spined Loach	*Cobitis taenia*		X		

(2005) for Eagle Lane. Overall, the total numbers of taxa recorded for each site were 29, 25 and 15, respectively.

Fish and the fishery

Fish began to return to the lowest reaches of the Tame near Chetwynd Bridge in the 1970s (Lester 1975), though three-spined sticklebacks (*Gasterosteus aculeatus*) were recorded on biological data sheets from Lea Marston over the same period (unpublished archive data). Fish surveys over the past 15 years have now recorded a diversity of species from both Lea Marston and the Chetwynd Bridge reaches (Table 13.1). Chub, dace, roach pike and perch are common and abundant and salmonids (trout) have also been recorded. In the Chetwynd and Elford reaches in the lower part of the river, the spined loach (*Cobitis taenia*) is now common. This is an Annex II species designated by the EU Habitats Directive. Salmonids have also been recorded on occasions indicating good water quality. In the middle and upper parts of the Tame, fish mortalities can still occur as a result of episodes of poor water quality, caused mainly by runoff from urban surfaces or storm-water outflows during heavy rainfall.

Relevant events and trends affecting ecological recovery

Over the past 40 years, many changes in the law, technologies, infrastructure and the economy have occurred, some or all of which have had repercussions for river pollution. For example, the Rivers (Prevention of Pollution) Act 1961, fully implemented in 1963, introduced retrospective control over effluents of all types. In 1966 natural gas began to replace coal gas. The coal-based gas works, with their grossly polluting effluents containing phenols, ammonia, coal tars and cyanides, closed, and the discharges ceased over a few years after about 1964. Also, in the mid 1960s, heavy industries, particularly blast furnaces, coke ovens and metal finishing industries began to decline, many moving operations overseas. Regional GDP (RGDP) expressed as a per capita percentage relative to the national GDP declined from over 100% to around 90% of its long-term average between 1965 and 1981. In the same period, unemployment rose relative to the national average from 48% to almost 120%. There is a highly significant direct relationship between the RGDP and ammonia values in the river (Fig. 13.8) which can be expressed by the equation $(y = 9.32\ Ln\ x + 86.3\ (R^2 = 0.93,\ p = < 0.001))$ where y is the average annual ammonia concentration and x is the RGDP (expressed as a percentage of the national GDP per capita). By about 1970, old sewage works in the Black Country were being closed and the sewage directed to the main treatment works at Minworth. Finally, in 1980 large artificial lakes through which the river was partly directed to improve settlement of solids and oxidation became fully operational following trials beginning in 1968. However, in a long-established industrial conurbation with a complex history of change, numerous localised events and actions will also have occurred, many of which may be unrecorded. Thus, the presence and location of hidden or culverted outfalls, contaminated areas of land and the sources of diffuse pollution may not all be known even at the present time. Further, some untraced pollution sources may have ceased because of industrial and economic changes.

The separate effects of the 1961 Act, the recession and gas works closures are difficult to distinguish at this stage. By the time the treatment lakes were fully opened, water quality was already greatly improved in the river downstream. There was little immediate improvement at Chetwynd Bridge following the opening of the lakes though ammonia values declined from around 2.5 to less than 1 mg L^{-1} over 15 years following the opening. However, Heng (2000) recorded a longitudinal reduction in ammonia concentrations of from 2.5 mg L^{-1} to 0.8 mg L^{-1} from the inlet to the outlet of the lakes. This change does not appear to have been detectable as far down as Chetwynd Bridge, probably as a result of dilution and effects of other tributaries.

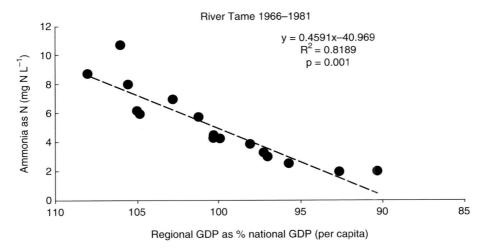

Figure 13.8. Relationship between average annual ammonia concentrations of (as N) in the River Tame at Chetwynd Bridge and the Annual Regional Gross Domestic Product (Birmingham/Black Country) expressed as a percentage of the National Gross Domestic Product (per capita).

At Lea Marston, just downstream of the main urban-industrial conurbations but upstream of the treatment lakes, the 1961 Act, industrial recession and gas works closures appear to have been the main significant events, though improvements in sewage treatment may also have contributed. This site is upstream of the treatment lakes; therefore, these would have had no effect at this site. The River Blythe which enters the Tame upstream of this site would be a source of cleaner water taxa.

At Eagle Lane, in the uppermost reaches of the river, the 1961 Act did not appear to be followed by an immediate decline in ammonia values but the closure of two large gas works in 1965 about 1–2 km upstream of Eagle Lane was followed by a decline from in excess of $10\,mg\,L^{-1}$ to less than $5\,mg\,L^{-1}$. The reasons for the subsequent rise to values of 8–$12\,mg\,L^{-1}$ in the 1970s and 1980s at Eagle Lane have not yet been investigated. Average BOD values (EA data sources) did not increase over the same period but stayed between 10 and $18\,mg\,L^{-1}$ from 1970 to 1990 when they declined to around $5\,mg\,L^{-1}$. As we have seen, the three sites showed differences in both the rates of recolonisation and succession and the composition of the invertebrate communities in the years after 2000.

Conclusions

Though this is an initial investigation of the recovery of the Tame using the long-term biological data and limited to three sites, a number of salient points have emerged. First, although the law may have been instrumental in stimulating improvements in water quality, it was most likely economic recession and the consequent closures of large industries together with the change from coal gas to natural gas that were initially the major factors in the rapid improvement of water quality in the river. Whether the recession itself was exacerbated by increasingly stringent pollution controls and their costs is unknown, but the improvements required were costly and other countries did not have the constraints. It is also clear that the Tame treatment lakes were not immediately significant factors in the reduction of ammonia or invertebrate recolonisation of the lower river, though further investigation may show that recovery was improved immediately downstream of the lake outfalls. The treatment would also contribute to the improvement in the longer term.

Recolonisation rates and succession by invertebrates clearly varied with location. At both Chetwynd Bridge and Lea Marston, the sites were downstream of major tributaries, which could contribute cleaner water species. However, at Lea Marston, despite the potential source of recolonisers from the River Blythe, no clean-water taxa were recorded, whereas at Chetwynd Bridge at least four clean-water taxa were recorded. By 2006, ammonia levels were very similar at both sites but it is likely that episodic pollution from diffuse sources of proximal urban runoff was more influential at Lea Marston than at the lower site. At Eagle Lane, Tipton, ammonia levels were not as low as at the other sites but were well below the threshold at which rapid recolonisation could occur. Though chemical recovery was slower, a major problem is the lack of any sources of clean-water fauna within or even near the system to provide a source of colonisers (Milner 1996; Langford et al. 2009). Biological surveys of the numerous small streams in the upper parts of the catchment show a complete absence of clean-water fauna (EA BIOSYS and unpublished archive data), and this is particularly true for the uppermost tributaries of the Oldbury branch of the Tame on which Eagle Lane is sited. All of the taxa that have recolonised at Eagle Lane, and many of those at the other sites, are not obligate running water taxa but are also found in ditches, ponds and pools. Examples are, apart from the Oligochaeta, snails such as *Lymnea* sp. and *Physa* sp. and leeches such as *Erpobdella* sp. and *Glossiphoniai sp.* Leeches, molluscs and oligochaetes are also likely to be associated because the leeches feed on Oligochaeta and gastropod molluscs (Elliott & Mann 1998). Sources of obligate clean river taxa are remote with the nearest being to the west in the Severn catchment 5–10 km distant.

The significance of the geomorphological condition of the Tame to recovery is equivocal, though there is some good evidence that more natural, weedy channels support a slightly more diverse invertebrate fauna than the heavily engineered channels (Beavan et al. 2001). In the most channelised reaches with concrete or bricked margins, there would be, predictably, a restriction of those clean-water species which require natural marginal habitat such as trailing vegetation or weeds (Armitage *et al.* 2001). Also, where riffle-pool sequences and instream-vegetation stands have been physically destroyed or removed there would be some suppression of invertebrate diversity even under clean-water conditions. However, the physical condition of the channel is unlikely to prevent colonisation by 'mid-stream' clean-water, lotic taxa such as species of Ephemeroptera, Plecoptera and Trichoptera if they were available to colonise. Therefore, water quality variations combined with the lack of animals for recolonisation remain the major constraints on recovery in the upper Tame. Morphological constraints are relatively small. The effects of biological improvement in relation to further chemical improvement are therefore difficult to predict in view of the poor availability of clean-water colonisers. Colonisation by fish may be more dependent on physical heterogeneity, particularly at high discharges (Booker 2003).

From this and other studies of the long-term recolonisation of rivers polluted to the source (e.g., Langford *et al.* 2009), it is clear that the problems of ecological recovery in many previously polluted streams can be a combination of residual pollution problems and biological processes such as the proximity and mobility of relevant taxa. Whilst it may be possible to improve chemical quality over a relatively short time, the dispersal of colonisers from clean streams and rivers may take much longer, possibly more than 30 years in the upper Tame (Langford *et al.* 2009). This has cost implications for effluent and water quality control in that to achieve biological recovery it may not be necessary to aim for continuous chemical improvement once a certain threshold has been reached, though the threshold quality must be maintained. Further, ecological quality improvement will not be directly related to chemical improvement but may be significantly delayed. Thus, though fish may recolonise quickly, invertebrate recolonisation may take many years and this may be very relevant for the upper Tame and rivers in similar condition in other industrial conurbations.

There are also implications for monitoring in that, though the chemistry of the water may indicate improvement, the fauna does not. Whether this is a result of other chemical variables not used in this analysis or of the lack of colonisers is not yet fully investigated. However, if the problem is biological rather than water quality, the validity of the biological assessment as a monitor of water quality is called into doubt. This may also have implications for classifications determined by the Water Framework Directive, which we hope further analysis of the long-term datasets will clarify.

Acknowledgements

The authors are indebted to Jean Langford for copy editing and for collating and checking the bibliography. Tom Worthington of Southampton University has helped considerably with data gathering and useful discussion. We would also like to thank all the staff of the Archives and Special Collections at Southampton University who maintain the Historic River Data Archive for their expertise and patient cooperation. Among them, Karen Robson has been our main contact and advisor. Thanks also to various sections of the Environment Agency who responded to our requests for data and to EA biological staff at Colden Common near Winchester for their help with data acquisition. The earliest raw data records from the 1950s to the 1980s, retained in the archive, originate from the work of Mr Frank Woodiwiss, Biologist to the Trent River Board and his staff for which we are eternally grateful. The data are retained and archived in the Archives and Special Collections of Southampton University Library. Our grateful thanks also go to the editors of this book and especially to Dr Lesley Batty for her help at the conference and with amendments to this chapter.

References

Alabaster, J. S. and Lloyd, R. (1980) *Water Quality Criteria for Freshwater Fish*. Butterworths, London & Boston.

Armitage, P. D., Lattmann, K., Kneebone, N. and Harris, E. (2001) Bank profile and structure as determinants of macroinvertebrate assemblages – seasonal changes and management. *Regulation of Rivers: Research and Management* **17**, 543–556.

Beavan, L., Sadler, J. and Pinder, C. (2001) The invertebrate fauna of a physically modified urban river. *Hydrobiologia* **445**, 97–108.

Bird, V. (1970) *Portrait of Birmingham*. Robert Hale, London. 239 pp.

Bird, V. (1973) *Warwickshire*. Batsford, London and Sydney.

Bird, V. (1974) *Staffordshire*. Batsford, London.

Booker, D. J. (2003) Hydraulic modelling of fish habitat in urban rivers during high flows. *Hydrological Processes* **17**, 577–599.

Box, J. D. and Walker, G. J. (1994) Conservation of the Blythe, a high-quality river in a major urban area in England. *Aquatic Conservation – Marine and Freshwater Ecosystems* **4**, 75–85.

Bracegirdle, C. (1973) *The Dark River*. John Sherratt & Son, Altrincham, UK.

Butcher, R. W. (1946) A brief account of a biological investigation of the headwater streams of the River Tame. In: *Annual Report and Statement of Accounts of the Year 1946, Appendix I*, 1946–50. Trent Fishery Board, Birmingham, UK.

Davenport, A. J., Gurnell, A. M. and Armitage, P. D. (2004) Habitat survey and classification of urban rivers. *River Research and Applications* **20**, 687–704.

De Pauw, N. and Vanhooren, G. (1983) Method for biological quality assessment of watercourses in Belgium. *Hydrobiologia* **100**, 153–168.

Dilworth, D. (1976) *The Tame Mills of Staffordshire*. Phillimore & Co. London and Chichester, UK.

Elliott, J. M. and Mann, K. H. (1998) *A Key to the British Freshwater Leeches*. Freshwater Biological Association, Scientific Publication No. 40.

Farrimond, M. S. and Martin, J. R. (1996) Improving water quality in the River Tame, Birmingham, UK. Paper to Rivertech 1996, 8 pp. IWRA Seminar, Chicago, 22–25 September 1996.

Gore, J. A. (1985) Mechanisms of colonization and habitat enhancement for benthic macroinvertebrates in restored river channels. In: *The Restoration of Rivers and Streams* (ed. J. A. Gore), pp. 81–101. Butterworths, Boston.

Gore, J. A. and Milner, A. M. (1990) Island biogeographical theory: can it be used to predict lotic recovery rates? *Environmental Management* **14**, 737–753.

Harkness, N. (1982) The River Tame – a short history of water pollution and control within an industrial river basin. *Water Science and Technology* **14**, 153–165.

Haslam, S. M. (1991) *The Historic River*. Cobden of Cambridge Press. 324 pp.

Hawkes, H. A. (1956) The biological assessment of pollution in Birmingham streams. *Journal of the Institute of Municipal Engineers* **82**, 452.

Hawkes, H. A. (1962) Biological aspects. In: *Aspects of River Pollution, Vol. 2, Causes and Effects* (ed. L. Klein), pp. 311–432. Butterworths, London.

Hawkes, H. A. (1975) River zonation and classification. In: *River Ecology* (ed. B. A. Whitton), pp. 312–374. Blackwell Scientific Publications, Oxford, UK.

Heng, H. (2000) Management of the Lower River Tame. Proceedings of the URGENT Annual Meeting 2000. http://urgent.nerc.ac.uk/ Meetings2000/2000Proc./water/heng/htm.

Hillier, C. (1976) *The Western Midlands*. Victor Gollancz, London.

Holland, D. G. and Harding, J. P. C. (1984) Mersey. In: *Ecology of European Rivers* (ed. B. A. Whitton), pp. 113–144. Blackwell Scientific Publications, Oxford, UK.

Hynes, H. B. N. (1960) *The Biology of Polluted Waters*. Liverpool University Press, Liverpool, UK.

Kinvig, R. H. (1950) The Birmingham district in Domesday times. In: *Birmingham and its Regional Setting*. Meeting 30 August – 6 September 1950, pp. 113–134. British Association for the Advancement of Science, Birmingham, UK.

Klein, L. (1962) *River Pollution, 11. Causes and Effects*. Butterworths, London, UK.

Langford, T. E. L. and Frissell, C. A. (2009) Restoration and conservation: evaluation of potential sites. In: *Assessing the Conservation Value of Freshwater* (eds. P. Boon and C. Pringle) Chapter 6. Cambridge University Press, Cambridge, UK.

Langford, T. E. L., Shaw, P. J., Ferguson, A. G. P. and Howard, S. R. (2009) Long-term recovery

of microinvertebrate biota in grossly polluted streams: recolonisation as a constraint to ecological quality. *Ecological Indicators* **9**, 1064–1077.

Lester, W. F. (1975) Polluted river: River Trent, England. In: *River Ecology* (ed. B. A. Whitton), pp. 489–513. Blackwell Scientific Publications, Oxford, UK.

Martin, J. R. (1993) Costs and benefits of chemical monitoring. In: *Freshwater Europe Symposium, River Water Quality Monitoring and Control*. 15, pp. 21–23. February 1993, Birmingham, UK.

Martin, J. R. (1997) *Water Quality Improvements in the Rivers Tame and Trent UK: Progress and Prospects*. Severn Trent Water, Birmingham, UK.

Metcalf-Smith, J. L. (1996) Biological water-quality assessment of rivers: use of macroinvertebrate communities. In: *River Restoration* (eds. G. Petts and P. Calow), pp. 17–43. Blackwell Scientific Publications, Oxford, UK.

Milner, A. M. (1996) System recovery. In: *River Restoration* (eds. G. Petts and P. Calow), pp. 205–226. Blackwell Scientific Publications, Oxford, UK.

NRA. (1996) *Tame Catchment Management Plan Consultation Report January 1996*. National Rivers Authority, Lichfield, UK.

Parsons, H. (1986) *The Black Country*. Robert Hale, London.

Pelham, R. A. (1950) The growth of settlement and industry c.1100–c.1700. In: *Birmingham and its Regional Setting*. Meeting 30 August – 6 September 1950, pp. 135–158. British Association for the Advancement of Science, Birmingham, UK.

Plot, R. (1686) *The Natural History of Staffordshire*. Oxford, UK.

Raistrick, A. (1973) *Industrial Archaeology*. Granada/Paladin, St Albans, UK.

Richardson, R. E. (1921) Changes in the bottom and shore fauna of the middle Illinois River and its connecting lakes since 1913–15 as a result of increase southward of sewage pollution. *Bulletin Illinois Natural History Survey* **14**, 33–75.

Sanderson, R. A., Eyre, M. D. and Rushton, S. P. (2005) The influence of stream invertebrate composition at neighbouring sites on local assemblage composition. *Freshwater Biology* **50**, 221–231.

Skerry, E. W. and Green, M. B. (1986) Integrated management – the Tame basin system. In: *IWPC Symposium 'Sewerage-value for Money' Paper 13*, pp. 137–148. 14–15 May, 1986. Heathrow, UK.

Spencer, K., Taylor, A., Smith, B., Mawson, J., Flynn, N. and Batley, R. (1986) *Crisis in the Industrial Heartlands: A Study of the West Midlands*. Clarendon Press, Oxford, UK.

Spicer, J. I. (1937) *The Ecology of the River Trent and Tributaries. A Scientific Survey of Nottingham and District*. British Association for the Advancement of Science. pp. 95–98. Burlington House, London.

Spicer, J. I. (1950) *The Fisherman's Struggle Against Pollution*. Trent Fishery Board, Nottingham, UK.

Sweeting, R. A. (1996) River pollution. In: *River Restoration* (eds. G. Petts and P. Calow), pp. 7–16. Blackwell Scientific Publications, Oxford, UK.

Twyman, E. S. (1950) Soils. In: *Birmingham and its Regional Setting*. Meeting 30 August – 6 September 1950, pp. 55–64. British Association for the Advancement of Science, Birmingham, UK.

Williams, D. D. (1981) Migrations and distributions of Stream Benthos.

In: *Perspectives in Running Water Ecology* (eds. M. A. Lock and D. D. Williams), pp. 155–208. Plenum Press, New York.

Williams, D. D. and Hynes, H. B. N. (1976) The recolonization mechanisms of stream benthos. *Oikos* **27**, 265–272.

Wills, L. J. (1950) Geology. In: *Birmingham and its Regional Setting*. Meeting 30 August – 6 September 1950, pp. 15–36. British Association for the Advancement of Science, Birmingham, UK.

Wise, M. J. (1950) The Cannock Chase region. In: *Birmingham and its Regional Setting*. Meeting 30 August – 6 September 1950, pp. 269–288. British Association for the Advancement of Science, Birmingham, UK.

Wohl, A. S. (1983) *Endangered Lives*. Methuen, London.

Woodiwiss, F. S. (1964) The biological system of stream classification used by the Trent River Board. *Chemistry and Industry* **14**, 443–447.

Wright, J. F., Sutcliffe, D. W. and Furse, M. T. (eds.) (2000) *Assessing the Biological Quality of Fresh Waters. Proceedings of an International Workshop, 16–18 September, 1997. Oxford*. Freshwater Biological Association, Ambleside, Cumbria, UK.

Yount, J. D. and Niemi, G. J. (1990) Recovery of lotic communities and ecosystems from disturbance – a narrative review of case studies. *Environmental Management* **14**, 547–569.

CHAPTER FOURTEEN

Manchester Ship Canal and Salford Quays: industrial legacy and ecological restoration

ADRIAN E. WILLIAMS, RACHEL J. WATERFALL,
KEITH N. WHITE AND KEITH HENDRY

Introduction

The upper reaches of the Manchester Ship Canal (MSC) and associated dock basins have been polluted by operational discharges, surface water runoff as well as upstream inputs from the River Irwell. The resulting poor water quality has been exacerbated by the deep (7 m) water column and limited water exchange. In this chapter, we describe the water quality management strategies put in place since the late 1980s to address poor water quality, specifically oxygenation of the water column of the MSC and isolation of the docks from the canal followed by destratification of the water column and habitat diversification. We then examine the effectiveness of these strategies in improving water quality, increasing biodiversity and enhancing the recreational potential of the enclosed dock basins and the MSC.

Establishing an inland port at Manchester

Long before the Industrial Revolution in the late eighteenth century Manchester was already at the centre of numerous industrial settlements located in the foothills of the southwestern Pennines (Gray 1993). The success of local industry rendered the use of traditional transport such as pack-horses and horse-drawn carts inadequate to meet its needs, leading to the increased use of river navigation. The main navigable rivers are the Irwell and Mersey, although water shortages, siltation and mudbanks were obstacles to the reliable transport of goods (Gray 1993). Obstructions had already been cleared in the Mersey estuary to allow vessels to reach Bank Quay, Warrington, and it was suggested that clearing and dredging the Rivers Mersey and Irwell would enable vessels to proceed as far as Manchester (Gray 1993). It was the Mersey and Irwell Navigation Company

Ecology of Industrial Pollution, eds. Lesley C. Batty and Kevin B. Hallberg. Published by Cambridge University Press. © British Ecological Society 2010.

that realised these plans, and by 1734, vessels were able to reach Manchester and Salford (Gray 1993) using a series of locks between Warrington on the River Mersey and Manchester on the River Irwell (Struthers 1993).

By the early nineteenth century, pressure on warehouse space at both Manchester and Liverpool Docks was becoming acute (Struthers 1993). From the late eighteenth century, however, the Mersey and Irwell faced competition from the nearby Bridgewater Canal, which offered a faster and more reliable route to the sea (Gray 1993). The Canal, together with the opening of the Liverpool and Manchester Railway in 1830, reduced the use of the Rivers Mersey and Irwell and the almost bankrupt Mersey and Irwell Navigation Company was taken over by the Bridgewater Trustees in the mid nineteenth century. The rivers were subsequently allowed to deteriorate owing to a lack of dredging as it was considered preferable to invest money in maintaining the Bridgewater Canal (Gray 2000).

The worldwide trade depression in the latter part of the nineteenth century resulted in a period of economic stagnation which was exacerbated by the high cost of importing raw cotton via the port of Liverpool and its transport to Manchester by rail (Gray 1993). As a result, the advantages of a deep, wide waterway from Manchester to the sea became increasingly economically attractive (Gray 1993). The idea of a ship canal to Manchester and Salford was promoted by Daniel Adamson, an owner of a local engineering firm, and after three Parliamentary Bills, the necessary powers were granted in 1885 (Struthers 1993).

Construction of the Manchester Ship Canal (MSC) between 1887 and 1894 was the means by which the inland city of Manchester became an inland seaport (Farnie 1980). At a cost of £15 million (equivalent to approximately £600 million today) the construction was a major undertaking; the MSC was the largest navigation in the world, wider at that time than the Suez Canal and deeper than any waterway in Europe (Struthers 1993). Indeed the achievement has since been hailed as 'a feat without precedent in modern history' (Farnie 1980). The Canal is 58 km long, linking Manchester to Eastham on the Mersey Estuary and comprises four locks. It utilises water from the Rivers Irwell and Mersey. Terminating at the four docks in Salford and four smaller docks at Pomona 1 km upstream, the canal transformed the economy of the area by providing impetus for the industrialisation of Trafford Park, the largest industrial estate in Europe employing at its peak 75 000 people (Struthers 1993). The docks prospered and became one of Britain's largest ports, reaching a peak in 1958 (Radway *et al.* 1988).

Waterway neglect and deterioration: the River Irwell

It is documented that, in the early part of the nineteenth century, the River Irwell was a 'clean river', teeming with fish (Struthers 1993); indeed the River Irwell was once a salmon river (Boult & Hendry 1995). A local anecdote supports the claim that salmon were extremely common in the catchment, as it suggest's that factory workers threatened strike action due to the monotony of being fed salmon on a daily basis by the owners. However industrial development around waterside locations usually proceeded without regard to the quality of water. Not only was the River Irwell utilised as a valuable means of transport but the factories clustered along its banks also released their waste products into the river. Water was often seen as a 'convenient dumping ground to carry away industrial and sewage effluent; out of sight was out of mind' (e.g., Hendry *et al.* 1993). The last salmon was caught in the Irwell in 1856 (cited in Boult & Hendry 1995), and during the 1870s regattas and rowing races were abandoned due to the condition of the Irwell (Struthers 1993). Canals, rivers and especially docks throughout the UK became progressively more polluted and the aquatic environment was severely degraded (White *et al.* 1993).

During the second half of the twentieth century, however, industrial use of waterways and docks declined. Changes in local manufacturing industries, shifts in patterns of trade, restrictive trade union practices and increasing competition on North Atlantic shipping routes all combined to cause a decline in the market for cargo handling services (Gray 2000). Ocean-going container ships and tankers also grew to such a size as could not be accommodated in the locks of the MSC. By the late 1970s, traffic fell to such an extent that the MSC Company seriously considered closing the upper reaches altogether (Gray 2000). Owing to its poor environmental quality, the River Irwell which feeds the MSC was neglected and allowed to deteriorate further. All that remained in 1984 following the closure of the docks at Salford was an abandoned 60-ha site comprising a polluted water course (Radway *et al.* 1988). The dramatic decline in dockland activity that began in the mid 1960s was not confined to Manchester, and the resulting dereliction was characteristic of many other UK ports with a previously strong industrial base (Kivell 1993).

Background to water quality problems

The Mersey Basin drains an area of 4570 km^2 in the North West of England and comprises three major waterways: the Rivers Mersey and Irwell and the MSC (e.g., Hendry *et al.* 1993, Fig. 14.1). Collectively these are among the most organically polluted in the UK (Hendry *et al.* 1993). Indeed the area of the MSC examined in this chapter, the Turning Basin located between Trafford Road Bridge and Mode Wheel Locks (Fig. 14.2), was recorded as Grade 4 (grossly polluted) by the Environment Agency in 1990 (EA 1990).

Such poor water quality in the upper MSC has been ascribed to discharges from waste water treatment works (WwTWs), storm sewage overflows, industrial effluent and agricultural activities (Webb 1993) from the Rivers Irwell, Medlock and Irk (Hendry *et al.* 1997). Mass flux analysis (APEM 1990a, b) revealed that the River Irwell in particular was a major source of organic loading into the MSC, with high levels of biological oxygen demand (BOD) and ammonia combined with low dissolved oxygen (DO). Such a scenario was

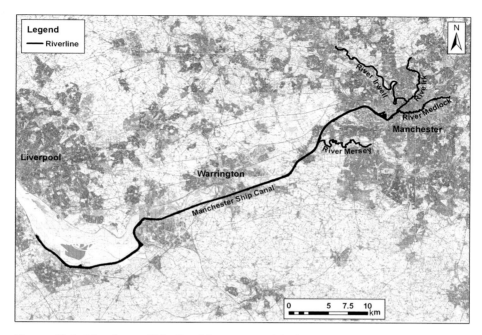

Figure 14.1. Manchester Ship Canal in context with the Mersey catchment.

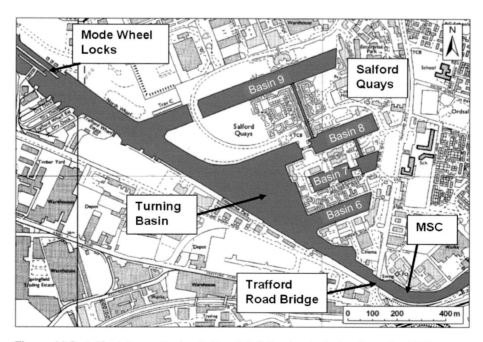

Figure 14.2. Salford Quays (Basins 7, 8 and 9) following isolation from the MSC and the Turning Basin.

an inevitable consequence of the fact that the Irwell flows through the cities of Manchester and Salford, from which large quantities of anthropogenic waste are generated. Indeed, in dry weather over 90% of the Irwell's flow comprises sewage effluent from the various WwTWs and industrial effluents along the river. Much of the organic pollution found in the Rivers Irk and Medlock can also be attributed to inadequately treated sewage and, until recently, farm waste.[1] The construction of the sewer network is such that localised storm events also led to frequent Combined Sewer Overflow (CSO) spill events (Rees & White 1993), particularly in the Rivers Irk and Medlock. Large volumes of organic pollutants would therefore enter the Irwell adding to the dry weather load.

The deep (up to 9 m), slow flowing water and steep vertical sides resulting from the canalisation of the River Irwell to form the MSC exacerbate the water quality problems (Hendry *et al.* 1997). Canalisation fundamentally changes the hydraulic regime of tributaries by slowing the flow, which has a profound effect upon water quality (Hellawell 1989). The consequential high residence time renders pollutants difficult to flush out and particulate contaminants settle readily onto the sediment layer, building up gradually to eventually create a thick layer of severely contaminated sediment. A non-uniform annual deposition of 0.5 m of fresh sediments occurs within the Turning Basin (APEM 1996). Sediment depth varies from 1.2 m to 4.2 m between Woden Street Footbridge and Mode Wheel Locks, and the total quantity of accumulated sediment is estimated at 460 000 m^3 (APEM 1996).

These sediments exert a demand for oxygen (Sediment Oxygen Demand – SOD) on the overlying water that, combined with the high retention time, facilitate stratification and bottom water anoxia during the warmer months. A confined anoxic layer in the lower depths of stratified water does not present a significant problem to pelagic fish populations. However under certain conditions, for example, following a CSO event coincident with warm dry weather, the deoxygenated layer can extend toward the surface and create total water column anoxia. Anoxia can occur in the absence of a CSO release during prolonged spells of hot, dry weather, particularly when there is little flow through the locks and minimal aeration from wind-driven mixing. In addition to the inadequate mixing and SOD, the high water column BOD and ammonia contribute to the oxygen deficiency (Hendry *et al.* 1997). Stratification prevents reaeration of the lower layer (hypolimnion) and when combined with a high BOD, also leads to anoxic sediments (Moss *et al.* 1986). Although the stratification eventually breaks down, the rate of atmospheric oxygen diffusion may still be exceeded by the BOD and/or SOD.

Sediment cores taken at various sites in the upper MSC and analysed for gas release and oxygen demand exerted on the overlying water highlighted the importance of SOD to the overall oxygen budget of the MSC (APEM 1990a, b).

---

[1] For a time, intensive pig farming in the Medlock Valley formed a significant contribution to the organic pollution load.

SOD in the MSC is spatially variable and changes according to sediment mobilisation during high flows and dredging activities, but peak values of over $1000\,mg\,m^{-2}\,hr^{-1}$ (*in situ*) have been recorded (HR Wallingford 1999). This is far in excess of what would be expected in unimpacted sites and is representative of heavily organically polluted sediments. Further research by Teesdale (2002) examining the spatial variation of SOD between Pomona Docks and Mode Wheel Locks revealed that SOD was highest within the Turning Basin and attributed it to greater particulate deposition due to the reduction in flow as the Irwell enters the MSC. The effect of newly deposited particulates on the sediment composition with depth was also highlighted in sediment core analysis in the upper MSC (APEM 1996). Surface sediment BOD averaged $1200\,mg\,kg^{-1}$ and reached peaks of almost $3\,000\,mg\,kg^{-1}$ in the Turning Basin, whilst at the majority of sites bottom sediment BODs averaged $630\,mg\,kg^{-1}$. The differences with depth were ascribed to the presence of recently deposited easily digested sewage-derived particulates at the surface whilst more inert carbon compounds, such as oil, would be prevalent in the lower and older (<100 years old) sediments. Surface carbon content was found to be around 20% in the Turning Basin (APEM 1996), twice that expected in an uncontaminated sediment (Donze 1990). Bottom sediments from areas near dock entrances at Pomona and Salford had carbon content in excess of 40%, which probably reflects past shipping activity.

The docks and channels upstream of Mode Wheel Locks provide an efficient settling area for sediment load brought down by the River Irwell. Low flow velocities are ensured by the maintenance of a standard operational water level in the dock area and by the need for deep water for navigation. Dredging is undertaken for flood defence and navigation purposes. Dredging used to be undertaken to maintain a water depth of 8.5 m, although in the past decade this has been reduced to 7 m. The location of the dredging varies from year to year and is partly dependent upon the flow conditions of that year, indeed some areas are not dredged for several years whilst other areas receive deposits of over 0.5 m each year. Dredging usually takes place between January and May with an estimated $60\,000\,m^3$ to $90\,000\,m^3$ of sediment being removed each year.

In addition to contributing to water column anoxia, the high organic content of the sediment is thought to play a role in sediment rafting, noxious gas generation and metal mobilisation (White *et al.* 1993; Boult & Hendry 1995) as well as providing a continual source of nutrients for potential algal growth. Sediment dredging is undertaken in the Turning Basin in response to flood defence requirements and navigational needs, a process that inevitably re-suspends sediment-bound nutrients. Recent total phosphorus concentrations in the water column ($530\,\mu g\,L^{-1}$, 2003–2007 average) are over an order of magnitude greater than required for algal proliferation. According to Reynolds[2]

---

[2] Professor Colin S. Reynolds – Institute of Freshwater Ecology.

(personal communication) a phosphorus concentration of less than $35\,\mu g\,L^{-1}$ is required to limit the algal population and less than $15\,\mu g\,L^{-1}$ is required to exert real control. Indeed phosphorus concentrations over $35\,\mu g\,L^{-1}$ are indicative of eutrophic water (OECD 1982). Presently, however, shading resulting from the turbid conditions prevailing in the MSC severely limit photosynthesis and therefore algal growth is currently constrained.

Sediment rafting and noxious gas generation was a major problem within the MSC in the late 1980s and was investigated by APEM between 1989 and 1990 (Webb 1993). The majority of sewage-derived organic sediments were thought to be deposited in the Turning Basin over the winter. As the water warmed in spring, microbial and fungal fibres grew through the newly deposited sediment layer binding it together (Boult & Hendry 1995). Due to the highly labile, organic nature of the sediments, microbial activity generated gases such as methane and hydrogen sulphide. These gases would accumulate within the cohesive sediment layer, making it buoyant and eventually causing it to slough off and erupt at the water surface. During the late 1980s, sediment rafts were so prolific that on occasion they could cover the entire surface of the 28-ha Turning Basin, giving the impression of solid ground. Anecdotal evidence suggests that during 1 week alone, three dogs drowned after jumping onto the surface of the rafts assuming they were solid.

Metal levels were found to be high throughout the sediment column in the upper MSC (APEM 1996) and exceeded the target environmental guidelines (Dutch Guidelines 1994). Concentrations of many metals increased with depth, reflecting a decrease in contaminating industrial discharges. Although the high concentrations of metals at the surface imply that the pollution entering the MSC is still relatively highly contaminated, alternative explanations include movement and reworking by resident organisms (bioturbation) and dredging which exposes old sediments. Certain metals including the so-called EC 'Black List' substances mercury and cadmium (EC Dangerous Substances Directive 76/464/EEC) were present at extremely high levels; maximum concentrations are given in Table 14.1.

The level of contamination can be put into context by comparing with contaminated sediments throughout the UK using data generated from the National Dock Survey of 1987 (Hendry *et al.* 1988). Lead, chromium and nickel were present in the MSC at higher concentrations than those recorded in other sediments, whilst concentrations of copper, cadmium and zinc were exceeded by those at only a single other site (Newcastle, Glasgow and London, respectively). Other industrial- and agriculturally derived pollutants such as pesticides and organics were generally found at comparably low concentrations, though occasionally elevated concentrations were recorded. These included PCBs and cyanide; the latter increased with sediment depth.

Table 14.1. *Maximum concentrations of selected metals recorded in sediments within the MSC together with target and intervention values[a] (Dutch Guidelines 1994)*

Metal	Maximum concentration (mg kg^{-1})	Target value (mg kg^{-1})	Intervention value (mg kg^{-1})
Copper	981	28	150
Zinc	1698	89	459
Lead	1574	72	451
Chromium	1963	58	219
Cadmium	20.5	0.8	12.5
Arsenic	82	24	45
Mercury	6.3	0.2	8
Nickel	81.3	14	83

Note: [a]The intervention values indicate the concentration of sediment contaminants above which its use for human, plant or animal life is defined as being seriously impaired. Target values indicate the sediment quality required for the full restoration of the sediment's use for human, animal and plant life.

In summary, the overriding issues driving the poor water quality in the MSC are three-fold: first, that the Mersey catchment is highly populated, creating a high sewage derived load to the canal. Second, the industrial legacy of the MSC means that the sediments contain high levels of persistent contaminants. Third, the change in hydraulic regime arising from the structure of the MSC, which was created for navigation, exacerbates the situation. It is unsurprising, therefore, that the MSC has experienced poor water quality since its construction. Harper (2000) recognised the MSC as the 'sump' of the catchment; thus, it would not be expected to behave as a 'normal' river. In fact the opposite was anticipated and serious future problems were predicted, as indeed has been the case.

Water quality and ecology: the industrial legacy

A summary of key water quality parameters in the MSC (April 1989 to May 1990) is given in Table 14.2, highlighting the condition of the water prior to remediation. The highly variable water quality in the early 1990s restricted the ecology to only the most pollution-tolerant organisms.

For many years, the macroinvertebrate diversity of the upper MSC was characterised by four or five pollution-tolerant detritivorous taxa. typical of poor water environments. The invertebrate community was dominated by a limited number of detritivorous, pollution-tolerant species, including worms (Oligochaeta), leeches (*Erpobdella* spp.), midge larvae (Chironomidae) and the water hog-louse (*Asellus aquaticus*) (White *et al.* 1993).

Table 14.2. *Mean and range (parentheses) for selected water quality parameters measured monthly from[a] – April 1989 to May 1990 at seven equidistant sites or[b] – August 1990 to July 1991 at two equidistant sites between Pomona Docks and Mode Wheel Locks (adapted from Hendry* et al. *1993)*

Parameter	Mean (range)
Dissolved oxygen (mg L^{-1})[a]	6.4 (0.0–13.1)
BOD (mg L^{-1})[a]	9.6 (2.5–36.0)
Ammonia (mg L^{-1})[a]	4.2 (0.1–9.3)
Suspended solids (mg L^{-1})[b]	16.3 (4.0–68.0)

Fish were absent from the MSC until the 1950s and only returned to some areas during the 1980s. Even then, only roach (*Rutilus rutilus*) and stickleback (*Gasterosteus aculeatus*) were found in the upper MSC (Hendry 1991). These species are able to withstand a high degree of pollution stress and are typical of eutrophic conditions (Jones 1964). Fish were absent during the summer months however presumably having migrated into the lower River Irwell to avoid asphyxia.

Indirect effects of bottom water anoxia include an alteration of the redox potential at the water-sediment interface which, in turn, can lead to mobilisation of phosphorus and some metals (Mortimer 1941). However there is little evidence of impacts of heavy-metal water and sediment contamination on fish condition through bioaccumulation up the food chain or direct toxicity. In a study of metal accumulation in biota in the lower Irwell (Critchley 1998), sediment bound metals appeared to be passed onto the invertebrate community. Accumulation of higher quantities of copper and zinc by *A. aquaticus* and *Erpobdella octoculata,* respectively, were attributed to different feeding habits resulting from differences in trophic level, metal detoxification strategy and the secretion of extracellular polysaccharides by *E. octoculata*. Results from analysis of sticklebacks found no evidence of biomagnification of metals through the food chain. Findings from a similar recent study of the lower Irwell and upper MSC supported this conclusion as sediments contained significantly more metals than macroinvertebrates, whilst macroinvertebrates such as *A. aquaticus* and *E. octoculata* contained higher levels than gudgeon (*Gobio gobio*) (Bassett 2005).

Notwithstanding periodic poor water quality influxes, principally due to CSO operation exacerbated by warm weather, gradual improvements to water quality and subsequently to ecology have occurred in the upper MSC during recent years. The improvements result from investments in WwTW infrastructure via the AMP[3]

---

[3] Asset Management Plan (AMP): a five-yearly investment cycle of maintenance, infrastructure, quality and environmental improvements undertaken by UK water companies.

process, although a substantial proportion of the progress is attributable to the remediation schemes, which are discussed below.

Salford Quays: the docks reborn

Following the closure of Manchester Docks in 1984 and in common with other dockland sites (Law 1988), the potential real estate value of the disused docks at Salford became highly attractive. However poor water quality precluded the initiation of such development due to aesthetically unpleasant floating rafts of sediment, bubbling gases and associated unpleasant odours. As a result, initiatives to improve conditions in the MSC began in the 1980s (Hendry et al. 1993) as part of an overall development plan supported by central government. Salford City Council (SCC) bought the land surrounding Salford Quays in the early 1980s (Law & Grime 1993), and regeneration was initiated in the mid 1980s.

SCC decided to construct bunds across the entrance to Docks 7, 8 and 9 and reroute surface water drainage; construction took place between 1986 and 1989 (Fig. 14.2). Warehouse demolition material was used to form the bunds which were capped to resemble the original dock walls (Hendry et al. 1993). A double lock into Basin 8 is the only remaining link into the MSC. As well as facilitating improved access around for construction vehicles, the immediate effect of impoundment was to eliminate the polluting load from the MSC and so allow improvements to the water quality to be considered. The isolation of some 20 ha of enclosed water created what is known today as Salford Quays.

Being deep and still waterbodies, the enclosed dock basins remained susceptible to stratification, as are similarly shaped natural waterbodies (e.g., Moss 1980; Horne & Goldman 1994). Temperature and DO profiles undertaken in Basin 7 in the summer of 1986 showed marked variations with depth, with the thermocline commonly occurring between 2 and 4 m (Montgomery 1988). Bottom water temperatures remained at 12–13° C, while towards the surface, temperatures of up to 18° C (occasionally $> 20°$ C) were observed. Anoxia was recorded in the bottom 4 m of water, whilst the surface water remained well oxygenated (80% saturation). Where algal blooms were present, supersaturation was recorded (Montgomery 1988).

To prevent the anticipated stratification and associated bottom water anoxia, Helixor mixing systems were installed into each basin shortly (between 6 and 16 months depending upon basin) after isolation. In total, the installed system comprises 16 Helixors: six in Dock 6 (open to the MSC), three in Basin 7, two in Basin 8 and five in Basin 9 (Walker et al. 1993). Water movement is achieved by compressed air from a land-based unit entraining water at the base of the Helixor tube, so enabling atmospherically oxygenated air to circulate to the bottom water (Fig. 14.3). Reviews of water quality in Manchester Docks before and after creation of Salford Quays by both Montgomery (1988) and Hendry (1991) found

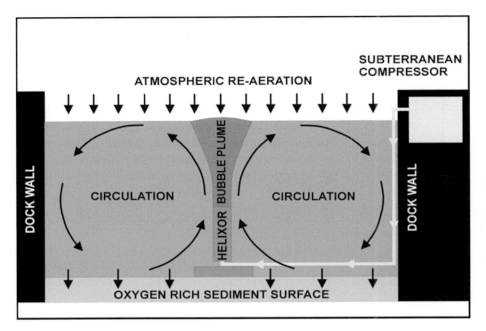

Figure 14.3. Diagrammatic representation of a Helixor mixing system.

a marked reduction in pollutants since isolation and demonstrated that the mixing system prevented bottom water anoxia and eradicated stratification. The Helixors created a uniformly high DO (>90% saturation) throughout the water column (Hendry *et al.* 1993); a notable achievement compared to the episodic bottom water anoxia experienced before and immediately following isolation (Fig. 14.4). Maintenance of the aerobic layer at the sediment-water interface is also demonstrated by the elimination of sulphide in the sediments (Walker *et al.* 1993).

The well-oxidised sediment–water interface reduced the release of phosphorus from the sediments and, over time, eliminated release altogether (e.g., as observed by Sas 1989). A general reduction in orthophosphate levels was observed after isolation, which continued following the installation of the Helixor mixing system. Concentrations generally remained below $0.05 \, \mathrm{mg \, L^{-1}}$ (Walker *et al.* 1993), whilst over the same period $1.67 \, \mathrm{mg \, L^{-1}}$ were present in the MSC. Recent 5-year average figures (2003–2007) for the Quays and the MSC are $0.023 \, \mathrm{mg \, L^{-1}}$ and $0.453 \, \mathrm{mg \, L^{-1}}$, respectively. Ammonia concentrations were reduced to $0.12 \, \mathrm{mg \, L^{-1}}$ in 1988 following isolation (Walker *et al.* 1993) and rarely approached $1 \, \mathrm{mg \, L^{-1}}$ (White *et al.* 1993), whilst levels prior to impoundment could be in excess of $9 \, \mathrm{mg \, L^{-1}}$ (Table 14.2). Further reductions have occurred, as illustrated in the most recent 5-year average (2003–2007) concentration of $0.035 \, \mathrm{mg \, L^{-1}}$ (APEM unpublished data).

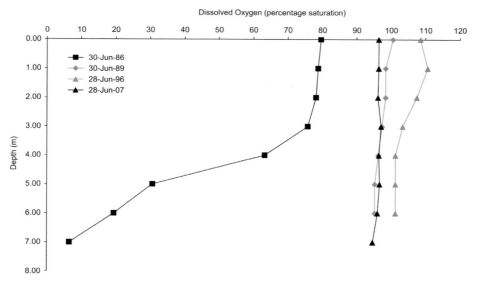

Figure 14.4. Typical dissolved oxygen profiles in Basin 7 shortly after impoundment (1986), following installation of the Helixors (1989), during the mid 1990s (1996) and recently (2007) (APEM unpublished data).

Dramatic reductions in ammonia have also occurred in the MSC, however, with a recent 5-year average of 1.46 mg L^{-1} (APEM unpublished data). Suspended solids concentrations decreased initially to around 3 mg L^{-1} from the previous levels of around 16 mg L^{-1} (Table 14.2), though development of the phytoplankton biomass disguised the reduction in suspended solids derived from the Irwell catchment (White *et al.* 1993). Owing to the low phytoplankton biomass and lack of polluting load, current suspended solids concentrations in the Quays are low, as indicated in the most recent 5-year average of 1.47 mg L^{-1} compared to 7.14 mg L^{-1} in the MSC during the same period (APEM unpublished data). The figure for the MSC nonetheless represents a substantial reduction from historic values provided above, albeit based on monthly sampling and, therefore, neglecting storm events which raised levels to >50 mg L^{-1} (APEM 2008a).

Isolation from the MSC and the resultant rapid decrease in suspended solids inevitably led to a concomitant increase in Secchi transparency. Initially, algal biomass in the Quays was low but increased rapidly as photic depth increased in the nutrient rich waters. The blue-green algae *Oscillatoria* dominated, increasing with both frequency and intensity over the next 4 years. The most severe blooms were recorded in the summer of 1992, when densities peaked at 5 m (filament length) ml^{-1} (Hendry *et al.* 1993). Throughout the following 5 years, the algal biomass remained high but showed a gradually decreasing trend, and by 1998 a stable and low biomass was achieved. It was originally anticipated

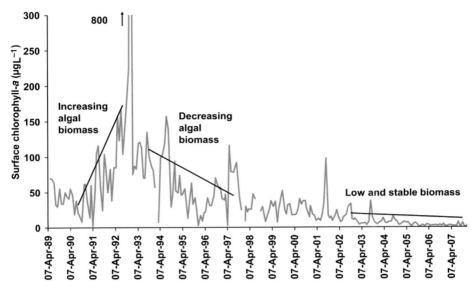

Figure 14.5. Chlorophyll-*a* concentrations in Salford Quays (Basin 7) between 1989 and 2007.

that the action of the Helixors would directly limit the growth of algae by intermittently circulating the cells into the aphotic zone at the lower depths (Radway *et al.* 1988; Horne & Goldman 1994). However Bellinger *et al.* (1993) suggested that the cells do not spend long enough in the aphotic zone to greatly reduce photosynthesis in the Quays. In the absence of nutrient additions to the enclosed waters, it was suggested that internal recycling initially played an important role in sustaining the algal blooms (Bellinger *et al.* 1993). Over the longer term, nutrients gradually became fixed in the oxygenated sediments and were, therefore, unavailable for algal growth. A linear regression equation on logged data from Basin 9 illustrated a significant correlation ($p < 0.01$) between *Oscillatoria* densities and total phosphorus concentrations over time (1989–2007). Whilst 'top-down' control mechanisms on algal biomass are commonly the most important in shallow lakes (Moss *et al.* 1994), other studies have found that 'bottom-up' control via alterations to nutrient availability and physical factors (e.g., mixing) can be most influential in deep waterbodies (McQueen 1990; Williams & Moss 2003).

Phytoplankton dynamics since isolation over time are mirrored closely by chlorophyll-*a* concentrations, as would be expected (Fig. 14.5). Stepwise multiple regression revealed that *Oscillatoria agardii*, and to a lesser extent the blue-green *Anabaena circinalis*, were the most significant contributors to algal biomass, the former having a correlation coefficient with chlorophyll-*a* of 0.610 ($p < 0.001$) (Hendry 1991). Chlorophyll-*a* concentration peaked at $800\,\mu g\,L^{-1}$ in

1992, after which a decreasing trend was apparent until low and stable conditions were reached and maintained post-2002. A population of *Dreissena polymorpha* (zebra mussel) has been present within the enclosed basins since the early 1990s; however it is unknown whether the filtering properties of this species played a part in reducing the phytoplankton biomass. Over the past few years, a diverse phytoplankton assemblage has been recorded composed of several groups, though dominance by chlorophytes has been maintained since 2000. The blue-green species *Microcystis* sp. has been recorded in the Quays during autumn since 2004, and the propensity for the cells to aggregate at the leeward end of the basins has caused densities to exceed the Environment Agency guideline thresholds of 20 colonies ml^{-1} (90 μm diameter) and 1.5 colonies ml^{-1} (200 μm diameter) (EA 2004) in these locations. However the densities are low and often undetectable at the routine sampling sites. There is therefore no evidence that current levels signify a reversion to the situation observed during the mid 1990s. Maintenance of the oxic sediment layer through Helixor mixing is crucial in retaining the 'trapped' nutrients and thus the continued successful control of algal biomass and protection of the water sports amenity within Salford Quays.

The rapid improvement in water quality, in particular the reduction in ammonia and suspended solids plus increased DO, facilitated diversification of other components of the Quays ecosystem. Zooplankton abundance initially increased in the enclosed basins from around 1 L^{-1} in Basin 9 in May 1988 to 500 L^{-1} by July due to a rise in copepod and to a lesser extent cladoceran populations (White *et al.* 1993). Similar seasonal changes occurred in the following 2 years, although the peak in 1990 was much lower. Total zooplankton density in all basins was positively correlated (stepwise multiple regression, $p < 0.05$) with algal biomass (as chlorophyll-*a* concentration; Hendry 1991). Although total zooplankton were significantly correlated with some algal groups ($p < 0.05$), no correlation were found with the blue-green algae *Oscillatoria* and *Anabaena*. Cladocerans, in particular *Daphnia longispina*, were however significantly correlated with blue-green algae ($p < 0.05$). The initial increase and subsequent relative decline in zooplankton abundance may have been a response to the shift in algal group dominance (White *et al.* 1993). It is widely recognised that blue–green species often form large inedible colonies, and this can be promoted by a reduction in other algal species by the grazing of large zooplankton (Lynch & Shapiro 1981; Williams *et al.* 2002). Therefore, 'top-down' control on the algal biomass from zooplankton was absent in the Quays.

Overpredation from stickleback also affected the zooplankton population (Hendry *et al.* 1997). Depletion of this resource and lack of an alternative food source caused the condition of stickleback to deteriorate, rendering them susceptible to secondary fungal infection by *Saprolegnia* spp. Periodic mass mortality of stickleback adversely affected the aesthetics of the water as fish

Table 14.3. *Average zooplankton group abundance in Salford Quays for two selected periods (number L^{-1}) (APEM unpublished data)*

	Basin 7		Basin 8		Basin 9	
	1990–94	2003–07	1990–94	2003–07	1990–94	2003–07
Copepods	5	12	6	11	6	8
Cladocerans	3	16	4	10	4	10
Rotifers	2	2	3	1	7	3

cocooned in fungus at the water surface suggested polluted water to the general public, although in actuality water quality was improving (Hendry *et al.* 1997).

Recent zooplankton abundance (average 2003–2007) reveals an increase in both copepod and cladoceran densities throughout all enclosed basins compared to the average in 1990–1994 (Table 14.3). Concurrently, rotifer species showed a decline in Basins 8 and 9. The increase in copepod and cladoceran densities, particularly in Basin 7, is likely due to the diverse habitat afforded by the natural colonisation and growth of macrophytes (Irvine *et al.* 1989) within the Quays (see below). The importance of aquatic plants as refuges for cladocerans against predaceous fish has been reported elsewhere (Moss *et al.* 1994). Significantly higher numbers of zooplankton were found in water lily beds (67 L^{-1}) compared to open water (6 L^{-1}) in a shallow lake (Beklioglu & Moss 1996).

Diversification of the macroinvertebrate community provided a further indication of the water quality improvements. The invertebrate community in the enclosed basins in the early years following isolation was dominated by species characteristic of organically polluted waters, and remained similar to those in the MSC. Whilst the pollution-tolerant species continued to dominate, species less tolerant to pollution such as the snail *Lymnaea peregra* became common (White *et al.* 1993). Various other species also colonised the basins, including mayfly (*Baetis* sp.), water bugs (*Corixa* sp.) and caddis flies (*Agraylea* sp. and *Phryganea bipunctata*). It was suggested that water quality would no longer restrict the invertebrate community but the homogeneous nature of the basins and lack of refuges and macrophytes would likely inhibit future diversification (White *et al.* 1993). This issue has subsequently been addressed (see below), and there has since been a further increase in diversity, with over 50 macroinvertebrate taxa being recorded in recent years, compared to just nine immediately following isolation of the dock basins (APEM 2008b).

The apparent instability in the fish community due to the very limited diversity, as demonstrated by the 'cocooned' sticklebacks described above, indicated that there was a need to ascertain whether the Quays could support a more diverse fish community. A small stocking trial was initiated in 1988, with 500 fish comprising roach, rudd (*Scardinius erythrophthalmus*), bream (*Abramis brama*) and carp (*Cyprinus carpio*) of varying age classes being introduced. The fish were monitored over a 12-month period for growth, diet, condition and contamination with heavy metals. Contamination by heavy metals released from the sediments was not marked in the stocked fish (Hendry 1991), though concentrations of zinc reached 138 mg kg^{-1} in indigenous stickleback (Hendry *et al.* 1997). This finding corroborates the view that maintaining an oxidised surface sediment layer segregates contaminants from the overlying water, reducing availability to biota. High growth rates were recorded, particularly of carp, roach and rudd. Due to the success of the pilot stocking, 12 000 coarse fish were stocked into the Quays in November 1989. Species added were similar to those in the trial, with the addition of chub (*Leuciscus cephalus*) and dace (*Leuciscus leuciscus*) to increase diversity, and also perch (*Perca fluviatilis*) principally as a predator. Stocking was deliberately restricted to the comparatively low density of 100 kg ha^{-1} compared with a typical carrying capacity of northern temperate lakes (300 kg ha^{-1}) to allow for expansion of the population. Fish were recaptured during subsequent years to enable assessment of their growth and condition. Growth of all fish was rapid during residence in the Quays due to the abundance of food, and roach growth rates were amongst the highest in the UK (Hendry *et al.* 1997). A Fulton Condition Factor (Fulton 1911) of 1.65 was calculated for fish in the enclosed basins, which is considered to compare favourably with normal populations elsewhere (White *et al.* 1993).

The prevailing homogeneity at the Quays, originating from its use as a port (deep basins, vertical walls and featureless sediment), was considered to severely limit the overall ecological stability of the system (Hendry *et al.* 1997). A programme of habitat diversification was therefore devised, with the aim of creating a more complex food web, promote colonisation and provide refuges and spawning substrates. Artificial structures intended to provide additional habitat and spawning substrates included tyre reefs, frayed ropes and discarded Christmas trees; the latter to mimic fronds of macrophytes. Submerged raised platforms were installed containing a number of aquatic macrophytes including bullrush (*Scirpus albescens*) and the common water lily (*Nymphaea alba*). The success of the introduction of macrophytes inspired further initiatives, including the use of submerged gabions to provide a base for additional macrophytes. In subsequent years, floating islands were introduced, planted with a combination of species including yellow iris (*Iris pseudocorus*), common reed (*Phragmites australis*) and purple-loosestrife (*Lythrum salicaria*).

These efforts to increase heterogeneity were necessary at a time when the prospect of natural colonisation by macrophytes was not considered a possibility, principally due to the basin depth. In recent years however water clarity within the Quays has increased rapidly and to a degree that has facilitated the

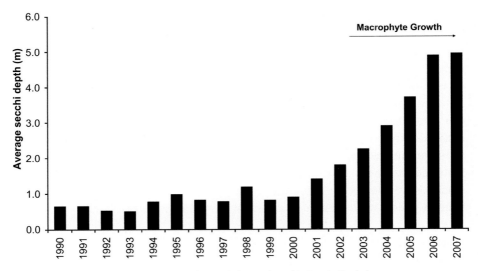

Figure 14.6. Annual average secchi depth in Basin 7 (St Louis Basin).

development of macrophyte stands (Fig. 14.6). Macrophyte biomass is pro-
foundly influenced by photic depth as it influences photosynthesis (Moss
1980; Carvalho *et al.* 2002). During a survey in 2007 (a combination of visual
assessment, bathyscope surveys and grapnel samples across a grid system)
seven macrophyte species and four filamentous algal species were recorded
(APEM 2008b). Percentage frequency analysis revealed the community to be
dominated by Nutall's water-weed (*Elodea nuttallii*) and to a lesser extent rigid
hornwort (*Ceratophyllum demersum*). A nationally scarce charophyte (*Nitella mucronata*) was also identified for a second year. Although *N. mucronata* is nationally
scarce, the species seems to be increasing in canal systems and in particular
favours restored canals (Nick Stewart[4] personal communication). Management
of the plant community is essential to optimise the recreational potential
of the Quays, and the annual weed cutting programme also aims to protect
N. mucronata from competition and excessive growth of other more prolific
species such as *E. nuttallii* and *C. demersum*.

Revitalising the MSC Turning Basin
Concurrently with the developments under way at Salford Quays initiated by
SCC, regeneration of the upper MSC was also being considered by the Mersey
Basin Campaign (MBC). The MBC was launched in 1985 and is coordinated by
the Department of the Environment, Transport and the Regions (DETR) (MBC

---

[4] Nick Stewart – UK charophyte referee for the Botanical Society of the British Isles and
author of the Red Data Book on stoneworts.

2000). It is an organisation composed of representatives from local authorities, local development corporations, North West Water (now United Utilities) and the Environment Agency (e.g., Webb 1993). The objective is to upgrade all watercourses in the region to at least an Environment Agency Class 2 classification by 2010 with the aim of supporting a diverse fish community, meeting European targets under the Water Framework Directive and encouraging waterside regeneration (MBC 2000). A 25-year programme was devised at an estimated total cost of £4 billion; £2 billion of United Utilities investment to improve the aesthetic and chemical quality of the waterways and £2 billion of further investment to develop the region's watersides (MBC 2000). Following a detailed water quality monitoring programme, a series of recommendations were proposed as part of an integrated management framework (Struthers 1993). However it was recognised that a 25-year regeneration programme is outside the generic development timescale and pressure from the local authorities resulted in short-term ameliorative measures to improve water quality (Hendry *et al.* 1993). Improving conditions adjacent to potential development sites between Woden Street Footbridge (subsequently altered to Trafford Road Bridge) and Mode Wheel Locks on the MSC was of particular interest to the MBC (APEM 2000) (Fig. 14.2). In this area of the MSC, the short-term measures were based around satisfying the oxygen requirements of the sediments and the water column (Rees & White 1993).

In a baseline study carried out by APEM in 1989, a semi-quantitative scale (0 to 3) was used to quantify the observed degree of bubbling caused by low DO. Regression analysis found a negative correlation with bottom water DO concentration ($p < 0.05$; Webb 1993). Reference to recorded DO concentrations for locations at which slight, moderate or severe gassing was noted indicated that severe gas generation was less likely to occur when DO concentrations were $4 \, \mathrm{mg \, L^{-1}}$ or higher; a minimum bottom water quality objective of $4 \, \mathrm{mg \, L^{-1}}$ was thus set (APEM 1989). However the elevated SOD and BOD combined with summer stratification created such a high oxygen demand that a mixing system (as in Salford Quays) would not be capable of raising the oxygen levels sufficiently to prevent bottom water anoxia. In addition, the canal's drainage and navigational functions prevent isolation as was possible with Salford Quays (Rees & White 1993). As a result of these factors, it was considered necessary to artificially oxygenate the water using pure oxygen.

A 3-month trial using a venturi oxygen injection (VITOX) device (Fig. 14.7) was undertaken during the summer of 1990 (APEM 1990b). Oxygen is injected via a venturi system into water drawn into each unit through mechanical screens. Fine bubbles are formed which readily dissolve under pressure and the oxygen-rich liquor is returned to the water body via a twin nozzle sparge system. The system was successful in elevating DO concentrations in the MSC, in common with other polluted waterbodies such as the River Thames

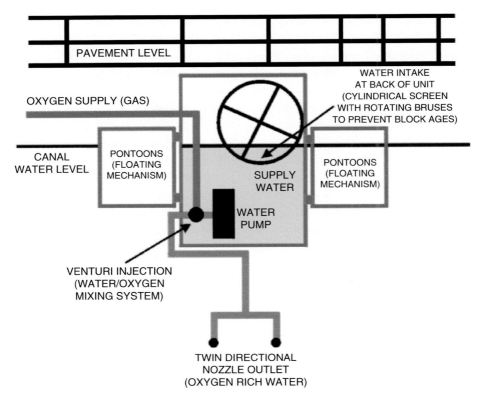

Figure 14.7. Schematic representation of an oxygenation unit.

(Lloyd & Whiteland 1990). A near continuous supply (>85%) of oxygen injection is required in the MSC during the summer, based on the frequency of DO concentrations falling to less than $4\,mg\,L^{-1}$ at temperatures of over $16°C$ as measured at Trafford Road Bridge (APEM 1990b). The extremely high SOD and associated eruption of sediment rafts meant that a certain amount of dredging was also necessary, as it was unlikely that oxygen enriched water would be able to penetrate the 0.5 m deep sewage-derived sediment layer.

Following the successful trial, designs for the upper MSC oxygenation system were finalised in February 2000 (APEM 2000). It was planned that the system would be operated for a total of 10 years during the summer months when the poorest water quality in the MSC occurs. Several configurations were considered but, based upon technical, planning, security and navigational issues, the five units were located as shown in Fig. 14.8. The system comprises a pipeline which distributes oxygen from a single bulk storage tank located near Mode Wheel Locks to five oxygen injection units situated at strategic locations in the Turning Basin (Fig. 14.8). Each oxygen injection unit is capable of supplying 3 Te (tonnes) O_2 per day ($125\,kgO_2\,hr^{-1}$), with an overall maximum

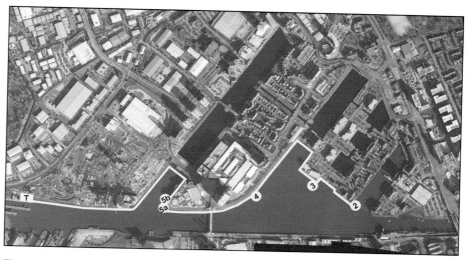

Figure 14.8. Oxygen injection unit locations (circles), pipeline and storage tank (T).

output of 15 TeO$_2$ per day. The oxygen supply infrastructure was completed in April 2001, and artificial oxygen injection commenced the following June. Major modifications to the injection nozzles were undertaken during the summer of 2001 to position the oxygen stream in the optimal vertical and horizontal trajectory, which was determined via experimentation. Field trials involved taking O$_2$ measurements from the surface and down through the water column at 1-m intervals. These profiles were taken at 10-m intervals across a horizontal grid. The number of nozzles (one or two) and their depth was altered as was the vertical and horizontal angle of the nozzles. Overlap between the units was minimised where possible. In addition the prevailing clockwise circulating current within the Turning Basin was utilised to further aid dispersion and mixing. Three-dimensional oxygen concentration contour plots for each nozzle configuration were created and visually assessed. It is believed that the current nozzle configuration maximises mixing and, therefore, dissipation of oxygen within the application area. A single injection unit was observed to increase bottom water oxygen concentrations to 4 mg L^{-1} within a 180-m radius with a secondary zone extending for up to 360 m (to the opposite bank of the Turning Basin) where an increase in oxygen concentration was seen (APEM 2000); water circulation patterns were also enhanced.

SOD in the MSC upstream of Mode Wheel Locks has, in general, reduced since 1989 (APEM 1999, 2000). However extremely high peak values still occur within the MSC and are continuing to contribute substantially to the overall oxygen demand. It is not known what effect the oxygen injection system will have on SOD over the long term, though we suggest that little reduction will be observed due to the continual 'rain' of organic material and the anoxic nature of the interstitial waters within the sediment itself. Nevertheless, the oxygenation system has been successful in elevating bottom water oxygen levels above the $4\,mg\,L^{-1}$ target value throughout the whole Turning Basin. A set of Helixors is used during the summer months to deliver compressed air and thus to help dissipate the oxygen enriched water within Basin 6, which is still connected to the MSC. The importance of a constant delivery of oxygen to the Turning Basin was highlighted during a period of inevitable oxygenation shutdown for maintenance work in 2003. Two days after switching the units off, bottom water oxygen fell to below the $4\,mg\,L^{-1}$ target and within a week was depleted still further to anoxia in some areas (Fig. 14.9). A similar occurrence happened in 2005, although for the first time the Helixor system alone was able to maintain the minimum DO target of $4\,mg\,L^{-1}$ within Basin 6, indicating that background conditions have improved (APEM 2006a).

Although the primary objective is to improve the aesthetic value of the MSC, the water quality improvements have also allowed the ecology within the system to develop and diversify. In addition, continual improvements are occurring in the area upstream of the artificial oxygenation, mirroring the long-term decline in suspended solids, BOD and ammonia described earlier. During the year of installation and commissioning (2001), similar BMWP[5] scores of around 15 were recorded in both the oxygenated and non-oxygenated (control) regions. Recent BMWP scores for the oxygenated and comparison regions were 28.7 and 21.6, respectively, representing increases of 91% and 44% from the initial score in 2001. Although at present statistically insignificant (t-test, $p > 0.09$), if the trend continues it would indicate that artificial oxygenation has allowed improvements to occur at a faster rate than would otherwise be possible.

In relation to the actual taxa present, for many years the macroinvertebrate diversity of the upper MSC has been very low, characterised by four or five pollution-tolerant species typical of poor water environments. Whilst these species remain, they are accompanied by an annually increasing number of species indicative of cleaner water. Taxa now recorded include caddis fly (*Phryganea bipunctata* and *Ecnomus tenellus*), freshwater shrimp (*Crangonyx pseudogracilis*), river limpet (*Ancylus fluviatilis*), flatworms (e.g., *Dugesia* sp. and *Dendrocoelum lacteum*) and damselfly (Coenagriidae). Particular increases in species diversity were recorded following the start of the artificial oxygenation (Fig. 14.10). In recent years, a peak of 41 taxa were identified in the oxygenated area of the upper MSC.

---

[5] Biological Monitoring Working Party Score: benthic macroinvertebrate families are allocated a score based on the tolerance to (organic) pollution and the score totalled (National Water Council 1981).

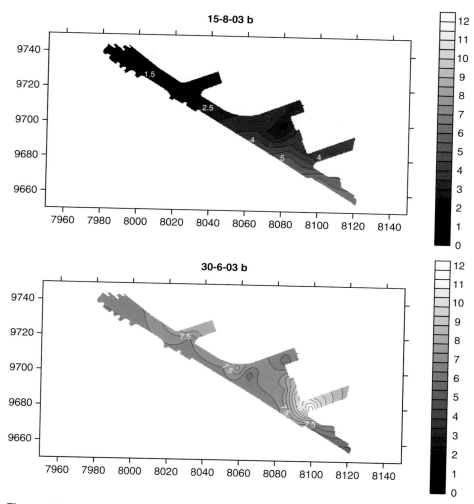

Figure 14.9. Typical bottom water dissolved oxygen contour plots used for oxygen dosing management comparing typical plots when the units are off (top) and on (bottom).

A. *aquaticus* continues to be the most abundant species within the upper MSC, though the number of individuals recorded has generally decreased since the commencement of oxygenation trials in 1998. This observation concurs with general ecosystem dynamics as conditions improve (e.g., Hawkes & Davies 1971). The result is that a greater number of species are represented, and there is a 'levelling out' in terms of relative abundance. As such, gradual changes in the invertebrate community composition are well illustrated by examining the shift in relative abundance of A. *aquaticus* and C. *pseudogracilis* (Fig. 14.11). The

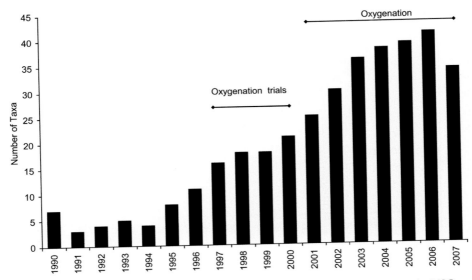

Figure 14.10. Total number of invertebrate taxa recorded in the Turning Basin MSC from 1990 to 2007.

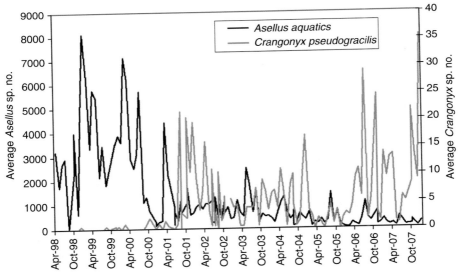

Figure 14.11. Numbers of *Asellus aquaticus* and *Crangonyx pseudogracilis* in the MSC (sampled using standard colonisation units (SCU)).

reduction in *A. aquaticus* and increase in *C. pseudogracilis* in 2001 coincided with the oxygenation system commissioning and initial operational phase.

Fish have also benefited from improvements to summer conditions resulting from the oxygenation remediation, enabling them to survive in the Turning

Basin all year round. Furthermore, the fish community has diversified into one that incorporates benthic species that, prior to oxygenation, could not have survived the anoxic conditions in these deeper parts of the water column. Recent netting data have revealed roach, perch, bream, chub and gudgeon (*Gobio gobio*), and a survey in 2004 revealed a marked shift in species compos-ition, with gudgeon dominating within the oxygenated region. The continued dominance of this species throughout the Turning Basin is significant as previous surveys undertaken between 1998 and 2000 (Nash *et al.* 2003) recorded extremely low numbers of gudgeon. Gudgeon are known to show plasticity in their habitat requirements, though they are described as being 'largely a fish of running water and only occasionally occurring in lakes . . . its oxygen require-ments are similar to those of grayling and bullhead' (Maitland & Campbell 1992). Clearly this benthic species is taking advantage of the improved condi-tions in the Turning Basin attributable to the summer oxygenation scheme.

Despite the enhanced oxygen levels in the Turning Basin, the depth and lack of spawning habitat renders this area unsuitable for reproduction and rearing of juvenile fish to sustain the current population. However length–frequency calculations on roach data from 2007 showed a comparatively wide range of lengths in the upper MSC compared to the lower sections of the Canal, having between three and five age cohorts (APEM 2008a). A distinct 0+ age class in the area shows recent recruitment and suggests spawning activity is present either within the Canal or in its feeder streams. Indeed results from electric fishing undertaken in the River Irwell in 2005 revealed important nursery areas for juvenile fish in the shallower margins immediately downstream of the entrance to Wilburn Street Basin 2.5 km from the Turning Basin (APEM 2005). Over 300 predominantly juvenile fish were caught, including chub, gudgeon, perch, roach and stickleback. It is suggested that, as these fish age, they will move downstream in search of new habitats and thus help sustain the fish populations in the oxygenated Turning Basin. The presence of juvenile fish also highlights the importance of generic water quality improvements within the catchment in helping to sustain and, in the future, enhance these nursery areas. Investments by United Utilities in tandem with catchment management research by the Mersey Basin Campaign are gradually resulting in tangible water quality improvements through reduced BOD, suspended solids, ammo-nia and nutrient concentrations (APEM 2006a).

Crucially, the MSC has been designated under the Freshwater Fisheries Directive (FFD) (78/659/EEC) as a cyprinid fishery from the River Irwell near Salford University to the freshwater limit of the Canal at Latchford Locks (22 km of the MSC). Despite the infrastructure investments and scientific research described above, the quality of water in the MSC does not currently meet the EC FFD standards, which was the impetus behind a recent intensive study undertaken by APEM (APEM 2008a). Results from fish sonar surveys

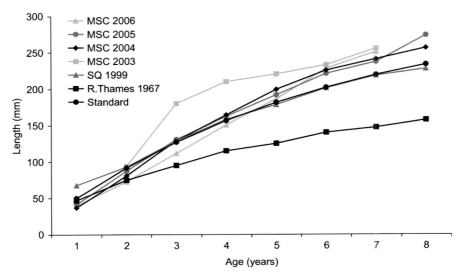

Figure 14.12. Growth rates for roach in the MSC, Salford Quays and the River Thames (after Williams 1967). Also shown is the Hickley Growth Standard Curve.

during winter months revealed shoals to be present throughout the Turning Basin. It seems likely that, during the winter months, fish may favour such areas in order to take advantage of the reduced flows as a way of maximising bioenergetic efficiency. During May however very few fish were recorded in the Turning Basin during the day, and it was suggested that the aggregation of spawning adults in this section of the MSC relocate to habitats upstream of the survey reach (i.e., the feeder streams) to spawn during the day. The free passage to the lower River Irwell provides greatly enhanced spawning and feeding potential due to increased habitat complexity available in the lower river. The fish retreated into the MSC overnight suggesting that selection of this area was due to the maintenance of oxygen levels through oxygenation, as it is unlikely to be for flow avoidance as flows would be relatively low in all stretches of the lower river at this time of year.

A measure of fish condition is provided by Fulton's Condition Factor ($K = W/L^3$, where W is weight and L is length) (Fulton 1911). The formula is based on the assumption that for a given length a heavier fish is in better condition (e.g., Bagenal & Tesch 1978). Average figures for roach and perch from the MSC in 2006 were 1.61 and 1.57, respectively, being comparable to recent figures obtained from the adjacent clean waters of Salford Quays (1.70 and 1.53) and thus indicating that the fish were in very good condition (APEM 2006b). Growth curves of MSC roach captured in 2006 followed the expected trend of rapid growth in younger fish, with growth becoming slower in adults (Fig. 14.12). Back-calculation revealed that fish at age 1+ were very similar in size to those in 2004 and 2005. Up to age 4+, roach were smaller than suggested by the Hickley Growth Standard (Hickley & Dexter 1979), which may indicate stunting through increased competition for food and spawning habitat. Of particular interest is that a consistent rate of growth (denoted by the slope of the line) of these younger fish was sustained, possibly due to the ample food supply, such that age 5 + fish were larger than the standard would suggest. It is important to note that the sample size of scaled fish in 2006 was smaller ($n = 17$) than previous years (in 2003 $n = 53$), which may affect the overall shape of the line. Nevertheless, the current trends in fish growth are likely to continue until the community stabilises and ultimately, if it is assumed that all individuals exhibit similar growth characteristics albeit with some degree of variability, the growth curve superimposes the Hickley Growth Standard (Hickley & Dexter 1979). A similar situation occurred within the comparatively clean waters of Salford Quays in 1999 following years of water quality improvements and 10 years after the introduction of the first fish. It is likely that long-term improvements in WwTW infrastructure through the AMP process would lead to a similar trend in the MSC, albeit over an extended time frame, although the oxygenation scheme has undoubtedly accelerated the rate of improvement in condition. As a measure of the achievement to date, also given in Fig. 14.12, is the curve derived from roach within the River Thames during the 1960s (Williams 1967) which shows markedly slower growth rates.

The sewage-derived effluents continuing to enter the MSC contain endocrine disrupting chemicals largely originating from anthropogenic oestrogen (the natural steroids oestrone and 17β-oestradiol) and synthetic steroids (ethinylestradiol), which can cause sex reversal of fish. This is an increasing problem in many inland waters running through heavily populated areas and leads to the feminisation of the fish population. Conlan et al. (2006) provides a comprehensive review of the subject and includes results from a study investigating intersex in roach and perch in the MSC and Salford Quays in 2000 (Nash et al. 2003). Results showed that the sex ratio in roach was biased towards females in the MSC and Salford Quays, though to a far lesser extent in the latter (Conlan et al. 2006). In addition, the incidence of intersex revealed that this condition was twice as common in males from the MSC as for those from Salford Quays. However, the discovery of intersex individuals in Salford Quays requires further investigation given that these waters have received no sewage derived inputs since isolation 20 years ago. It was suspected that potentially oestrogenic compounds may be present in the sediments originating prior to the isolation. The high level of intersex in the MSC is comparable with studies of roach in other polluted waterways (e.g., Jobling et al. 1998). It was postulated that the level of intersex and the highly skewed sex ratio may be due to oestrogenic effects of chemicals in the MSC (Conlan et al. 2006). The study was the first to identify intersex in male perch; in the MSC 100% of males had ovotestes.

In contrast, none of the perch sampled in Salford Quays had intersex gonads. The apparent difference in sensitivity between perch and roach in Salford Quays was ascribed to different routes of exposure through feeding. Perch feed on zooplankton and other fish whilst roach tend to consume a relatively large amount of detritus whilst foraging in the silt – thus increasing their exposure to potentially contaminated sediments (Conlan *et al.* 2006). The overall impacts of feminisation on fish populations are, as yet, poorly understood (Conlan *et al.* 2006).

Future challenges

Considering that the waterside development and ecological recovery at Salford Quays, in common with similar developments both nationally and internationally (Hoyle *et al.* 1988), is due to the water quality improvements, the importance of routine monitoring and management of the water itself cannot be understated. The success of Salford Quays since 1986 has been possible only through development and implementation of a holistic and dynamic monitoring and management plan. For example, the recent natural colonisation of macrophytes in the Quays was a prospect previously not envisaged. Nevertheless, the management strategy has been modified to account for this and other developments of the ecosystem such as the initiation of sampling the leeshore in response to the presence of *Microcystis* sp. from the early 2000s.

Implementation of oxygenation remediation within the Turning Basin in 2001 has resulted in a dramatic improvement in water quality, aesthetic value and ecology of the upper reaches of the MSC. The number of pollution sensitive macroinvertebrate species has increased and BMWP scores suggest that the rate of increase in the Turning Basin has been accelerated as a result of the elevated oxygen concentrations. Gudgeon which were absent from the MSC for many decades are now ubiquitous and in fact dominate recent catches in the oxygenated region. Condition of roach, as determined by Fulton's Condition Factor, is also excellent, being similar to recent figures for fish in Salford Quays. Despite improvements in water quality over the last few years however the MSC remains a borderline habitat for fish populations. Episodic pollution and major habitat limitations such as spawning habitat, nursery areas and food resources for juveniles in particular still restrict the potential for self-sustaining populations in the longer term.

A further factor affecting the future success of the MSC regeneration is the requirement for frequent dredging. Dredging is essential for navigation and flood control, though the stresses exerted on water column oxygen concentrations can be severe, particularly during the summer months. This leads to the possibility of widespread mortalities of macroinvertebrates and fish. Moreover, low oxygen concentrations could lead to the release of sediment bound phosphorous which, if matched with a reduced sediment load and thus turbidity, could in turn promote algal blooms. In addition, toxic heavy metals and persistent organics may be released into the water column under anoxic conditions and general dredging disturbance. Late spring and summer dredging operations have the potential therefore to lead to serious deleterious effects in the Turning Basin.

An additional consideration in the MSC is the potential for metals to be mobilised from sediments under oxidising conditions (Forstner 1993). In a laboratory investigation to elucidate the dynamics of metal transport from a metal-contaminated and eutrophic lake, Nguyen *et al.* (2007) found that zinc, copper and cadmium were removed from solution during the anoxic period and released to solution during oxic conditions. In contrast, iron was rapidly released to the water under anoxic conditions, but taken up during oxic conditions. Xue *et al.* (1997) found that hypolimnetic oxygenation in a eutrophic lake enhanced the release of copper from the sediment but also accelerated the entrapment and deposition of copper and zinc by freshly formed manganese and iron oxides. However the oxic surface layer was suggested as an effective retainer or 'lid' for arsenic in a saline lake (Lyons & Lebo 1997). Riedel *et al.* (1997) undertook microcosm experiments to investigate metal fluxes from sediment under variable water column oxygen concentrations (saturated, 10% saturation and anaerobic). Under saturated or 10% oxygen saturation, arsenic fluxes were negligible, whilst in contrast, copper fluxes out of the sediment increased with increasing oxygen concentrations.

Within the MSC, the effect of the artificial oxygenation remediation on the surface sediments is not fully understood at present, though the beneficial impact is thought to be limited by the constant 'rain' of organic material deposited to the sediments. Surface water arsenic, cadmium, chromium, lead and mercury concentrations have been monitored biannually since 2002 in Salford Quays in compliance with the EC Bathing Waters Directive (76/1260/EEC). Most metals have remained at extremely low concentrations, being below the limits of detection. Arsenic however is detectable, though it remains at very low concentrations (maximum of $3.8 \mu g \, L^{-1}$) and below the $50 \mu g \, L^{-1}$ stipulated in the EC Surface Water Abstraction Directive (75/440/EEC). Moreover, as discussed earlier, there has been no evidence to suggest an adverse effect of metal contamination on the biota within either the MSC or Salford Quays.

Currently, high turbidity in the MSC reduces the photic depth and hence, algal photosynthesis. However with the anticipated reduction in suspended solids in effluents following the AMP investment by United Utilities, together with the currently high nutrient levels (augmented by dredging operations), the situation is likely to be reversed in the future. Based on evidence from Salford Quays and also Preston Riversway Dock (APEM 1991), we suggest that severe blooms of blue-green algae may occur and will represent a serious threat to both the recreational and aesthetic value of riparian developments along the canal. Higher organisms may also be affected due to changes in water chemistry via pH-mediated toxicity of unionised ammonia (Alabaster & Lloyd 1980). Algal blooms in the upper MSC will also affect the oxygenation unit operation. First, night-time dissolved oxygen sags will become more pronounced due to the increased oxygen requirements of respiration. Second, death of the algal cells will probably cause the BOD and SOD to increase, thereby raising the amount of oxygen required to maintain the minimum DO level of $4\,mg\,L^{-1}$.

The Mersey Basin Campaign via its Science Group and Healthy Waterways Trust (HWT) is currently considering the long-term management of algal blooms in the MSC as part of a large catchment-wide strategic study. Management will inevitably entail a combination of both reductions in nutrient inputs into the Irwell catchment (phosphate stripping and reductions in agricultural inputs) as well as methods of sustaining oxygen-rich bottom waters. Longer term improvements to WwTW effluents in terms of nutrient reductions are anticipated through Water Framework Directive-driven initiatives.

In light of these pressures, it is apparent that some form of aeration/mixing will be required in the upper MSC for the foreseeable future. Evidence from earlier studies show that conditions have improved sufficiently to enable the minimum target of $4\,mg\,O_2\,L^{-1}$ to be achieved by Helixors in Basin 6 during periods of oxygenation shutdown, when other areas of the Turning Basin experienced depleted oxygen concentrations. There are many types of aeration and mixing systems available, and a thorough evaluation will be required to determine the most suitable technique based on oxygen demand and the influence of projected algal blooms on operating efficiency and cost. The partnership between the Manchester Ship Canal Company, United Utilities, APEM and HWT will be essential in developing future plans within the AMP5 budget.

Conclusions

The success of Salford Quays is illustrated by the balance achieved between encouraging a diverse and stable ecosystem whilst retaining a recreational facility and is testimony to the holistic programme of management. The water complies with the requirements of the EC Bathing Waters Directive; thus, the Quays are used as a major water sports centre for the North West and successfully hosted the triathlon in the Commonwealth Games of 2002 and subsequent World Cup events. The area contains over 200 businesses and 2 000 homes. It is clear that earlier initiatives to rectify the legacy of pollution in the MSC have also been highly successful. However there is still a long way to go in terms of sustaining and furthering the improvements in the upper reaches of the MSC. The success of Salford Quays shows that, with adequate research followed by imaginative and holistic management, a sustainable solution to poor water quality can be achieved.

References

Alabaster, J. S. and Lloyd, R. (1980) Ammonia. In: *Water Quality Criteria for Freshwater Fish*. Butterworths, London.

APEM. (1989) *Mersey Basin Campaign Central Catchment Group Water Quality Study. Historical Data Review and Engineering Appraisal. Volume II: Desk Study*. APEM, Manchester, UK.

APEM. (1990a) *Mersey Basin Campaign Central Catchment Group Water Quality Study. Volume I: Water Quality Review*. APEM, Manchester, UK.

APEM. (1990b) *Mersey Basin Campaign Central Catchment Group Water Quality Study. Volume II: Engineering Appraisal*. APEM, Manchester, UK.

APEM. (1991) *Water Quality Management Options at Preston Dock. Desk Study Report*. APEM, Manchester, UK.

APEM. (1996) *Sediment Accumulation and Contamination in the Manchester Ship Canal – Implications for Water Quality. Final Report*. APEM, Manchester, UK.

APEM. (1999) *Manchester Ship Canal Harbour Project – Sediment Oxygen Demand Re-visit*. APEM, Manchester, UK.

APEM. (2000) *Manchester Ship Canal Harbour Project – Historical Review*. APEM, Manchester, UK.

APEM. (2005) *Manchester Ship Canal Oxygenation Project – Ecological Improvements: 2005*. APEM, Manchester, UK.

APEM. (2006a) *Manchester Ship Canal Oxygenation Project: 2005*. APEM, Manchester, UK.

APEM. (2006b) *Manchester Ship Canal Oxygenation Project – Ecological Improvements: 2006*. APEM, Manchester, UK.

APEM. (2008a) *Manchester Ship Canal AMP4 Water Quality Investigation. Data Collection Report*. APEM, Manchester, UK.

APEM. (2008b) *Salford Quays Specialised Scientific Support Services: 2007*. APEM, Manchester, UK, In prep.

Bagenal, T. and Tesch, F. W. (1978) Age and growth. In: *Methods for Assessment of Fish Production in Fresh Waters* 3rd edn (ed. T. Bagenal), Chapter 5, IBP Handbook No. 3. Blackwell Scientific Publications, London.

Bassett, R. P. (2005) Metal pollution within sediment, invertebrates and gudgeon *(Gobio gobio)* tissue from the Manchester Ship Canal, Lancashire, England. Unpublished BSc Thesis, Manchester Metropolitan University, UK.

Beklioglu, M. and Moss, B. (1996) Existence of a macrophyte-dominated clear water state over a very wide range of nutrient concentrations in a small shallow lake. *Hydrobiologia* **337**, 93–106.

Bellinger, E. G., Hendry, K. and White, K. N. (1993) Nutrient cycling in a closed dock and its biological implications. In: *Urban Waterside Regeneration: Problems and Prospects* (eds. K. N. White, E. G. Bellinger, A. J. Saul, A. J. Symes and K. Hendry), Chapter 39. Ellis Horwood, London.

Boult, S. and Hendry, K. (1995) Gas generation in the Manchester Ship Canal. *Open University Geological Society Journal* **16**, 1–5 (Symposium Edition).

Carvalho, L., Bennion, H., Darwell, A. *et al.* (2002) *Physico-chemical Conditions for Supporting Different Levels of Biological Quality for the Water Framework Directive for Freshwaters*. R&D Technical Report. Environment Agency, Bristol, UK.

Conlan, K., Hendry, K., Fawell, J., Williams, A. E. and Hubble, M. (2006) *Endocrine Disrupters: A Review of the Science Underpinning the ED Research Programme (TX/04)*. UK Water Industry Research, London.

Critchley, B. (1998) An investigation into the release of trace metals from contaminated sediments and metal accumulation by the biota in a section of the lower reaches of the River Irwell, Manchester. Unpublished MSc Thesis, University of Manchester, UK.

Donze, M. (1990) *Aquatic Pollution and Dredging in the European Community*. Published at the Occasion of the Eleventh Lustrum of the Association of Dutch Dredging Contractors. Delwel, The Hague the Netherlands.

Dutch Guidelines. (1994) *Ministry of Housing, Spatial Planning and Environment*. The Hague, the Netherlands.

Environment Agency (EA). (1990) *River Quality 1990*. According to the NWC

Classification Scheme (Map). Environment Agency.

Environment Agency (EA). (2004) *Guidance for Cyanobacterial Monitoring and Management of Incidents.* National Centre for Ecotoxicology and Hazardous Substances, Wallingford, Oxfordshire, UK.

Farnie, D. A. (1980) The Ship Canal company and the corporation of Manchester. In: *The Manchester Ship Canal and the Rise of the Port of Manchester 1894–1975*, Chapter 1. Manchester University Press, Manchester, UK.

Forstner, U. (1993) Metal speciation – general concepts and applications. *International Journal of Environmental Analytical Chemistry* **51**, 2–23.

Fulton, T. W. (1911) *The Sovereignty of the Sea.* Blackwood, Edinburgh, UK.

Gray, E. (2000) *Salford Quays – The Story of the Manchester Docks.* Memories, Manchester, UK.

Gray, T. (1993) *A Hundred Years of the Manchester Ship Canal.* Memories, Manchester, UK.

Harper, E. (2000) *The Manchester Ship Canal Water Quality. The 20th Century in Perspective and Recommendations for the New Millennium.* Report for the Environment Agency, Manchester Ship Canal Company, North West Water and the Mersey Basin Campaign.

Hawkes, H. A. and Davies, L. J. (1971) Some effects of organic enrichment on benthic invertebrate communities in stream riffles. In: *The Scientific Management of Animal and Plant Communities for Conservation* (eds. E. Duffey and A. S. Watt). Blackwell Scientific Publications, London.

Hellawell, J. M. (1989) *Biological Indicators of Freshwater Pollution and Environmental Management.* Elsevier Applied Science, London.

Hendry, K. (1991) The ecology and water quality management of disused dock basins and their potential for alternative uses. Unpublished PhD Thesis, Department of Environmental Biology, University of Manchester, UK.

Hendry, K., Bellamy, W. M. and White, K. N (1997) Environmental improvements to

enhance and develop freshwater fisheries – Salford Quays, a UK case study. In: *Fisheries and the Environment: Beyond 2000* (eds. B. Japar Sidik, F. M. Yusoff, M. S. Mohd Zaki and T. Petr). Universiti Putra Malaysia, Serdang, Malaysia.

Hendry, K., Webb, S. F., White, K. N. and Parsons, A. N. (1993) Water quality and urban regeneration – a case study of the Central Mersey Basin. In: *Urban Waterside Regeneration: Problems and Prospects* (eds. K. N. White, E. G. Bellinger, A. J. Saul, A. J. Symes and K. Hendry), Chapter 31. Ellis Horwood, London.

Hendry, K., White, K. N., Conlan, K. *et al.* (1988) *Investigation into the Ecology and Potential for Nature Conservation of Disused Docks.* Nature Conservancy Council Contract Report, HF3–11–52 (3).

Hickley, P. and Dexter, K. F. (1979) A comparative index for quantifying growth in length of fish. *Fisheries Management* **10**, 147–151.

Horne, A. J. and Goldman, C. R. (1994) *Limnology.* 2nd edn. McGraw-Hill, London.

Hoyle, B. S., Pinder, D. A. and Husain, M. S. (1988) *Revitalising the Waterfront.* Bellhaven, London.

H. R. Wallingford (1999) *Manchester Ship Canal – Measurement of in-situ SOD between Latchford and Mode Wheel Locks.* Report EX4109.

Irvine, K., Moss, B. and Balls, H. (1989) The loss of submerged plants with eutrophication II. Relationships between fish and zooplankton in a set of experimental ponds, and conclusions. *Freshwater Biology* **22**, 89–107.

Jobling, S., Nolan, M., Tyler, C. R., Brighty, G. and Sumpter, J. P. (1998) Widespread sexual disruption in wild fish. *Environmental Science and Technology* **32**, 2498–2506.

Jones, J. R. E. (1964) *Fish and River Pollution.* Butterworths, London.

Kivell, P. (1993) *Land and the City – Patterns and Processes of Urban Change.* Routledge, London.

Law, C. M. (1988) Urban revitalisation, public policy and the redevelopment of redundant port zones – lessons from Baltimore and Manchester. In: *Revitalising the Waterfront*

(eds. B. S. Hoyle, D. A. Pinder and
M. S. Husain). Bellhaven, London.

Law, C. M. and Grime, E. K. (1993) Salford
Quays 1: the context. In: *Urban Waterside
Regeneration: Problems and Prospects* (eds. K. N.
White, E. G. Bellinger, A. J. Saul, A. J. Symes
and K. Hendry), Chapter 9. Ellis Horwood,
London.

Lloyd, P. J. and Whiteland, M. R. (1990) Recent
developments in oxygenation of the tidal
Thames. *Journal of the Institute of Water and
Environmental Management* **4**, 103–111.

Lynch, M. and Shapiro, J. (1981) Predation,
enrichment and phytoplankton
community structure. *Limnology and
Oceanography* **26**, 86–102.

Lyons, W. B. and Lebo, M. E. (1997) Observations
on the diagenetic behaviour of arsenic in a
saline lake: Pyramid Lake, Nevada. *International
Journal of Salt Lake Research* **5**, 329–335.

Maitland, P. S. and Campbell, R. N. (1992)
Freshwater Fishes of the British Isles. Harper
Collins, London.

McQueen, D. J. (1990) Manipulating lake
community structure. Where do we go
from here? *Freshwater Biology* **23**, 613–620.

Mersey Basin Campaign (MBC). (2000) *Progress
Report (1985–2000)*. Mersey Basin Campaign,
Manchester, UK.

Montgomery, T. M. (1988) The future of the
Manchester Ship Canal – implications of
land use and water quality in the upper
reaches. Unpublished MSc Thesis,
Department of Environmental Biology,
University of Manchester, UK.

Mortimer, C. H. (1941) The exchange of dissolved
substances between mud and water in
lakes. I. *Journal of Ecology* **29**, 280–329.

Moss, B. (1980) *Ecology of Freshwaters*. Blackwell
Scientific Publications, London.

Moss, B., Balls, H., Irvine, K. and Stansfield, J.
(1986) Restoration of two lowland lakes by
isolation from nutrient rich water sources
with and without removal of sediment.
Journal of Applied Biology **23**, 391–414.

Moss, B., McGowan, S. and Carvalho, L. (1994)
Determination of phytoplankton crops by

top-down and bottom-up mechanisms in a
group of English lakes, the West Midland
Meres. *Limnology and Oceanography* **39**,
1020–1029.

Nash, K. T., White, K. and Hendry, K. (2003) *The
Effect of Water Quality on Coarse Fish Productivity
and Movement in the Lower River Irwell and
Upper Manchester Ship Canal: A Watercourse
Recovering from Historical Pollution*. APEM,
Manchester, UK.

National Water Council. (1981) *River Quality –
The 1980 Survey and Future Outlook*. HMSO,
London.

Nguyen, L. T., Lundgren, T., Håkansson, K. and
Svensson, B. H. (2007) Influence of redox
cycle on the mobilization of Fe, Zn, Cu and
Cd from contaminated sediments: a
laboratory investigation. In: *Water Resources
Management IV* (eds. C. A. Brebbia and
A. G. Kungolos). WIT Press, Southampton, UK.

Organisation for Economic Cooperation and
Development (OECD). (1982) *Eutrophication of
Waters, Monitoring Assessment and Control*.
Organisation for Economic Cooperation
and Development, Paris.

Radway, A., Walker, L. S. and Carradice, I. (1988)
Water Quality Improvements at Salford
Quays. *Journal of the Institute of Water and
Environmental Management* **2**, 523–531.

Rees, A. and White, K. N. (1993) Impact of
combined sewer overflows on the water
quality of an urban watercourse. *Regulated
Rivers Research and Management* **8**, 83–94.

Riedel, G. F., Sanders, J. G. and Osman, R. W.
(1997) Biogeochemical control on the flux
of trace elements from estuarine sediments:
water column oxygen concentrations and
benthic infauna. *Estuarine, Coastal and Shelf
Science* **44**, 23–38.

Sas, H. (1989) *Lake Restoration by Reduction of
Nutrient Loadings: Expectations, Experiences,
Extrapolations*. Academia Verlag Richarz,
Sant Augustin.

Struthers, T. (1993) The Greater Manchester
experience. In: *Urban Waterside
Regeneration: Problems and Prospects* (eds. K. N.
White, E. G. Bellinger, A. J. Saul, A. J. Symes

and K. Hendry, K), Chapter 8. Ellis Horwood, London.

Teesdale, C. (2002) The spatial variability of sediment oxygen demand in the Manchester Ship Canal. Unpublished MSc Thesis, University of Manchester, UK.

Walker, L. S., Carradice, I. and Warren, R. S. (1993) Improvements in water quality at Salford Quays (1986–1990). In: *Urban Waterside Regeneration: Problems and Prospects* (eds. K. N. White, E. G. Bellinger, A. J. Saul, A. J. Symes and K. Hendry), Chapter 32. Ellis Horwood, London.

Webb, S. F. (1993) The evaluation of oxygen injection to effect water quality improvements in the Manchester Ship Canal. In: *Urban Waterside Regeneration: Problems and Prospects*, (eds. K. N. White, E. G. Bellinger, A. J. Saul, A. J. Symes and K. Hendry), Chapter 24. Ellis Horwood, London.

White, K. N., Hendry, K. and Bellinger, E. G. (1993) Ecological change as a consequence of water quality improvements at Salford Quays. In: *Urban Waterside Regeneration: Problems and Prospects* (eds. K. N. White, E. G. Bellinger, A. J. Saul, A. J. Symes and K. Hendry), Chapter 40. Ellis Horwood, London.

Williams, A. E and Moss, B. (2003) Effects of different fish species and biomass on plankton interaction in a shallow lake. *Hydrobiologia* **491**, 331–346.

Williams, A. E., Moss, B. and Eaton, J. (2002) Fish induced macrophyte loss in shallow lakes: top-down and bottom-up processes in mesocosm experiments. *Freshwater Biology* **47**, 2216–2232.

Williams, W. P. (1967) The growth and mortality of four species of fish in the River Thames at Reading. *Journal of Animal Ecology* **66**, 695–720.

Xue, H., Gächter1, R. and Sigg, L. (1997) Comparison of Cu and Zn cycling in eutrophic lakes with oxic and anoxic hypolimnion. *Aquatic Sciences* **59**, 176–189.

(a)

(b)

Figure 2.1. Metallophyte vegetation on ancient lead-mining sites in the UK. (a) Sparse cover of *Agrostis capilliaris* and *Silene uniflora* on acidic wastes at Goginan lead mine, central Wales; (b) Continuous metallophyte turf colonising superficial mine workings at Gang mines, near Matlock, Peak District. The calcareous substrate here and mosaic of metal contamination levels produce a rich assemblage of metallophytes including *Minuartia verna* in the most metal-contaminated areas. Photos: A. J. M. Baker.

(a)   (b)

Figure 2.2. (a) *Viola calaminaria, Festuca ovina* subsp. *guestphalica* and *Thlaspi caerulescens* at the ancient mining site of Schmalgraf, Belgium. (b) *Armeria maritima s.l., Viola calaminaria* and *Thlaspi caerulescens* at what was one of the richest metallophyte habitats of northwestern Europe at Rabotrath, Belgium. Most of the metallophyte communities have disappeared in the last 4 years since the land was taken into agricultural production. Photos: A. Van der Ent.

Figure 4.1. Acid mine drainage from Parys Mountain, Anglesey, UK, showing the characteristic orange colour of iron hydroxide deposits coating the riverbed. Photograph taken 3 km downstream of the main discharge.

Figure 4.2. *Habrophlebia fusca*, showing evidence of iron hydroxide deposits on its appendages.

Figure 8.3. The red waters of Psyche Bend Lagoon attest to the impact of recent (2004) acidification of the lagoon following its sacrifice in a broader management programme (Photo: Peter Gell).

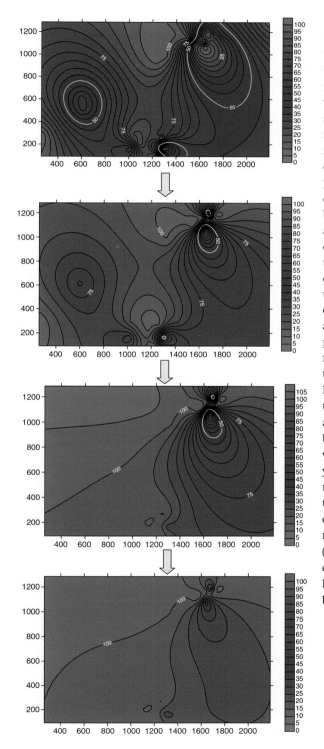

Figure 12.6. Tox-map (of the study site) constructed time course of bioremediation, leading from a highly toxic to a virtually non-toxic site. Each step was linked to use of the manipulations coupled to microbial biosensor testing detailed in Fig. 12.1 to progressively identify constraints to bioremediation and alleviate them through site engineering. Figure 12.6 therefore represents a time course of bioremediation at the site as various constraints were alleviated and the bioremediation proceeded progressively further. At the start of the time course, the tox-map highlights that there were three particularly toxic areas of the site (designated by increasingly bright red) which are delineated by a yellow boundary representing site material that is too toxic for degradation by micro-organisms (predominantly BTEX degrading bacteria) and hence for any intrinsic bioremediation to occur.

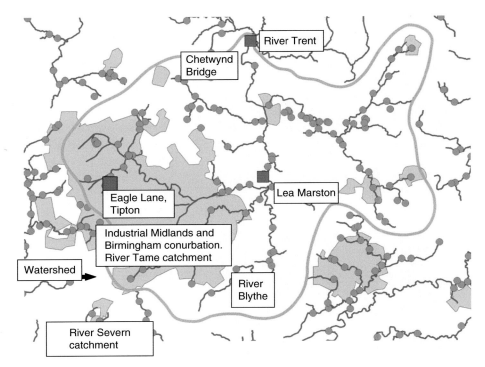

Figure 13.1. River Tame system showing relative positions of the three sampling sites used for this analysis. Green spots indicate EA biological sampling stations. Red squares, the sites used in this analysis.

Figure 13.2. Black Country landscape in the 1960s. Mine spoil heaps and pit pools such as these were common in the seventeenth century and stocked with coarse fish to provide food. The large block in the background is the remains of a slag-heap from a blast-furnace demolished in the 1940s. Today the area is all housing. (© T. E. L. Langford.)

During mining – soil is stored in stockpiles, ideally the topsoil and subsoil is separated

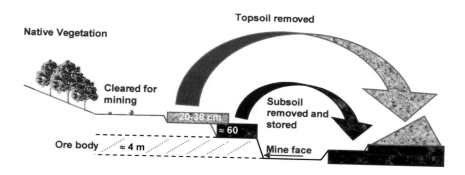

After mining – stockpiled top- and subsoil are replaced in pedological order

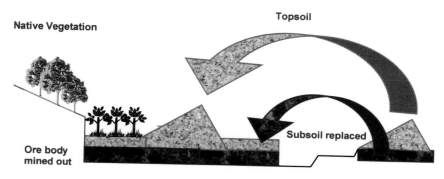

Figure 15.2. Conventional model for soil handling in mine site restoration schemes. Ideally, the topsoil and subsoil are separated into separate or sequential stockpiles.

**Direct return of topsoil avoids the need for stockpiling.
It has the minimum economic cost for the maximum
ecological benefit**

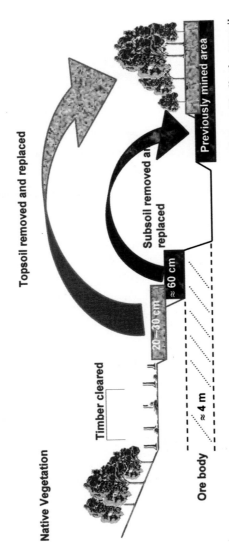

Figure 15.3. Direct return model for soil handling in mine site restoration schemes. Ideally the topsoil and subsoil are placed directly onto the newly prepared landscape (see text). If necessary, the soils are stockpiled only for a few hours or days at most. This is the most effective manner of retaining soil fertility.

CHAPTER FIFTEEN

Large-scale mine site restoration of Australian eucalypt forests after bauxite mining: soil management and ecosystem development

MARK TIBBETT

Introduction

Mining is essential to provide the resources for modern industrial societies but can result in a catastrophic destruction of pre-mining ecosystems. In Australia, these are often natural ecosystems, commonly pristine and with significant endemism in the flora and fauna. In all cases, the mining of bauxite ore in Australia occurs in areas covered by eucalypt forests or open woodland. Five major bauxite mines are in operation in Australia and these are (from NE to SW) Weipa (Rio Tinto) where the natural vegetation is an open *Eucalyptus tetradonta* (Darwin Stringybark) woodland, Gove (Rio Tinto), a mixed *E. tetradonta* and *E. miniata* woodland (both adjacent to the Gulf of Carpentaria) and Huntly, Willowdale (Alcoa) and Boddington (Worsley) mining in the unique *E. marginata* (Jarrah) forest region of Western Australia (Fig. 15.1).

Bauxite mining is an important economic activity for Australia, and it is an industry in which it leads the world. Australia is the world's largest bauxite producer, mining 40% of the world's bauxite ore. Australia's aluminium industry is worth over $7.8 billion (in 2004) in export earnings and employs over 16 000 people directly and many more in associated service industries. Australia has, therefore, made the decision to sacrifice some of its unique forested areas in order to maintain economic prosperity for its people.

In order to minimise the negative effect of bauxite mining, typically a form of strip mining or open cast mining, stringent measures (including financial instruments) have been put in place by the state and federal governments whereby the mining companies are required to restore the natural forest into sustainable ecosystems that reflects the original forest prior to mining as much as possible. As mining results in such a catastrophic destruction of the entire terrestrial ecosystems, including the regolith and associated hydrology, there is

Ecology of Industrial Pollution, eds. Lesley C. Batty and Kevin B. Hallberg. Published by Cambridge University Press. © British Ecological Society 2010.

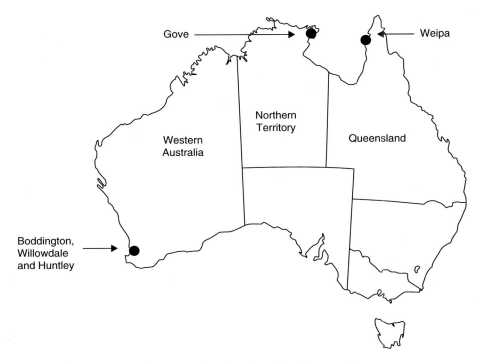

Figure 15.1. Location of five operating bauxite mines in Australia.

a tacit acceptance that complete restoration on human time scales is not achievable. Therefore the term 'rehabilitation' is often used to describe the restoration processes and targets in the local literature.

Australia has developed some world leading practices in mine site restoration after bauxite mining (Bell 2001; Mulligan *et al.* 2006). Restoration techniques are underpinned by two key practices: (i) incremental rehabilitation, restoring land progressively to forest after it has been mined out (Fourie & Tibbett 2006; Koch 2007a) and (ii) integrating mining with restoration, a practice that requires joint planning by both ecological and mining engineers (Hinz 1992; Koch 2007a).

This is particularly evident in improvements in soil handling techniques in bauxite mining, examples of which are outlined below.

Soils and bauxite ore

Australia has one of the most ancient landscapes in the world where the same land surface has often been subject to weathering and pedogenesis for hundreds of millions of years (McKenzie *et al.* 2004). Bauxite ore bodies found in Australia and elsewhere are formed through long-term pedogenic activity where the eluviation of Fe and Al released from the chemical weathering of secondary (clay) minerals in the upper soil horizons has illuviated in subsoil horizons. Over millennia, subsoil concentration of Al can reach significant concentrations that constitute ore grade sufficient for mining. In effect, the mining of bauxite ore is the mining of ancient subsoil horizons in which Al has accumulated.

In terms of classification, the soils in which Australian bauxite has formed approximate to Red Kandosols (Australian Soil Classification or Ferralsols, World Reference Base) or Oxisol (Soil Taxonomy). They are ancient and highly weathered soils (palaeosols) with limited horizon differentiation. They have little macro-structural development and are micro-aggregated (to c. 300 μm) with a significant proportion of piezolites and are thus highly permeable to water. The fine earth (less than 2-mm fraction) is dominated by kaolinite and is rich in amorphous Al and Fe oxide minerals which fix much of the phosphorus, rendering it unavailable to plants. Much of the cation exchange capacity of the soil is associated with organic matter and is pH dependent. In the northern mines, a typical profile has soil c. 80 cm over bauxitic ferricrete; in the western mines, lateritic cap rock may be found even at the surface and is of substantive nature that requires blasting. In the northern mines, the landscape is flat and most of the soils are mined out where as in the west between one third to two thirds of the forest is left unmined, depending on ore grade.

Soil handling and the restoration process

Although the geomorphology of the landscapes being mined for bauxite in Australia is somewhat different (hilly in the southwest and flat in the north), the fundamental processes of forest restoration are similar between regions and mine sites (e.g., Hinz 1992; Koch 2007a). These restoration processes, particularly with regard to soil handling, are now considered state of the art (Mulligan et al. 2006).

Topsoil is a valuable resource for mine site restoration as it is often stripped and stored separately for use after mining (Bell 2001). This conventional method for topsoil handling is shown in Fig. 15.2 as a generalised model for bauxite mining in Australia, but applies broadly to other surface- or strip-mining operations such as coal and mineral sands. The process shown is often referred to as 'double stripping' where the topsoil and subsoil are removed and stored separately (Bell 2001; Mulligan et al. 2006). The topsoil is of particular importance as it retains a valuable source of nutrients, micro-oganisms, organic matter and seeds (Jasper 2007; Koch 2007b; Tibbett 2008). In order to retain biological activity in the topsoil, locally provenanced plant species are sometimes sown on the stockpiles to develop a green cover.

Although the recognition that the original soils provide a valuable resource to the mine, stripping and storage of soil has some disadvantages. First, it requires double handling of the soil; that is, the soil needs to be moved

During mining – soil is stored in stockpiles, ideally the topsoil and subsoil is separated

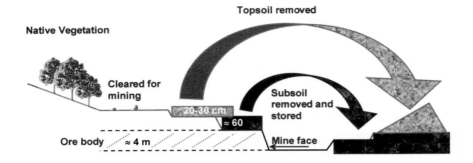

After mining – stockpiled top- and subsoil are replaced in pedological order

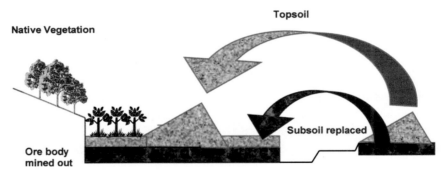

Figure 15.2. Conventional model for soil handling in mine site restoration schemes. Ideally, the topsoil and subsoil are separated into separate or sequential stockpiles. See colour plate section.

twice: into and out of storage. Second, space is required to store the soil. Herethere is a trade-off between the depth of soil dump and the land area required to store the soil. As a general rule, the deeper the soil store (and hence the smaller the footprint) the poorer is the soil's retention of favourable biophysical properties.

Stockpiling of any sort may reduce seed viability and soil physical, chemical and biological fertility (Johnson *et al.* 1991; Schaffer *et al.* 2007). To address this, soil stripped from land prior to bauxite mining is often moved directly to an area of land that has previously been mined and is ready to receive soil

Direct return of topsoil avoids the need for stockpiling. It has the minimum economic cost for the maximum ecological benefit

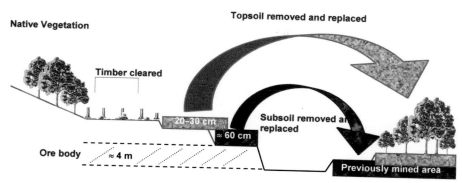

Figure 15.3. Direct return model for soil handling in mine site restoration schemes. Ideally the topsoil and subsoil are placed directly onto the newly prepared landscape (see text). If necessary, the soils are stockpiled only for a few hours or days at most. This is the most effective manner of retaining soil fertility. See colour plate section.

prior to vegetative restoration (Hinz 1992; Vlahos *et al.* 1999; Koch 2007b) (Fig. 15.3). This process has become known as 'direct-return' soil handling and provides the maximum ecological benefit for the smallest financial cost on the ground.

Depending on the mine, before and/or after soil is spread on the new landscape the mine floor is deep-ripped (subsoiled) in order to break up compacted materials due to the nature of the regolith materials and the trafficking of heavy machinery (Fig. 15.4) (Croton & Ainsworth 2007; Kew *et al.* 2007; Szota *et al.* 2007). The intention is to move a straight or winged tine (shown in Fig. 15.4) through the subsoil in order to create physical conditions more amenable to root growth and penetration in the reconstructed soil profile (Schaeffer *et al.* 2008).

The diversity of the natural forest is typically high and therefore substantial seed collection operations have to take place, normally exploiting locally provenanced material representing over 100 plant species (Hinz 1992; Vlahos *et al.* 1999; Koch 2007b). Sowing may be by hand or mechanical seed spreader and tends to take place during the dry season as the time of sowing has little overall effect on seedling emergence of these well-adapted plant species (Fig. 15.5) (Worthington *et al.* 2006). Those species that do not take from seed are propagated in nurseries and planted out by hand (Koch 2007a).

After the soil and seed handling processes are complete, soil fertility is addressed through chemical fertilisation. This is thought necessary in order

Figure 15.4. Dozer with winged tine (arrowed) used for deep-ripping (subsoiling) of bauxite mine pits after the return of soil layers.

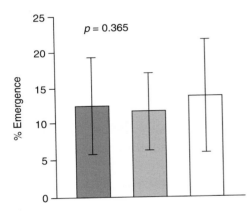

Figure 15.5. Total emergence of 18 Jarrah forest seedlings following sowing in mid dry season (dark grey bar), late dry season (light grey bar) and at the break of season (clear bar). Values are means ± SE. (n = 5) (after Worthington *et al.* 2006).

to 'kick-start' the incipient ecosystem and replace losses of nutrient that occurred due to plant removal and the physical mixing of the soil. Fertilisation application tends to be ground based in the northern mines, while the south-western mines rely on airborne (helicopter) delivery in order to avoid re-compaction of the soil. Typically, the fertilisers used are phosphorus based, as this is the major limiting nutrient in the Jarrah forest soils (Chen *et al.* 2008). Single super phosphate and diammonium phosphate are typically used as the major nutrient sources (Hinz 1992; Vlahos *et al.* 1999; Koch 2007b).

The development of ecosystems and their successional processes after mining is a contentious and complex matter. A true understanding of ecosystem development must include above- and below-ground components of the developing systems and consider a range of biotic and abiotic parameters.

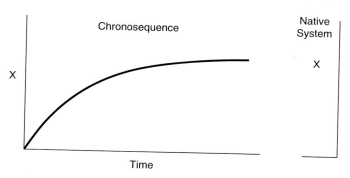

Figure 15.6. Idealised model for biotic or abiotic parameter (x) used to assess the success of individual restoration measures benchmarked against a native, undisturbed or other appropriately selected target systems. Time may be in years, decades or centuries.

It must also consider generational change such as seeding and recruitment; and the development of positive and negative functional attributes such as pollination, herbivory and pathogens. Ultimately, these must be viewed in terms of the ecosystem's resilience to perturbation such as storms and fire (e.g., Smith *et al.* 2004). An overview of some of the fundamental characteristics of the developing forest ecosystems after bauxite mining is discussed in the following sections. As well as above- and below-ground development of biotic parameters, some key abiotic measurements will also be described and discussed. Many of these are shown as progressing towards a benchmark value determined at a native or reference site (or sites) to give an indication of putative success (or otherwise) of each criterion (Fig. 15.6).

The capacity to monitor the development of post-mining ecosystems is apparent from the presence of mining companies alongside the restored sites for many decades (e.g., Gardner & Bell 2007). The restored forests after mining are effectively potential long-term ecological research sites. Unfortunately, while vegetation monitoring often takes place on the same sites over many years, most of the true research completed on these systems has relied on space-for-time substitution analysis (Pickett 1989). While simple and convenient (many sites of different age can be visited on the same day), this approach suffers from several weaknesses in post-mining systems and elsewhere (Fleming 1999). Primarily, these include the variability of soil, the differences in initial growing seasons by restored land unit, the variability of available seed in a given (previous seed collection) season, and final position in the landscape (in relation to aspect, hydrology and so on). This effectively amounts to pseudo-chronosequences, not true sequences through time at the same site; although these are often treated the same in the literature (Johnson & Miyanishi 2008). Therefore, in this chapter a chronosequence of sites is in reality a space-for-time substitution. This is not perfect, but is serviceable for analysis and inference of ecosystem development across the relatively large land areas restored.

Plant communities

After taking the measures to optimise soil handling as described above, it is the development of vegetation that is usually considered the primary target for restoration success (e.g., Norman *et al.* 2006). Detailed analysis of plant community succession after bauxite mining has been described elsewhere (Grant & Loneragan 2003; Grant 2006), and a brief overview of some key observations will be discussed here.

Restored *E. marginata* and *E. tetrordonta* forests both go through a series of predictable changes that include increases in tree height and stem densities to a levels comparable to native forest after 20 years or more (Spain *et al.* 2006). Structurally, the northern *E. tetrordonta* seems to follow a conventional successional pathway (Figs. 15.7a and 15.7b) (Spain *et al.* 2006); there is evidence to

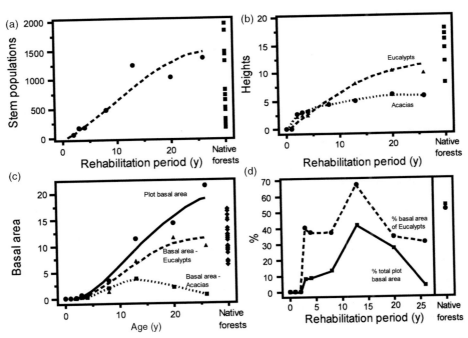

Figure 15.7. Selected plant community characteristics in a chronosequence of restored *Eucalyptus tetrodonta* sites and in nearby unmined native *E. tetrondonta* forests. (a) Population stem densities (stems ha^{-1}); (b) Mean heights (m); (c) Total basal areas and basal areas of eucalypts and acacias (m^2 ha^{-1}); (d) Basal areas of *E. tetrodonta* as a percentage of total plot basal area and as a percentage of the basal area of eucalypts. Data presented are based on individuals taller than 1.5 m and with woody stems greater than 1 cm in diameter at 1.5 m (after Spain *et al.* 2006).

suggest that the floristic composition of the southwestern *E. marginata* can reflect the initial vegetation complex for many years after restoration (Koch & Ward 1994; Grant & Loneragan 2001; Norman *et al.* 2006). That said, both cases (the northern and southwestern forests) have a predicable dominance of acacias, that are short-lived and produce copious seed, for the first decade or so of restoration. After this, the overstorey eucalypts begin to dominate the forest and the acacias decline in their relative to the basal area (Figs. 15.7b and 15.7c). The overstory vegetation is typically dominated by just a few standard species (Figs. 15.7c and 15.7d) (Spain *et al.* 2006; Koch & Samsa 2007) and it is in the understory where the majority of floristic diversity occurs and needs to be re-established (Koch 2007a).

Accumulation and quality of litters

Litter accumulation is considered the major accession route for both organic matter and most plant nutrient elements to the soil (Spain 1973) and is also an important habitat for invertebrate colonisation of restored lands with direct implications for soil development and nutrient cycling (Abbott 1989; Ward *et al.* 1991). It is therefore measured as a key indicator of ecosystem development (Majer *et al.* 2006; Grant *et al.* 2007) and is now also of interest as a precursor to soil carbon as a component of carbon accounting (Tibbett 2008).

Litters tend to develop as a function of above-ground primary productivity, so litters tend to accumulate slowly in the first few (up to 8) years and it takes about a decade of ecosystem development to reach almost concurrent canopy and litter closure (Fig. 15.8). However, within a decade after this, litters accumulate masses greater than that found in natural undisturbed reference sites. Nonetheless, litters are evidently decomposing in these restored systems (Grant *et al.* 2007) as is apparent from the lower mass of litters after the wet season (Fig. 15.8).

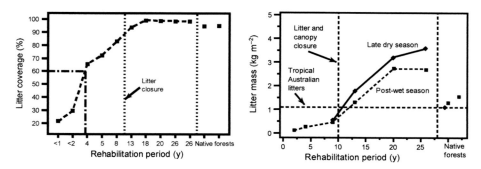

Figure 15.8. Litter properties in the rehabilitated and unmined native forest soils: (left panel). Percentage surface coverage of litter, late dry season 2002 (right panel). Seasonal changes in dry mass (after Spain *et al.* 2006).

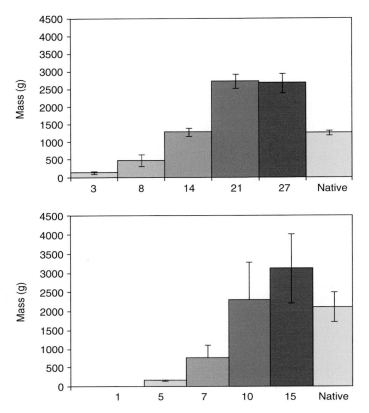

Figure 15.9. Surface litter masses (grams per square metre) of land rehabilitated with *Eucalyptus tetradonta* forest at Gove (top panel) and *E. marginata* (Jarrah) forest at Boddington (bottom panel) and comparative litter masses on unmined (native) forest soil. Bars = standard error of the mean (after Tibbett 2008).

Stocks of litter on the forest floor are typically greater in restored sites than native, unmined forest sites. Litter accumulation has been reported as being almost four times greater than the surrounding Jarrah forest (Ward & Koch 1996). Litter masses that accumulated at two contrasting bauxite mines, one from the northern forests and the other in Western Australia, are both greater in the restored than the native forests after around 15 years (Fig. 15.9). In the western Jarrah forests, litter accumulation matched the native systems in only 10 years, while in the northern forests, this took a little longer, at 14 years. At the now closed Jarrahdale mine site (Western Australia) the amount of carbon held in the litter layer increased by 50% between 8-year-old and 15-year-old rehabilitated sites (Sawada 1999). This time frame is in keeping with the natural senescence of many of the reseeding, short-lived *Acacia* species that tend to dominate the canopy of the rehabilitated forest until this time. At Gove,

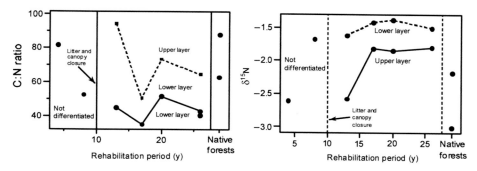

Figure 15.10. Litter properties in the rehabilitated and unmined native forest soils. Litter C:N ratio (left panel) and $\delta^{15}N$ values in litter layers (‰) (right panel) (after Spain *et al.* 2006).

for example, the *Acacia* tree basal area and the phyllode component of the litter tends to decrease from around 14 years (Spain *et al.* 2006). From this, it seems likely that a rapid increase in litter stocks occurs as part of the natural successional cycle as (primarily legumous) reseeders are replaced by the standard resprouting trees that make up the overstory in a mature Eucalypt forest canopy.

A study of litter quality at the Gove mine site in the Northern Territory provided some interesting findings (Spain *et al.* 2006). Within a few years after litter closure, litters accumulated to such depths that the litter differentiated into two distinct horizons, forming an upper and lower litter layer; not normally found in the native forest. These litter layers had distinct C:N ratios and $\delta^{15}N$ signatures (Fig. 15.10) (Spain *et al.* 2006). The lower litter layer had a narrower C:N ratio, indicating a loss of readily oxidisable C. Carbon to nitrogen ratios are good indicators of litter decomposability (Taylor *et al.* 1989; Tian *et al.* 1992), and by such a measure these lower litters should be more decomposable at this stage, but seem to accumulate nonetheless. The $\delta^{15}N$ values of litters were in the range commonly measured for plant tissues (-5 to $+2$) (Fry 1991). The lower litter layer had an enriched $\delta^{15}N$ signature compared to the upper litter layer. This characteristic is typical of enhanced litter decomposition (Connin *et al.* 2001; Gioacchini *et al.* 2006) and is thought to be a result of microbial mediated fractionation and preferential retention of $\delta^{15}N$ enriched residual material.

Litter masses, C:N ratios and $\delta^{15}N$ signatures in the restored sites are all different from the native forest at the Gove site. This may be simply because the restored sites have not been burnt in an area where fire is a common factor in forest systems; however, native sites that have not been fired for over a decade showed similar trends to the burnt native forest for these parameters. It is therefore possible that there is some unidentified impediment to litter decomposition in the restored system that may be related to a biotic dysfunction in litter decomposition.

Soil fertility and management

Soil development is the improvement that occurs in physical, chemical and biological properties, as the soil reorganises and acquires fertility with increasing restoration period. As with recently restored soils elsewhere, those of post-mining environments are incipient at best and are therefore likely to be poor in the stocks of nutrient elements available, at least initially (Odum 1969). Continuing soil development is, therefore, necessary to conserve ecosystem nutrient balances and is thus essential for self-sustainability.

The restored soils at most bauxite mines are developing on are permeable pisolitic, gravel-rich and highly weathered soil materials (Spain et al. 2006). Physical aspect of pedogenic activity seems to progress surprisingly rapidly and parameters such as soil hydraulic conductance can develop at a remarkable pace (Fourie & Tibbett 2007). Chemical and biological fertility, however, seem to develop more slowly and these aspects may need intervention to allow ecosystem restoration to succeed in human time scales.

Chemical fertilisation is common to all Australian bauxite mines as studies have shown that natural fertility after soil movement is too low to secure significant seedling establishment and vigour, important in the initial stabilisation of soils (Bradshaw & Chadwick 1980; Hinz 1992; Ward 2000; Standish et al. 2008). Recently, detailed studies have been undertaken to establish the effect of fertiliser application (particularly phosphorus) and soil management on soil fertility in the long term and short term (George et al. 2006; Standish et al. 2008). These have found significantly enhanced concentration of soil phosphorus (up to five times that of unfertilised soil and natural forest soils) 18 months after fertilisation (George et al. 2006) and significant residuals that remain after 13 years (Standish et al. 2008). This may be particularly important in forests where low nutrient status is considered to have contributed to the high floristic diversity (Myers et al. 2000; Hopper & Gioia 2004). The management of soil and placement of fertiliser also seems to be important in determining the availability of soil phosphorus in what are typically P-fixing soils (Fig. 15.11) (George et al. 2006). Management strategies such as soil scarification and deep ripping (and the generation of ridges and furrows) and the timing of fertiliser application can have a profound effect (again of up to five times the difference) on phosphorus availability and determine the position of the available phosphorus in the soil profile (Fig. 15.11).

Biological fertility is more difficult to measure, predict and manage. A wide range of functional groups of organisms must be considered here such as invertebrates, heterotrophic bacteria and fungi, mycorrhizal fungi, pathogens, rhizobia, frankia and blue-green algae (Jasper 2007; Majer et al. 2007). Certain

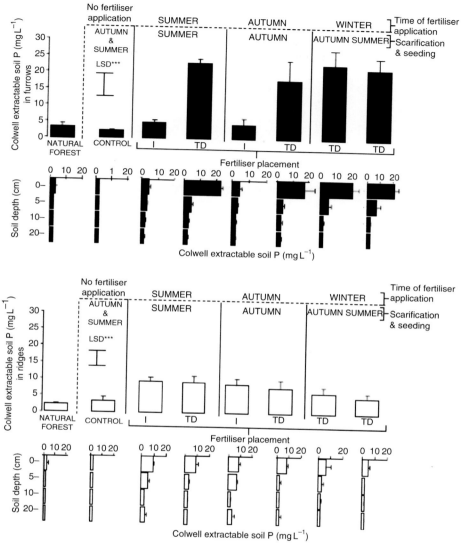

Figure 15.11. Collwell extractable soil phosphate comparisons between fertiliser placement (incorporated (I) and top-dressed (TD)) and timing in furrows ■ and ridges □ (formed following deep ripping) (***P < 0.01) (after George et al. 2006).

aspects of biological fertility return quite nicely to the levels found in pre-mining systems; for example, enzymes important in nitrogen cycling such as chitinase (1,4-beta-poly-N-acetylglucosaminidase – EC 3.2.1.14) can be found to match the range in natural forests very well (Fig. 15.12) (Spain et al. 2006). Other aspects of soil biology, however, seem to fall short of natural systems, even after a quarter of a century of forest regrowth.

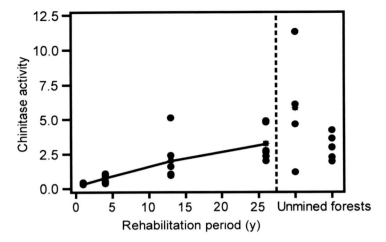

Figure 15.12. The development of chitinase activity (1,4-beta-poly-N-acetylglucosaminidase – EC 3.2.1.14) in restored and unmined native forest soils (after Spain *et al.* 2006).

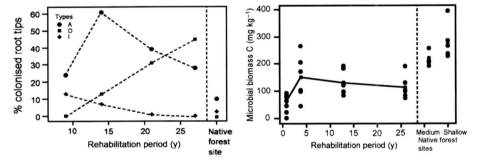

Figure 15.13. The development of biological fertility in restored and unmined native forest soils: Microbial biomass C (right panel) and changing pattern of representation of three ectomycorrhizal fungal morphotypes in the root tips of woody plants (left panel) (after Spain *et al.* 2006).

Ectomycorrhizal fungi are a major component of the biological fertility of (nearly) all forest systems (Smith & Read 1997). These fungi are important in tree nutrition (Tibbett & Sanders 2002) and are thought to undergo successional processes along with a forest ecosystem (e.g., Chu-Chou & Grace 1982; Frankland 1992). It takes quite a number of years (7–15) for these fungi to return to the reforested systems (Gardner & Malajczuk 1988; Glen *et al.* 2008). Evidence from molecular studies and mycorrhizal root tip morphotyping suggests that ectomycorrhizal communities may remain different from the native forest for decades (Spain *et al.* 2006; Glen *et al.* 2008) (Fig. 15.13). While this has real implication for soil fungal diversity, the functional significance of these different communities is not clear.

A major target that is often considered for below-ground restoration success is the return of the soil microbial biomass (the standing crop of microbes) (Sawada 1999; Harris 2003). In some cases after bauxite mining, the soil microbial biomass has remained consistently lower (by about one third) than the equivalent standing crop of microbes found in natural sites (Fig. 15.13) (Spain et al. 2006). It is noteworthy that the same sites in Gove in the Northern Territory had low microbial biomass and correspondingly large litter masses, indicative of low rates of decomposition. It is possible that this accumulating litter and reduced microbial biomass represents some form of soil microbiological dysfunction; however, further study is required to unravel the processes leading to these conditions in the restored forests.

Conclusions

Reconstruction of functional forest ecosystems are major operations for bauxite mining companies in Australia. Given the size and complexity of the task, it is generally considered to be well executed and among the better restoration schemes after mining (Koch & Hobbs 2007). This is in part due to the intelligent soil management methods employed, although issues still remain with artificial chemical fertilisation (George et al. 2006; Standish et al. 2008).

The above-ground parts of the ecosystem that are more easily observed show the greatest parity with natural systems. Plant populations are on track to match native systems fairly well, with discrepancies largely accounted for by the restored forest age. The below-ground parts of the ecosystem are harder to track and also show some similarities to natural systems, at least after decades of growth. However, it is in aspects of soil fertility, particularly biological fertility, that some differences are evident. The true long-term effect of these differences may only become clear over much longer time scales or after a major perturbation such as a cyclone or wildfire.

References

Abbott, I. (1989) The influence of fauna on soil structure. In: *Animals in Primary Succession: The Role of Fauna in Reclaimed Lands* (ed. J. D. Majer), pp. 39–50. Cambridge University Press. Cambridge, UK.

Bell, L. C. (2001) Establishment of native ecosystems after mining – Australian experience across diverse biogeographic zones. *Ecological Engineering* **17**, 179–186.

Bradshaw, A. D. and Chadwick, M. J. (1980) *The Restoration of Land*. Blackwell Scientific Publications, Oxford, UK.

Connin, S. L., Feng, X. and Virginia, R. A. (2001) Isotopic discrimination during decomposition of organic matter: a theoretical analysis. *Soil Biology Biochemistry* **33**, 41–51.

Chen, Y., Worthington, T., George, S., Walker, E., Wallrabenstein, H. and Tibbett, M. (2008) *Characterisation of Materials Planned for Use in the Closure of BRDA1 II. Nutrient Assessment and Constraints to Plant Growth*. Report No. CLR-08-04-C. Centre for Land Rehabilitation, University of Western Australia.

Chu-Chou, M. and Grace, L. J. (1982) Mycorrhizal fungi of eucalyptus in the North Island of New Zealand. *Soil Biology and Biochemistry* **14**, 133–137.

Croton, J. T. and Ainsworth, G. L. (2007) Development of a winged tine to relieve mining related soil compaction. *Restoration Ecology* **15**, S48–S53.

Fleming, R. A. (1999) Statistical advantages in, and characteristics of, data from long-term research. *Forestry Chronicle* **75**, 487–489.

Fourie, A. B. and Tibbett, M. (2006) Minimising closure costs by integrating incremental rehabilitation into mining operations. In: *Proceedings of the 2nd International Seminar on Strategic Versus Tactical Approach to Mining*, Section 17 (ed. Y. Potvin), pp. 1–12. Australian Centre for Geomechanics, Perth, Australia.

Fourie, A. B. and Tibbett, M. (2007) Post-mining landforms: engineering a biological system. In: *Proceedings of the Second International Seminar on Mine Closure* (eds. A. B. Fourie and M. Tibbett). Australian Centre for Geomechanics, Perth Australia.

Frankland, J. C. (1992) Mechanisms in fungal succession. In: *The Fungal Community – Its Organization and Role in the Ecosystem*, 2nd edn (eds. J. C. Carrol and D. T. Wicklow), pp. 383–402. CRC Press, Baton Rouge.

Fry, B. (1991) Stable isotope diagrams of freshwater food webs. *Ecology* **72**, 2293–2297.

Gardner, J. H. and Bell, D. T. (2007) Bauxite mining restoration by Alcoa World Alumina Australia in Western Australia: social, political, historical, and environmental contexts *Restoration Ecology* **15**, S3–S10.

Gardner, J. H. and Malajczuk, N. (1988) Recolonisation of rehabilitated bauxite mine sites in Western Australia by mycorrhizal fungi. *Forest Ecology and Management* **24**, 27–42.

George, S. J., Braimbridge, M. F., Davis, S. G., Ryan, M., Vlahos, S. and Tibbett, M. (2006) Phosphorus fertiliser placement and seedling success in Australian Jarrah Forest.

In: *Proceedings of the First International Seminar on Mine Closure* (eds. A. B. Fourie and M. Tibbett), pp. 341–350. Australian Centre for Geomechanics, Perth, Australia.

Gioacchini, P., Masia, A., Canaccini, F., Boldreghini, P. and Tonon, G. (2006) Isotopic discrimination during litter decomposition and δ 13 C and δ 15 N soil profiles in a young artificial stand and in an old floodplain forest. *Isotopes in Environmental and Health Studies* **42**, 135–149.

Glen, M., Bougher, N. L., Colquhoun, I. J., Vlahos, S., Loneragan, W. A., O'Brien, P. A. and Hardy G. E. (2008) Ectomycorrhizal species in fungal communities of rehabilitated bauxite mine sites in the Jarrah Forest of Western Australia. *Forest Ecology and Management* **255**, 214–225.

Grant, C. D. (2006) State-and-transition successional model for bauxite mining rehabilitation in the Jarrah Forest of Western Australia. *Restoration Ecology* **14**, 28–37.

Grant, C. D. and Loneragan, W. A. (2001) The effects of burning on the understory composition of rehabilitated bauxite mines in Western Australia: community changes and vegetation succession. *Forest Ecology and Management* **145**, 255–279.

Grant, C. D. and Loneragan, W. A. (2003) Using dominance-diversity curves to assess completion criteria after bauxite mining rehabilitation in Western Australia. *Restoration Ecology* **11**, 103–109.

Grant, C. D., Ward, S. C. and Morley, S. C. (2007) Return of ecosystem function to restored bauxite mines in Western Australia. *Restoration Ecology* **15**, S94–S103.

Harris, J. (2003) Microbiological tools for monitoring and managing restoration progress. In: *Land Reclamation* (eds. H. M. Moore, H. R. Fox and S. Elliot), pp. 201–206. A. A. Balkema Publishers, Lisse.

Hinz, D. A. (1992) Bauxite mining and Walyamirri, the return of the living environment, paper two, the rehabilitation programme. Australian Mining Industry Council,

Seventeenth Annual Environmental Workshop, Yeppoon, Queensland, 5–9 October, 1992, pp. 102–114.

Hopper, S. D. and Gioia, P. (2004) The southwestern Australian floristic region: evolution and conservation of a global hotspot of biodiversity. *Annual Review of Ecology, Evolution, and Systematics* **35**, 623–650.

Jasper, D. A. (2007) Beneficial soil microorganisms of the Jarrah Forest and their recovery in bauxite mine restoration in Southwestern Australia. *Restoration Ecology* **15**, S74–S84.

Johnson, D. B., Williamson, J. C. and Bailey, A. J. (1991) Microbiology of soils at opencast coal sites. 1. Short-term and long-term transformations in stockpiled soils. *Journal of Soil Science* **42**, 1–8.

Johnson, E. A. and Miyanishi, K. (2008) Testing the assumptions of chronosequences in succession. *Ecology Letters* **11**, 419–431.

Kew, G. A., Mengler, F. C. and Gilkes, R. J. (2007) Regolith strength, water retention, and implications for ripping and plant root growth in bauxite mine restoration. *Restoration Ecology* **15**, S54–S64.

Koch, J. M. (2007a) Restoring Jarrah forest understory vegetation after bauxite mining in Western Australia. *Restoration Ecology* **15**, S26–S39.

Koch, J. M. (2007b) Alcoa's mining restoration process in South Western Australia. *Restoration Ecology* **15**, S11–S16.

Koch, J. M. and Hobbs, R. J. (2007) Synthesis: is Alcoa successfully restoring a Jarrah Forest ecosystem after bauxite mining in Western Australia? *Restoration Ecology* **15**, S137–S144.

Koch, J. M. and Samsa, G. P. (2007) Restoring Jarrah Forest trees after bauxite mining in Western Australia. *Restoration Ecology* **15**, S17–S25.

Koch, J. M. and Ward, S. C. (1994) Establishment of understorey vegetation for rehabilitation of bauxite-mined areas in the Jarrah Forest of Western-Australia. *Journal of Environmental Management* **41**, 1–15.

Majer, J. D., Brennan, K. E. C. and Moir, M. L. (2007) Invertebrates and the restoration of a forest ecosystem: 30 years of research following bauxite mining in Western Australia. *Restoration Ecology* **15**, S104–S115.

Majer, J. D., Orabi, G. and Bisevac, L. (2006) Incorporation of terrestrial invertabrate data in mine closure completion criteria. In: *Proceedings of the First International Seminar on Mine Closure* (eds. A. B. Fourie and M. Tibbett), pp. 709–717. Australian Centre for Geomechanics, Perth, Australia.

Mulligan, D., Scougall, J., Williams, D. J. *et al.* (2006) *Leading Practice Sustainable Development for the Mining Industry: Mine Rehabilitation.* Department of Industry Tourism and Resources, Canberra, Australia.

McKenzie, N., Jacquier, D., Isabell, R. and Brown, K. (2004) *Australian Soils and Landscapes.* CSIRO Publishing, Collingwood, Australia.

Myers, N., Mittermeier, R. A., Mittermeier, C. G., da Fonseca, G. A. B. and Kent, J. (2000) Biodiversity hotspots for conservation priorities. *Nature* **403**, 853–858.

Norman, M. A., Koch, J. M., Grant, C. D, Morald, T. K. and Ward, S. C. (2006) Vegetation succession after bauxite mining in Western Australia. *Restoration Ecology* **14**, 278–288.

Odum, E. P. (1969) The strategy of ecosystem development. *Science* **164**, 262–270.

Pickett, S. T. A. (1989) Space-for-time substitution as an alternative to long-term studies. In: *Long-term Studies in Ecology: Approaches and Alternatives* (ed. G. E. Likens), pp. 110–135. Springer-Verlag, New York.

Sawada, Y. (1999) Microbial indices for assessing the progress of rehabilitation of mined land and mine residues. Unpublished PhD Thesis, University of Western Australia.

Schaffer, B., Eggenschwiler, L., Suter, B. *et al.* (2007) Influence of temporary stockpiling on the initial development of restored topsoils. *Journal of Plant Nutrition and Soil Science* **170**, 669–681.

Schaeffer, B., Schulin, R. and Boivin, P. (2008) Changes in shrinkage of restored soil caused by compaction beneath heavy agricultural machinery. *European Journal of Soil Science* **59**, 771–783.

Smith, M. A., Grant, C. D., Loneragan, W. A. and Koch, J. M. (2004) Fire management implications of fuel loads and vegetation structure in Jarrah Forest restoration on bauxite mines in Western Australia. *Forest Ecology and Management* **187**, 247–266.

Smith, S. E. and Read, D. J. (1997) *Mycorrhizal Symbiosis*. Academic Press, London.

Spain, A. V. (1973) A preliminary study of spatial patterns in the accession of conifer litterfall. *Journal of Applied Ecology* **10**, 557–567.

Spain, A. V., Hinz, D. A., Ludwig, J., Tibbett, M. and Tongway, D. (2006) Mine closure and ecosystem development: Alcan Gove bauxite mine, NT, Australia. In: *Proceedings of the First International Seminar on Mine Closure* (eds. A. B. Fourie and M. Tibbett), pp. 299–308. Australian Centre for Geomechanics, Perth, Australia.

Standish, R. J., Morald, T. K., Koch, J. M., Hobbs, R. J. and Tibbett, M. (2008) Restoring Jarrah Forest after bauxite mining in Western Australia — the effect of fertilizer on floristic diversity and composition. In: *Proceedings of the Third International Seminar on Mine Closure, Johannesburg, South Africa* (eds. A. Fourie, M. Tibbett, I. Weiersbye and P. Dye), pp. 717–725. Australian Centre for Geomechanics, Perth, Australia.

Szota, C., Veneklaas, E. J., Koch, J. M. and Lambers, H. (2007) Root architecture of Jarrah (*Eucalyptus marginata*) trees in relation to post-mining deep ripping in Western Australia. *Restoration Ecology* **15**, S65–S73.

Taylor, B. R., Parkinson, D. and Parsons, W. F. J. (1989) Nitrogen and lignin content as predictors of litter decay rates: a microcosm test. *Ecology* **70**, 97–104.

Tian, G., Kang, B. T. and Brussaard L. (1992) Biological effects of plant residues with contrasting chemical composition under humid tropical conditions – decomposition

and nutrients release. *Soil Biology and Biochemistry* **24**, 1051–1061.

Tibbett, M. (2008) Carbon accumulation in soils during reforestation: the Australian experience after bauxite mining. In: *Proceedings of the Third International Seminar on Mine Closure, Johannesburg, South Africa* (eds. A. Fourie, M. Tibbett, I. Weiersbye and P. Dye), pp. 3–12. Australian Centre for Geomechanics, Perth, Australia.

Tibbett, M. and Sanders, F. E. (2002) Ectomycorrhizal symbiosis can enhance plant nutrition through improved access to discrete organic nutrient patches of high resource quality. *Annals of Botany* **89**, 783–789.

Vlahos, S., Bastow, B. B. and Rayner, G. A. (1999) Bauxite mining rehabilitation in the northern Jarrah Forest. In: *Fifth International Alumina Quality Workshop*, Vol. 2, pp. 559–569. Bunbury, Western Australia.

Ward, S. C. (1995) Rehabilitation and revegetation. In: *Best Practice Environmental Management in Mining Series*. Environment Protection Agency, Canberra, Australia.

Ward, S. C. (2000) Soil development on rehabilitated bauxite mines in south-west Australia. *Australian Journal of Soil Research* **38**, 453–464.

Ward, S. C. and Koch, J. M. (1996) Biomass and nutrient distribution in a 15.5 year old forest growing on rehabilitated bauxite mine. *Australian Journal of Ecology* **21**, 309–315.

Ward, S. C., Majer, J. D. and O'Connell, A. M. (1991) Decomposition of eucalypt litter on rehabilitated bauxite mines. *Australian Journal of Ecology* **6**, 251–257.

Worthington, T., Braimbridge, M. F., Vlahos, S. and Tibbett, M. (2006) When to sow your seed for optimal forest rehabilitation: lessons from the Jarrah Forest of south western Australia. In: *Proceedings of the First International Seminar on Mine Closure* (eds. A. B. Fourie and M. Tibbett), pp. 319–328. Australian Centre for Geomechanics, Perth, Australia.

CHAPTER SIXTEEN

Sustaining industrial activity and ecological quality: the potential role of an ecosystem services approach

LORRAINE MALTBY, ACHIM PAETZOLD AND
PHILIP WARREN

Introduction

Modern societies benefit greatly from the products of industry and ecosystems provide the raw materials and energy required to produce them. Societies also benefit from a wide range of other ecosystem services including the supply of food, fuel, fibre and water, the regulation of disease and climate, recreational opportunities and aesthetic enjoyment. However, there is a potential conflict between these two types of benefits as increased industrialisation is often associated with increased release of hazardous chemicals or habitat modification, which have the potential to degrade ecosystem services, including those required for continued industrial production and development.

Here we consider the relationship between industrialisation and environmental impact using the development of Sheffield's metal industries as a case study. We then go on to explore the broader question of ecological quality and how it is defined, before outlining a quality assessment framework based on ecosystem services that may provide a tool for managing ecosystems for the optimal delivery of services. Finally, we consider the global aspects of economic development, industrialisation and environmental degradation.

Environmental impacts of industry: Sheffield metal industries and the River Don

Sheffield is synonymous with steel and metal has been worked here since at least the Middle Ages. Early metal workers capitalised on the environmental resources provided by the region: fast flowing rivers for water power, oak woodlands for charcoal, iron ore for smelting and grit stone for grinding. Later the locally abundant coal became an important source of power. The invention

Ecology of Industrial Pollution, eds. Lesley C. Batty and Kevin B. Hallberg. Published by Cambridge University Press. © British Ecological Society 2010.

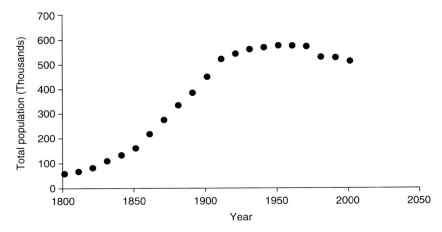

Figure 16.1. Population of Sheffield from 1801 to 2001. Pre-1961 data are from census parish tables, whereas 1961–2001 data are compiled by the Office of National Statistics (data from: A vision of Britain (through time), http://www.visionofbritain.org.uk/index.jsp).

of crucible steel and Sheffield plate in the eighteenth century and the Bessemer process and stainless steel in the nineteenth and twentieth centuries, all contributed to the development of the metal industry in Sheffield (Tweedale 1995). The metal trades established themselves along the Sheaf, Don, Rivelin, Loxley and Porter, the five major rivers of Sheffield, and by the late eighteenth century there was an average of between three (River Don) and six (River Rivelin) water-powered metal-working sites per mile of river (Wray *et al.* 2001). The steel industry peaked in the nineteenth century when the Lower Don Valley became the city's major industrial heartland, although there were also steelworks in the Upper Don.

The history of the River Don has been documented in the Environment Agency publication '*Domesday to the dawn of the new millennium: 900 years of the Don Fishery*' (Firth 1997) and the following provides a brief overview based primarily on this source. Industrial development in the Don catchment was associated with major changes to the region's rivers including altering the natural course of the rivers, constructing weirs to harness the flow and reservoirs to supply the industrial and drinking water needs of Sheffield. These physical modifications and impoundments changed the character of the river, which was reduced to a series of pools when the compensation flows from reservoirs were stopped at the end of the working day. The population of Sheffield grew rapidly to provide the labour force needed to sustain the developing industries, increasing from 60 100 in 1801 to 451 200 in 1901 (Fig. 16.1). Most of the workers lived in poor-quality housing with inadequate sewage systems. Untreated sewage was washed into rivers and in 1860 the River Don was little more than an open sewer and was described as being black and foul smelling. The numerous mines that provided the coal, coke and ganister needed by the metal industries pumped untreated metal-rich mine water into local rivers resulting in them turning an ochreous yellow and the industries themselves discharged a cocktail of pollutants (e.g., metals, oil, acid) into surface waters, reducing water quality further.

The River Don and its tributaries used to support a diverse fishery with 31 fish species, including salmon and trout, being recorded pre- 1850. Fish populations began to decline rapidly from the beginning of the nineteenth century, initially due to physical modifications, but then compounded by declining water quality. Weirs and dams impeded the movement of fish and changes in water flow altered habitat characteristics. Fish were unable to reach upstream spawning sites. Sedimentation occurring upstream of weirs produced conditions suited for roach, bream and perch whereas scouring downstream of weirs produced habitat more suited to dace, chub, barbel and grayling. Fish populations continued to decline as water quality deteriorated, and by 1860 the River Don flowing through Sheffield and downstream to Rotherham was virtually fishless. Fish populations persisted in the headwaters and tributaries of the Don, although several of these received waste water from mines that had, and still have, a negative impact on fish, invertebrate, algal and microbial communities (e.g., Bermingham *et al.* 1996; Ashton 1997; Amisah & Cowx 2000a). By 1974, the Don fishery had been reduced to a small number of brown trout in a few upstream reaches.

The effects of Sheffield's industrial development on the River Don were not confined to aquatic species. During the height of the manufacturing industry, the River Don flowing through Sheffield ran at a constant 20° C, the water having been used for cooling purposes. This warm water provided suitable conditions for the germination of wild fig (*Ficus carica*) seeds that entered the river via sewage works. As a result, more than thirty 80- to 90-year-old fig trees can be found growing along the River Don in the eastern end of the city (Gilbert 1990).

Industrialisation in the Don catchment changed the river from a provider of power, food (fish), drinking water and recreational space to a recipient of domestic and industrial waste that presented a serious risk to human health. The abundant local resources (oak woodlands, coal, iron ore and ganister) that underpinned the development of the metal industries were overexploited resulting in mine closures and importing of raw materials. By the mid twentieth century, the population of Sheffield had peaked at 577 050 (1951 census) and the River Don had obtained a reputation for being one of the most polluted rivers in Europe (Firth 1997).

Industrial decline and environmental improvement

Sheffield experienced a massive deindustrialisation in the 1980s, beginning with the steel workers' strike in 1980, and followed 4 years later by the miners' strike. Between 1975 and 2005, the UK production of coal and steel decreased

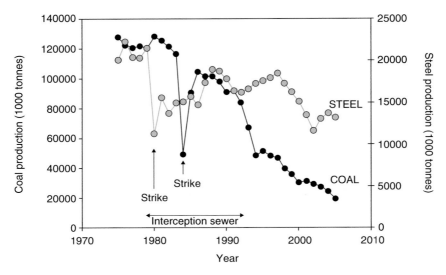

Figure 16.2. Change in UK coal and steel production between 1975 and 2005. Arrows indicate the steel workers' strike in 1980 and the miners' strike in 1984 (data from DTI and ISSB).

by 84% and 34%, respectively (Fig. 16.2) and 70 000 jobs (25% of the total) were lost from Sheffield's economy between 1979 and 1987. Several steel companies closed in the Lower Don Valley and the manufacturing workforce reduced by 35 000 between 1980 and 1990 (DiGaetano & Lawless 1999). The severe reduction in industrial activity, coupled with the installation of a new interception sewer (1979–1993), reduced the inputs of pollutants into local rivers, particularly the River Don, although contaminated sediments remained. But has water quality also improved?

River water quality in England and Wales has traditionally been assessed using a combination of chemical and biological measures. Biological assessments have focused on benthic macroinvertebrates and the principle index used is the Biological Monitoring Working Party (BMWP) score or its derivative, the Average Score per Taxon (ASPT). The BMWP score ascribes a value to macroinvertebrate families present at a site according to their assumed pollution sensitivity; sensitive taxa (e.g., stoneflies) are given a high value, insensitive taxa (e.g., chironomid larvae) a low value. These values are then summed to give the BMWP score for the site and the higher the score, the higher the water quality (Hawkes 1997). There has been a marked increase in BMWP score along the River Don over the last 20 years indicative of an improvement in water quality (Fig. 16.3), which coincides with the demise in the steel and coal industries and improvements in the sewer system. The Environment Agency's current classification of river quality for the River Don is 'good' to 'fairly good' for chemical quality, but only 'fairly good' to 'poor' for biological quality (http://www.environment-agency.gov.uk/maps/info/river/). There has been a

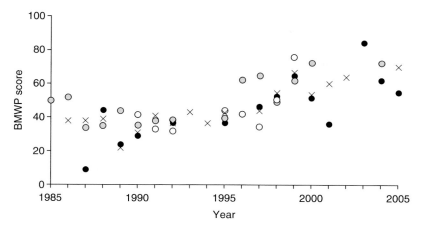

Figure 16.3. Change in BMWP score along the River Don, Sheffield between 1985 and 2005. Moving northwest of Sheffield downstream to the Lower Don Valley, the four sites are at Oughtibridge (black circle), Hillsborough (grey circle), Corporation Street (white circle), Weedon Street (cross).

programme of fish stocking in the River Don since 1975. In the 1990s, although there was a healthy brown trout population in the upper reaches of the Don, further downstream, fish abundance and diversity were still impoverished, with limited numbers of roach, gudgeon, dace, perch, brown trout and grayling being sampled during fish surveys (Amisah & Cowx 2000b).

The improvement in water quality observed in the River Don has been mirrored in other industrial areas including Liverpool, Manchester, Leeds, Newcastle and the Midlands (see Chapter 13). Further improvements in water quality are expected as a result of the implementation of the EU Water Framework Directive (WFD). The WFD is an important piece of environmental legislation that has the aim of achieving 'good ecological status' for surface waters by 2015. However, this aspiration raises some interesting issues about what we consider to be an improvement in quality and how these judgements can be made. While it would be hard to argue that changes in Sheffield's rivers from lifeless, or virtually so, to their present state is not an improvement, what would constitute further improvements and are these realistic and desirable?

Reference conditions and environmental management

The WFD defines ecological status in terms of derivations from 'undisturbed conditions' and, hence, requires the identification of reference states to which the current state can be compared. Reference condition for the WFD has been defined as 'a state in the present or in the past corresponding to very low pressure, without the effects of major industrialisation, urbanisation and intensification of agriculture, and with only very minor modification of physico-chemistry, hydromorphology and biology' (UKTAG 2007). The question is, does such a state occur in the present UK landscape and if not, how far into the past should we go, and do we have the appropriate data on past environments?

Two concepts frequently discussed when considering the quality of ecosystems are biological integrity and ecosystem health. Biological integrity in an ecosystem has been defined as 'the capability of supporting and maintaining a balanced, integrated, adaptive community of organisms having species composition, diversity, and functional organization comparable to that of natural habitats of the region' (Karr & Dudley 1981). Common definitions of 'natural' are, in essence, either being without human influence or without human technology (Hunter 1996; Angermeier 2000). The terms 'balanced, integrative and adaptive' are not further specified in the definition and without further clarification cannot provide consistently applicable criteria for assessing ecosystem state. This leaves the concept resting mainly on naturalness. The principle of biological integrity indices is to select and integrate indicators of the structure and functioning of ecosystems (e.g., species richness, tolerant taxa, trophic structure, individual health of organisms) that can detect divergence from natural reference conditions attributable to human activities (Karr & Chu 2000). Consequently, high biological integrity is an attribute of ecosystems unaffected by humans, and it provides a benchmark by which to measure the extent of human impacts on ecosystems (Angermeier 2000). Because integrity is determined relative to a regionally defined natural reference state, it implies an existing or historical target.

It can be very difficult to characterise the conditions of an ecosystem without human influence or technology, given the long history and dominance of human activities in many regions of the world (Kareiva et al. 2007; Nilsson et al. 2007). Many ecosystems, such as large temperate lowland rivers, no longer exist in a 'natural' state in many regions of the world. Alternative methods of defining natural reference states include the use of historical or paleo-ecological data (Gillson & Willis 2004). However, historical data of appropriate quality are often scarce and paleo-ecological data are essentially limited to a selected set of organisms and habitats, such as diatoms in lakes (Swetnam et al. 1999). Even if good historical data are available, many landscapes have been altered by humans for hundreds to thousands of years and therefore historical references would describe other degrees and types of human influence rather than a natural state without human influence (Pickett & Ostfeld 1995). And, of course, both natural long-term change and temporal variability in the biological structure of ecosystems, as a result of natural disturbances or internal ecological mechanisms, further complicates the separation of human induced alteration from other processes (Bunn & Davies 2000; Milner et al. 2006).

Apart from the problems of defining the reference state, there is also a fundamental practical issue, which is that return to the natural state would require the removal of most, if not all, human activity in an area and this is simply not feasible for ecosystems in regions of the world that are dominated by humans (Vitousek *et al.* 1997; Palmer *et al.* 2004). Therefore, pristine, or even pre-industrial, reference states can be an exclusive management goal only for a relatively small part of the world, where human activity is restricted either naturally (e.g., inhospitable terrain) or by legislation (e.g., National Parks).

Ecosystem health and resilience

A healthy ecosystem is defined as 'being stable and sustainable, maintaining its organization and autonomy over time and its resilience to stress' (Rapport *et al.* 1998). Rapport *et al.* (1998) and Constanza and Mageau (1999) proposed that vigour (metabolism or productivity), organisation (biodiversity, number of trophic interactions) and resilience to stress (the ability to resist or recover from stress) are key attributes of ecosystem health. The key issue that emerges here is that ecosystem health is a time-dependent concept. All integral parts of the definition ('stable', 'sustainable', 'maintaining autonomy over time' and 'resilience') have a time component, and satisfying these criteria requires limited change over time. Thus, health is essentially defined by the lack of change in system organisation and functioning. Such a definition might seem to avoid the problem of needing definition of a reference state, but this is not entirely so, since under the definition of health above any change in state could be interpreted as a loss of health and, even adding in the more directional markers of 'key attributes', the health of an already degraded system is essentially undefined unless its history is known. In other words, either the direction of change needs to be determined or a normal state/target needs to be defined. One important difference from integrity, however, is that this target can be aspirational and does not necessarily have to have existed.

A key component in the definition of ecosystem health is the concept of resilience. Resilience has been defined as the rate at which a system approaches steady state after perturbation (Pimm 1984), or more recently for systems that operate in a non-linear fashion 'as the capacity of a system to absorb disturbance and reorganize while undergoing change so as to retain essentially the same function, structure, identity and feedbacks' (Folke *et al.* 2004). Ideally, resilience must be assessed in the context of a specific disturbance, i.e., a system that is highly resilient to one type of disturbance might not be resilient to another type of disturbance (Walker *et al.* 2004). The problem of using resilience as a measure, when defined as the capacity to absorb/recover from a perturbation, is that this is only known for the largest disturbance yet experienced for specific ecosystems. Because of our limited ability to predict the dynamics of an ecosystem under stress, it is in practice extremely difficult to quantify the resilience of an ecosystem (Constanza & Mageau 1999; Karr 1999).

Consequently, resilience is of limited use for management aimed to protect ecosystems. Apart from the challenges associated with defining and measuring resilience, it might not always be desirable to achieve high resilience of an ecosystem because it reduces the capability of the system to move to another, possibly more desirable ecosystem state (Walker *et al.* 2004).

The concept of health for an ecosystem, by analogy with human health, suggests that optimal ('healthy') states for ecosystems could be defined (Scrimgeour & Wicklum 1996). Critics of the ecosystem health concept argue that there are no intrinsically defined optimum states of an ecosystem because the interactions between parts are not optimised for the functioning of the whole system (Scrimgeour & Wicklum 1996; Calow 2000). Applying the proposed key attributes (vigour, organisation and resilience) of ecosystem health would mean, for instance, that oligotrophic lakes (low productivity, low biodiversity) would be less healthy than productive (higher vigour) and diverse (higher organisation) eutrophic lakes (Karr 1999). We agree with Karr (1999) that such definitions of health are hard to defend on scientific grounds and are of limited use in practical ecosystem management. The concept of ecosystem health is evolving, and the role of socioeconomic values of ecosystems including their effect on human health is emphasised to various degrees in more recent definitions (Meyer 1997; Rapport *et al.* 1999). However, no consistent and integrative definition of this meaning of ecosystem health has been developed so far.

When we reduce biological integrity and ecosystem health to their essential parts, two principal concepts emerge: we can either define the quality of an ecosystem by its resemblance to a reference state with little human influence (integrity) or by the capacity of an ecosystem to maintain its organisation over time (ecosystem health). Closeness to a natural state, sometimes explicitly defined as integrity, has been adopted by various implementations of major environmental legislation, including the US Clean Water Act and the EU Water Framework Directive, despite the practical problems associated with the establishment of the natural reference conditions (Karr 1991; Lackey 2001).

The ecosystem health concept requires a definition of a 'normal' or desired state to be operational. If the natural state is used as benchmark, then the definition of ecosystem health essentially merges into the concept of ecological integrity, with a stronger emphasis on stability and resilience. We argue that there is no objective ecological basis that could justify the sole focus on a natural reference in the definition of ecological quality, and we concur with the argument of Freyfogle and Lutz Newton (2001) that subjective values of society are needed to decide whether a particular ecosystem state or ecological process is good or bad. We do not argue against 'naturalness' as a quality criterion for ecosystems but we emphasise that the ascription of value to 'naturalness' is a domain of society and cannot be inferred solely from ecological theories (Freyfogle & Lutz Newton 2001). Consequently, at an operational level, it is a societal decision whether the aspiration to 'naturalness' has primacy to other values and demands that humans place on ecosystems.

Sustainable ecosystems and sustainable societies?

Humans place both direct and indirect demands on ecosystems, whether for basic provisioning (food, water and fuel), or the regulating and renewing processes that maintain suitable environmental conditions for life. The conditions and processes through which ecosystems sustain human life have been termed ecosystem services (Constanza et al. 1997) and have been classified as provisioning services (ecosystem goods, such as food and timber), regulating services (e.g., regulation of climate, floods and water quality), cultural services (e.g., recreation, aesthetic enjoyment and spiritual fulfillment) and supporting services (e.g., soil formation, nutrient cycling and photosynthesis), which are necessary for the production of all other services (Millennium Ecosystem Assessment 2005). In addition to ecosystem services influencing human well-being, human well-being influences ecosystem services through direct and indirect drivers of change (Fig. 16.4). Economic development, often associated with industrialisation, is generally associated with improved standards of living, reduced infant mortality and increased life expectancy. However, economic development requires energy, generates waste and often results in social inequalities. Energy demands are supplied by extracting fossil fuels or harvesting timber, and waste material is released into waterways, disposed of in landfill or burnt; all of which have environmental impacts (Hodgson et al. 2007).

Ecosystem services are likely to be interdependent and the management or optimisation of one service may have negative consequences on others (Rogriguez et al. 2006). As human influence on the world's ecosystems increases, understanding and managing potential trade-offs resulting from the various demands society places on ecosystems becomes ever more important. Because existing ecological quality definitions do not really allow us to address these issues, we are challenged to develop alternative ways of defining ecological quality that can accommodate the different demands that society places on ecosystems, account for potential trade-offs between them and be applied at the scales at which management operate. Deriving such definitions will allow us to explore the robustness and generality of those definitions in principle, and then the challenges in making them operational. Here we suggest one such definition, explore some of the attributes such a measure might have and consider some of the practicalities of its implementation. As illustration of the ideas involved, we draw on some of the issues of quality and service provision that arise in the context of river, or catchment, management.

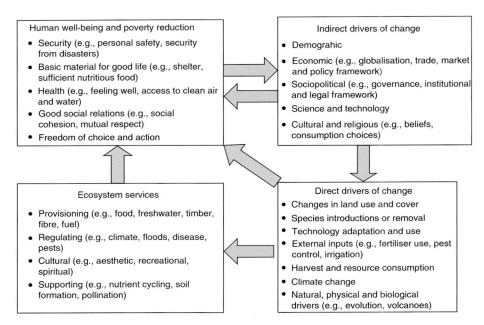

Figure 16.4. Interactions between ecosystem services, human well-being and drivers of change. Based on the MEA and reported from Hodgson *et al.* (2007).

Towards a quality assessment framework based on ecosystem services

First, we suggest that there are certain features that might be desirable, or necessary, to generate a robust concept of ecological quality based around the idea of ecosystem services. These are as follows.

1. The flexibility to allow context dependent definition of quality. 'Good' quality need not be the same in all situations; what is a meaningful and practical aspiration for a river in a sparsely populated rural area may not be appropriate for one in an urban area. Both systems provide ecosystem services, but the balance of these may be rather different in the two situations and we want to recognise that each system might be different, yet might be functioning in a way very appropriate to the context it is in. We want to be able to recognise that this 'fit' of function to context may be an important element of whether a system is of good quality.

2. The ability to take account of more than one ecosystem service, and the trade-offs between services. There are few, if any, situations in which all the assessment of quality could, or should, be based on the delivery of a single function or service. Partly, this is because there will almost always be explicit demand for multiple services in any system, and partly because even if one element is taken to be paramount, services are not independent. Alteration of one process or function may have negative

consequences for other aspects of the system which were not explicitly recognised or managed for, but when changed become a significant issue (e.g., flood regulation).

3. The criterion that the functions and services in a system are assessed in terms of their sustainability, not exploitative potential. Clearly, there are some aspects of ecosystems where potential short-term delivery rates of goods and services may be high, but have no prospect of being maintained in the long term. Any assessment of service provision – and hence any measure of quality depending on it – must work as far as possible with sustainable levels of service provision.

4. The ability to incorporate existing elements of ecological quality where these are relevant, as a component, but not the whole, of the assessment. These existing criteria may be quite different; for example, one important group of criteria used in measuring ecological quality in freshwaters are thresholds of chemical concentration, or biological toxicity, which may be quite precise quantitative measures, encapsulated in legislation such as that for drinking water supply. At the other end of the spectrum, criteria such as those discussed earlier, of 'naturalness' or closeness to some pre-human state, may still be criteria that we wish to take account of, but need to be incorporated in a broader assessment, not the sole target.

Several elements emerge here: the inclusion of multiple services, the facility to include existing criteria and the notion of demand for services as being an element of the context in which quality is evaluated. Taking these, we propose that one way of assessing quality, which could capture all the features above, is to use the match between the demand for a range of services, and the extent to which that demand is met sustainably, as our index of quality: a system that can support, in the long term, the range and magnitudes of services we desire from it is, in some sense, of good quality. We term this approach of matching requirement and provision across a number of ecosystem services the Ecosystem Service Profile (ESP), and we feel that such a measure has some potentially useful characteristics.

The first is that, although services are delivered and measured in many different ways, the matching of requirement and provision for each service is likely to be in the same units, and it may avoid the problematic job of translating services into common units (monetary other otherwise).

Second, if for a range of services, the match between demand and provision can be assessed, then strong mismatches in which demand exceeds provision draw attention to the places where management might be directed to improve quality. Clearly, improving such a mismatch might involve the increase in provision, or decrease in demand, both valid management options. This makes the important point that quality improvement might come from reducing demand for something. Whilst this might initially seem a little counterintuitive, it is something that already happens (albeit without the outcome being framed in 'quality' terms) in management: for example, domestic water supply management involves both increase of supply and regulation of demand.

A third feature that emerges from such a definition of quality is that it is not absolute. There is no single target for what constitutes good quality, but existing targets can be incorporated into it: if there is a strong demand for features of a natural riverine environment, for example, as a component of the recreational, conservation or aesthetic services a system provides – then that becomes one of the criteria on which the quality of the system might fall down.

Constructing an ESP and its use in environmental management

There are clearly a number of stages involved in generating such a service profile and the challenges at each stage are significant.

First of all, some set of services must be defined, which are appropriate to the system (e.g., in the case of a river, fish production for angling, flood mitigation, water supply and aspects of biological conservation might all feature). Whilst in some cases the list may be fairly obvious, we do not underestimate the practical difficulty of generating an agreed list appropriate to a system. This might have to be based on standard 'core' lists defined for particular types of system, or derived on a system-by-system basis. Stakeholder involvement is probably useful but one important issue is that attention should be given to services which might not naturally have a high 'visibility' in the current state of the system, but have the potential to be important. Again, flood regulation would be a good example: if flood regulation is currently provided very effectively by a system, its consequence – lack of serious flooding – may mean that most people do not realise is it a function delivered by the current system, so might not identify it as a service they 'use'.

The next stage of the process would be to identify indicators for the services. This is necessary (a) because many services cannot easily be measured directly, and (b) because it will be necessary to work with data that are already available, or can be readily generated. So, for example, the extent of flood regulation – in essence the extent to which flooding has not happened – is hard to assess directly, but it may be more practical to assess related variables such as predicted maximum water retention capacity given land use and floodplain capacity in the area. Similar issues arise with measuring sustainable provision – an indirect approach may be needed. The constraints of data availability are also significant since practical use of such a measure would not easily accommodate the time and cost involved in detailed long-term data gathering. However, these issues apply to many current management scenarios, and the use of indicators is already established in a number of areas, such as biodiversity conservation (Karr 1999; Scholes & Biggs 2005).

For the chosen indicators, both provision and requirement must then be quantified. In some cases, common units may be self-evident (e.g., volume of water) but for others it may be difficult to derive common scales for both, and this may need to be taken into account when indicators are being chosen. For example, quantifying biodiversity in terms of species diversity may be achievable for many groups of organisms, but it is likely to be intractable to establish a meaningful figure in this metric for what people expect, or want, from the system. It may be more practical to work with the presence or absence of key taxa, or the occurrence of particular types of recognisable biodiversity. In some cases, the requirement may already be present in, for example, health standards, where the choice of units is already determined.

If all these issues can be resolved (albeit imperfectly), then the ESP can be constructed from the ratios of provision and demand. At this stage, we have a measure of how well the system is delivering what humans demand of it. This is useful in its own right, but it also provides a framework within which the quality outcomes of different management changes could be explored. To do this requires an understanding of the interrelationships between the ecosystem services concerned. Although a challenging task, such understanding need not be fully mechanistic but might be empirical, or even based on expert opinion. Equipped with such information the consequences of different management options can be explored, and potentially presented to stakeholders in an integrated way (e.g., Fig. 16.5).

Industry, development and pollution

Using Sheffield as an example, we have illustrated the environmental impact of industrialisation and how decline in heavy industry has been associated with improvements in ecological quality. Good environmental/ecological quality is desired by many societies in developed countries and is a stated goal of European legislation (i.e., WFD), and we have proposed a framework for assessing ecological quality based on ecosystem services. The same societies that desire good quality environments also require the products of heavy industry to sustain their lifestyle; a lifestyle aspired to by many developing and increasingly industrialised countries. In this section, we broaden our discussion to consider the global aspects of trade in industrial products and ecosystem services.

The environmental, economic and social developments of countries are inextricably linked. Economic development is associated with improvements in human well-being (e.g., reduced infant mortality, increased food security and enhanced life-span), but may also be associated with increased demand for ecosystem goods, and environmental degradation, including habitat loss and pollution. Left unabated, these will result in deterioration of ecosystem services and hence have detrimental effects on well-being (e.g., poor water and air

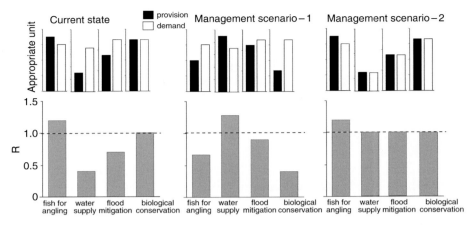

Figure 16.5. Upper panels illustrate the provision and demand profiles for four ecosystem services (fish for angling, water supply. flood mitigation and biological conservation) in a catchment either in its current state or under two different management scenarios: scenario 1 has a modified provision profile and scenario 2 has a modified demand profile. The ecosystem service profiles (ESP) are illustrated in the lower panels, where R is the provision:demand ratio. The catchment in its current state meets the demand (R > 1) for two of the four services (fish for angling and biological conservation). Management scenario 1 results in the demand for flood mitigation and water supply being met, but at the expense of fish for angling and biological conservation. Management scenario 2 enables demands for all four services to be met.

quality, reduced food quality and quantity and increased disease). The Millennium Ecosystem Assessment concluded that 'two-thirds of the ecosystem services it examined are being degraded or used unsustainably' and that 'the harmful effects of the degradation of ecosystem services are being borne disproportionately by the poor . . . and are sometimes the principal factor causing poverty and social conflict' (Millennium Ecosystem Assessment 2005). In addition to adverse effects on well-being, degradation of ecosystem services will affect business and industry by reducing the availability and increasing the cost of raw materials and by changing the regulatory, economic and social environment within which they operate.

Although developed countries remain the world's largest industrial producers, this is changing. Whereas in 1980, developed countries accounted for 82% of manufactured goods, by 2000 this had reduced to 69%, with the largest gains by Asia and Latin America (UNDESA-DSD 2006). As industrial economies have developed, there has been a shift in emphasis from heavy pollution-intensive industry (e.g., iron and steel, coal and mineral mining, industrial chemicals, pulp and paper), so-called 'dirty' industry (Mani & Wheeler 1998), to light manufacturing and service industries. As a consequence, developing countries have increased their production from heavy industry to meet the import demand of developed countries. Heavy industry tends to require a high input of labour, natural resources and energy and is, therefore, costly. Reduced resource costs are often cited as one of the reasons why there is a shift in heavy industry from developed to developing countries (Mani & Wheeler 1998). Another possible explanation is the 'Pollution Haven Hypothesis', which states that differences in the stringency of environmental regulation between developed and developing countries will provide the latter with a comparative advantage in pollution-intensive production and hence the relocation of heavy industry, and consequent transfer of pollution, from developed to developing countries (Cole 2004). For these and other reasons, a hump-backed relationship between national wealth (per capita income) and environmental degradation, known as the environmental Kuznets curve, has been proposed (Grossman & Krueger 1995). It has been hypothesised that, initially, there is a positive relationship between economic development and environmental degradation/pollution, but that as development progresses a point is reached beyond which environmental degradation reduces with increasing wealth generation.

The Pollution Haven Hypothesis and the environmental Kuznets curve have received a considerable amount of attention recently, although the evidence for both is equivocal. Cave and Blomquist (2008), for example, investigated whether environmental policy in the European Union was resulting in pollution havens. They found evidence to support the Pollution Haven Hypothesis for energy-intensive trade but not for toxic-intensive trade, although the toxic index used focused on carcinogens rather than environmental pollutants in general. Cole (2004) investigated the relationship between trade in 'dirty goods' (i.e., those from pollution-intensive industries) between developed and developing countries on air and water quality. There was evidence of a relationship between per capita income and most environmental quality indicators that was consistent with the environmental Kuznets curve. Further analysis indicated that there was a positive relationship between the share of 'dirty' exports and environmental quality indicators, but this was less strong when imports were considered. Cole (2004) concluded that the 'share of manufacturing output in gross domestic product generally has a positive statistically significant relationship with pollution' and that the 'downturn in emissions experienced at higher income levels appears to be a result of the increased demand for environmental regulations and increased investment in abatement technologies, trade openness, structural change in the form of declining share of manufacturing output, and increased imports of pollution intensive output'. Whereas increased investment in abatement technologies and reduced production could result in an overall decrease in pollution, increased reliance on imports simply displaces the pollution and may increase global pollution levels if overseas industries are less efficient and have less stringent pollution controls.

Conclusions

The ecosystem services that deliver support for the basic requirements of human life, and the wherewithal for the development of the complex industrial, and post-industrial societies we live in, are threatened by the demands we make on them. In seeking to manage these demands, we need quality criteria that take account of the fact that return of most ecosystems to pre-industrial, or pre-human states is not a practical, or necessarily desirable, goal.

The proposed framework for ecological quality assessments based on ecosystem services, unlike traditional approaches (e.g., general water quality standards), accounts for the fact that the value put on the state of a system will depend on the specific management context (land use, population density etc.) and associated societal demands. For instance, in catchments of low population density and land use intensity, conservation of biodiversity and aesthetical enjoyment might be primarily valued, whereas in urban dominated catchments the supply of drinking water, flood mitigation and attenuation of pollutants are likely to be highly desired ecosystem services. If the demand–supply balance is periodically reviewed, it also allows for changes over time, either through changes in societal demand and expectation, or changes in external drivers of function such as climate. By incorporating a suite of services, and their interdependencies, demand for one service has to take into account not just the provision of that service, but the effect of provision at that level on others – where the services may include supporting, provisioning, regulating and cultural services.

At the level we have outlined here, the ESP concept provides a framework for ecological quality assessments. There are significant research challenges ahead before we could actually apply the concept in practice, but our aim here is to stimulate critical debate about the idea. We suggest that the ESP framework provides an adaptable and robust approach, which has the potential to foster a more integrative approach to ecosystem assessment and management in the future, and to meet the broad challenges laid down by new and forthcoming environmental legislation.

Whereas the focus of environmental management tends to be local, regional or national, ecosystem services are traded globally. The decline in heavy industry in many developed countries has generally been associated with an improvement in environmental quality, hence addressing the desire of such societies for minimally impacted environments. However, societies in developed countries have a high demand for industrial products and meet this demand by importing ecosystem services and the products derived from them, often from countries where environmental legislation and management is less well-developed and enforced. Importing ecosystem services, or the products derived from them, does not mitigate the environmental impact of their production, it simply produces a spatial disconnect between well-being benefit and environmental cost: the importer gains the benefit whereas the exporter carries the cost.

References

Amisah, S. and Cowx, I. G. (2000a) Impacts of abandoned mine and industrial discharges on fish abundance and macroinvertebrate diversity of the upper River Don in South Yorkshire, UK. *Journal of Freshwater Ecology* **15**, 237–250.

Amisah, S. and Cowx, I. G. (2000b) Response of the fish population of the River Don, South Yorkshire to water quality and habitat improvements. *Environmental Pollution* **108**, 191–199.

Angermeier, P. L. (2000) The natural imperative for biological conservation. *Conservation Biology* **14**, 372–381.

Ashton, E. A. (1997) The effects of metals on stream benthic algal communities. Unpublished PhD Thesis. University of Sheffield, UK.

Bermingham, S., Maltby, L. and Cooke, R. C. (1996) Effects of coal mine effluent on aquatic hyphomycetes I. Field study. *Journal of Applied Ecology* **33**, 1311–1321.

Bunn, S. E. and Davies, P. M. (2000) Biological processes in running waters and their implications for the assesssment of ecological integrity. *Hydrobiologia* **422/423**, 61–70.

Calow, P. (2000) Critics of ecosystem health misrepresented. *Ecosystem Health* **6**, 3–4.

Cave, L. A. and Blomquist, G. C. (2008) Environmental policy in the European Union: fostering the development of pollution havens? *Ecological Economics* **65**, 253–261.

Cole, M. A. (2004) Trade, the pollution haven hypothesis and the environmental Kuznets curve: examining the linkages. *Ecological Economics* **48**, 71–81.

Constanza, R. and Mageau, M. (1999) What is a healthy ecosystem? *Aquatic Ecology* **33**, 105–115.

Constanza, R., D'Arge, R., de Groot, R. S. *et al.* (1997) The value of the world's ecosystem services and natural capital. *Nature* **387**, 253–260.

DiGaetano, A. and Lawless, P. (1999) Urban governance and industrial decline. Governing structures and policy agendas in Birmingham and Sheffield, England, and Detroit, Michigan, 1980–1997. *Urban Affairs Review* **34**, 546–577.

Firth, C. J. (1997) *Domesday to the Dawn of the New Millennium: 900 years of the Don Fishery* Environment Agency, UK.

Folke, C., Carpenter, S., Walker, B. *et al.* (2004) Regime shifts, resilience, and biodiversity in ecosystem management. *Annual Review of Ecology and Systematics* **35**, 557–581.

Freyfogle, E. T. and Lutz Newton, J. (2001) Putting science in its place. *Conservation Biology* **16**, 863–873.

Gilbert, O. L. (1990) Wild figs by the River Don, Sheffield. *Watsonia* **18**, 84–85.

Gillson, L. and Willis, K. J. (2004) 'As Earth's testimonies tell': wilderness conservation in a changing world. *Ecology Letters* **7**, 990–998.

Grossman, G. M. and Krueger, A. B. (1995) Economic growth and the environment. *Quarterly Journal of Economics* **110** 353–377.

Hawkes, H. A. (1997) Origin and development of the Biological Monitoring Working Party score system. *Water Research* **32**, 964–968.

Hodgson, S. M., Maltby, L., Paetzold, A. and Phillips, D. (2007) Getting a measure of nature: cultures and values in an ecosystem services approach. *Interdisciplinary Science Reviews* **32**, 249–262.

Hunter, M., Jr (1996) Benchmarks for managing ecosystems: are human activities natural? *Conservation Biology* **10**, 695–697.

Kareiva, P., Watts, S., McDonald, R. and Boucher, T. (2007) Domesticated nature: shaping landscapes and ecosystems for human welfare. *Science* **316**, 1866–1869.

Karr, J.R. (1991) Biological integrity: a long-neglected aspect of water resource management. *Ecological Applications* **1**, 66–84.

Karr, J.D. (1999) Defining and measuring river health. *Freshwater Biology* **41**, 221–234.

Karr, J.R. and Chu, E.W. (2000) Sustaining living rivers. *Hydrobiologia* **422/423**, 1–14.

Karr, J.R. and Dudley, D.R. (1981) Ecological perspectives on water quality goals. *Environmental Management* **5**, 55–68.

Lackey, R.T. (2001) Values, policy, and ecosystem health. *BioScience* **51**, 437–443.

Mani, M and Wheeler, D. (1998) In search of pollution havens? Dirty industry in the world economy, 1960 to 1995. *Journal of Environment and Development* **7**, 215–247.

Meyer, J.L. (1997) Stream health: incorporating the human dimension to advance stream ecology. *Journal of the North American Benthological Society* **16**, 439–447.

Millennium Ecosystem Assessment. (2005) *Ecosystems and Human Well-being: Synthesis.* Island Press, Washington, DC.

Milner, A.M., Conn, S.C. and Brown, L.E. (2006) Persistence and stability of macroinvertebrate communities in streams of Denali National Park, Alaska: implications for biological monitoring. *Freshwater Biology* **51**, 373–387.

Nilsson, C., Jansson, R., Malmqvist, B. and Naiman, R.J. (2007) Restoring riverine landscapes: the challenge of identifying priorities, reference states, and techniques. *Ecology and Society* **12**, http://www.ecologyandsociety.org/vol12/iss11/art16/.

Palmer, M., Bernhardt, E., Chornesky, E. *et al.* (2004) Ecology of a crowded planet. *Science* **304**, 1251–1252.

Pickett, S.T.A. and Ostfeld, R.S. (1995) The shifting paradigm in ecology. In: *A New Century of Natural Resource Management* (eds. R.L. Knight and S.F. Bates), pp. 261–277. Island Press, Washington, DC.

Pimm, S.L. (1984) The complexity and stability of ecosystems. *Nature* **307**, 321–326.

Rapport, D.J., Boehm, G., Buckingham, D. *et al.* (1999) Ecosystem health: the concept, the ISEH, and the important tasks ahead. *Ecosystem Health* **5**, 82–90.

Rapport, D.J., Constanza, R. and McMichael, A.J. (1998) Assessing ecosystem health. *Trends in Ecology and Evolution* **13**, 397–402.

Rodriguez, J.P., Beard, T.D., Jr, Bennett, E.M. *et al.* (2006) 'Trade-offs across space, time, and ecosystem services. *Ecology and Society* **11**. http://www.ecologyandsociety.org/vol11/iss1/ art28.

Scholes, R.J. and Biggs, R. (2005) A biodiversity intactness index. *Nature* **434**, 45–49.

Scrimgeour, G.J. and Wicklum, D. (1996) Aquatic ecosystem health and integrity: problems and potential solutions. *Journal of the North American Benthological Society* **15**, 254–261.

Swetnam, T.W., Allen, C.D. and Betancourt, J.L. (1999) Applied historical ecology: using the past to manage the future. *Ecological Applications* **9**, 1189–1206.

Tweedale, G. (1995) *Steel City. Entrepreneurship, Strategy and Technology in Sheffield 1743–1993.* Clarendon Press, Oxford, UK.

UKTAG. (2007) Recommendations on Surface Water Classification Scheme for the purposes of the Water Framework Directive. http://www.wfduk.org/UKCLASSPUB/ LibraryPublicDocs/sw_status_classification.

UNDESA-DSD. (2006) *Trends in Sustainable Development.* United Nations, New York.

Vitousek, P.M., Mooney, H.A., Lubchenco, J. and Melillo, J.M. (1997) Human domination of earth's ecosystems. *Science* **277**, 494–499.

Walker, B., Holling, C.S., Carpenter, S.R. and Kinzig, A. (2004) Resilience, adaptability and transformability in social-ecological systems. *Ecology and Society* **9**. http://www.ecologyandsociety.org/vol9/iss2/art5/.

Wray, N., Hawkins, B. and Giles, C. (2001) 'One Great Workshop': The Buildings of the Sheffield Metal Trades. English Heritage, UK.

Index

technetium
  biogeochemistry of Tc reduction in
    sediments, 230–232
  impact of bioreduction strategies on
    solubility, 230–232
  influences of nitrate and nitrate-reducing
    bacteria on solubility, 234–236
terbuthylazine, 214
terbutryn, 214
Thlaspi caerulescens, 11
threatened species
  in polluted environments, 3
transformation products
  degradation products (degradates), 111
  ecotoxicity of degradates, 112–114
  emerging contaminants, 111–114
  environmental fate of degradates, 111–112
  environmental occurrence of
    degradates, 111
  pesticide risk assessment, 111
Trent Biotic Index, 128, 256

uranium waste
  bioremediative stimulation of native
    U(VI)-reducing bacteria, 232–234
  fate and transport in the environment,
    232–234
  influences of nitrate and nitrate-reducing
    bacteria on solubility, 234–236
US Environmental Protection Agency (USEPA)
  ecological risk assessment (ERA) scheme,
    189–190
Usnea spp., 49

Vezdaea leprosa, 46
Violetalia calaminariae vegetation order, 12, 20
Violetum calaminariae, 9, 12

Water Framework Directive (EU), 2, 86, 87,
  126–127, 136, 156
  restoration of wetland systems, 164–165
wetland systems
  commitments to restore wetlands, 163–165
  diffuse pollution sources, 162–163
  establishing baseline conditions, 165–166
  identifying targets for restoration, 165–166
  palaeoecological records in sediments,
    166–167
  reference conditions, 165
  restoration under the Water Framework
    Directive, 164–165
  sediment records, 166–167
  South Australia's wetland strategy,
    163–164
wetland systems studies
  acidity/alkalinity, 179–180
  appropriate baseline for restoration,
    182–183
  eutrophication, 173–175
  metal pollution in sediments, 172–173
  methods, 169–170
  multiple drivers and symptoms, 180–182
  Murray Darling Basin study area, 167–168
  regional integration of site studies
    information, 182
  results and discussion, 170–183
  salinity, 175–179
  sediment pollution, 170–173
  sedimentation rates, 170–172
  study areas, 167–168
  turbidity and its biological impacts, 173
  upper Mississippi River Basin study area,
    167, 168

Xanthoria parietina, 44

Printed in the United States
by Baker & Taylor Publisher Services